D1032461

UCLA Symposia on Molecular and Cellular Biology, New Series

Series Editor, C. Fred Fox

RECENT TITLES

Volume 50
Interferons as Cell Growth Inhibitors and Antitumor Factors, Robert M. Friedman, Thomas Merigan, and T. Sreevalsan, *Editors*

Volume 51
Molecular Approaches to Developmental Biology, Richard A. Firtel and Eric H. Davidson, *Editors*

Volume 52
Transcriptional Control Mechanisms, Daryl Granner, Michael G. Rosenfeld, and Shing Chang, *Editors*

Volume 53
Progress in Bone Marrow Transplantation, Robert Peter Gale and Richard Champlin, *Editors*

Volume 54
Positive Strand RNA Viruses, Margo A. Brinton and Roland R. Rueckert, *Editors*

Volume 55
Amino Acids in Health and Disease: New Perspectives, Seymour Kaufman, *Editor*

Volume 56
Cellular and Molecular Biology of Tumors and Potential Clinical Applications, John Minna and W. Michael Kuehl, *Editors*

Volume 57
Proteases in Biological Control and Biotechnology, Dennis D. Cunningham and George L. Long, *Editors*

Volume 58
Growth Factors, Tumor Promoters, and Cancer Genes, Nancy H. Colburn, Harold L. Moses, and Eric J. Stanbridge, *Editors*

Volume 59
Chronic Lymphocytic Leukemia: Recent Progress and Future Direction, Robert Peter Gale and Kanti R. Rai, *Editors*

Volume 60
Molecular Paradigms for Eradicating Helminthic Parasites, Austin J. MacInnis, *Editor*

Volume 61
Recent Advances in Leukemia and Lymphoma, Robert Peter Gale and David W. Golde, *Editors*

Volume 62
Plant Gene Systems and Their Biology, Joe L. Key and Lee McIntosh, *Editors*

Volume 63
Plant Membranes: Structure, Function, Biogenesis, Christopher Leaver and Heven Sze, *Editors*

Volume 64
Bacteria–Host Cell Interactions, Marcus A. Horwitz, *Editor*

Volume 65
The Pharmacology and Toxicology of Proteins, John S. Holcenberg and Jeffrey L. Winkelhake, *Editors*

Volume 66
Molecular Biology of Invertebrate Development, John D. O'Connor, *Editor*

Volume 67
Mechanisms of Control of Gene Expression, Bryan Cullen, L. Patrick Gage, M.A.Q. Siddiqui, Anna Marie Skalka, and Herbert Weissbach, *Editors*

Volume 68
Protein Purification: Micro to Macro, Richard Burgess, *Editor*

Volume 69
Protein Structure, Folding, and Design 2, Dale L. Oxender, *Editor*

Volume 70
Hepadna Viruses, William Robinson, Katsuro Koike, and Hans Will, *Editors*

Volume 71
Human Retroviruses, Cancer, and AIDS: Approaches to Prevention and Therapy, Dani Bolognesi, *Editor*

Please contact the publisher for information about previous titles in this series.

Protein Purification

Micro to Macro

Protein Purification

Micro to Macro

Proceedings of a Cetus-UCLA Symposium
Held at Frisco, Colorado, March 29–April 4, 1987

Editor

Richard Burgess
Department of Oncology
McArdle Laboratory for Cancer Research
University of Wisconsin-Madison
Madison, Wisconsin

Alan R. Liss, Inc. • New York

Address all Inquiries to the Publisher
Alan R. Liss, Inc., 41 East 11th Street, New York, NY 10003

Printed in the United States of America

Library of Congress Cataloging-in-Publication Data

Protein purification.

(UCLA symposia on molecular and cellular biology ; new ser., v. 68)
Proceedings of the Cetus-UCLA Symposium on Protein
Purification—Micro to Macro.
1. Proteins—Purification—Congresses. I. Burgess, Richard. II. Cetus Corporation. III. University of California, Los Angeles. IV. Cetus-UCLA Symposium on Protein Purification—Micro to Macro (1987 : Frisco, Colo.) V. Series.
QP551.P69753 1987 660′.63 87-22736
ISBN 0-8451-2667-9

Contents

Contributors

Marie A. Abbott, Eli Lilly and Company, Indianapolis, IN 46285 **[239]**

Bharat B. Aggarwal, Department of Molecular Immunology, Genentech, Inc., South San Francisco, CA 94080 **[17]**

Geoffrey Allen, Department of Protein Chemistry, Wellcome Biotech, Beckenham, Kent BR3 3BS, England **[367]**

Paul C. Anderson, Biochemistry Department, Gortner Laboratory, University of Minnesota, St. Paul, MN 55108 **[131]**

Guido Antoni, Sclavo Research Centre, Siena, Italy **[421]**

Dennis N. Arvidson, Department of Microbiology and Molecular Biology Institute, University of California, Los Angeles, CA 90024 **[401]**

A.M. Athalye, Department of Chemical Engineering, University of Wisconsin, Madison, WI 53706 **[475]**

Janet M. Attwood, Department of Biochemistry, University of Wisconsin-Madison, Madison, WI 53706 **[197]**

Cosima Baldari, Sclavo Research Centre, Siena, Italy **[421]**

M. Belew, Repligen Corporation, Cambridge, MA 02139 **[491]**

S. Berenzenko, Sterling Organics R&D, Newcastle upon Tyne, NE3 3TT England **[247]**

Milan Bier, Center for Separation Science, University of Arizona, Tucson, AZ 85721 **[315, 329]**

M.G. Bite, Sterling Organics R&D, Newcastle upon Tyne, NE3 3TT England **[247]**

Michael D. Bonifacio, Royal North Shore Hospital, St. Leonards, NSW 2065, Australia **[225]**

M. Giuseppina Borri, Sclavo Research Centre, Siena, Italy **[421]**

Paola Bossù, Sclavo Research Centre, Siena, Italy **[421]**

Jeffrey Bruton, Becton Dickinson Research Center, Research Triangle Park, NC 27709 **[263]**

Richard R. Burgess, Department of Oncology, McArdle Laboratory for Cancer Research and U.W. Biotechnology Center, University of Wisconsin-Madison, Madison, WI 53706 **[xvii, 279]**

The numbers in brackets are the opening page numbers of the contributors' articles.

Frank P. Buxton, Biochemicals Division Allelix Inc., Mississauga, Ontario, Canada L4V 1P1 **[355]**

M. Cristina Casagli, Sclavo Research Centre, Siena, Italy **[421]**

Costante Ceccarini, Sclavo Research Center, Siena, Italy **[217]**

King-Lan Cheng, Cambridge BioScience Corporation, Worcester, MA 01605; present address: Department of Tumor Biology, University of Texas System Cancer Center, M.D. Anderson Hospital, Houston, TX 77030 **[443]**

Stephen Cockle, Connaught Research Institute, Willowdale, Ontario, Canada M2R 3T4 **[375]**

M.C.M. Cockrem, Department of Chemical Engineering, University of Wisconsin, Madison, WI 53706 **[475]**

Jerry R. Colca, Department of Gastrointestinal Diseases Research, The Upjohn Company, Kalamazoo, MI 49001 **[27]**

J. Michael Conlon, Clinical Research Group for Gastrointestinal Endocrinology of the Max-Planck-Gesellschaft at the University of Göttingen, Federal Republic of Germany **[255]**

Raffaella Conti, Sclavo Research Center, Siena, Italy **[217]**

Michael M. Cox, Department of Biochemistry, University of Wisconsin-Madison, Madison, WI 53706 **[197]**

E.L. Cussler, Department of Chemical Engineering and Materials Science, University of Minnesota, Minneapolis, MN 55455 **[307]**

R. Wayne Davies, Biochemicals Division, Allelix Inc., Mississauga, Ontario, Canada L4V 1P1 **[355]**

L. Derry, Sterling Organics R&D, Newcastle upon Tyne, NE3 3TT England **[247]**

Jeffrey R. Deschamps, Laboratory for the Structure of Matter, Naval Research Laboratory, Washington, DC 20375-5000 **[207]**

Cinzia D'Ettorre, Sclavo Research Centre, Siena, Italy **[421]**

Daryll B. DeWald, Departments of Biotechnology Research and Diabetes, The Upjohn Company, Kalamazoo, MI 49001 **[27]**

Anne S. Dillon, Department of Chemical Engineering, Massachusetts Institute of Technology, Cambridge, MA 02139 **[117]**

Ned B. Egen, Center for Separation Science, University of Arizona, Tucson, AZ 85721 **[315]**

Karin Ernst-Cabrera, Department of Biophysics, The Weizmann Institute of Science, Rehovot, 76100 Israel **[163]**

Julie M. Fagan, Department of Animal Sciences, Rutgers University, New Brunswick, NJ 08903 **[459]**

Millicent A. Firestone, Center for Separation Science, University of Arizona, Tucson, AZ 85721 **[329]**

Cesira Galeotti, Sclavo Research Centre, Siena, Italy **[421]**

Cynthia A. Gates, Department of Biochemistry, University of Wisconsin-Madison, Madison, WI 53706 **[197]**

Keith Gewain, Schering-Plough Corp., Bloomfield, New Jersey 07003 and DNAX, Palo Alto, CA 94304 **[383]**

Paolo Ghiara, Sclavo Research Centre, Siena, Italy **[421]**

S.J. Gibbs, Department of Chemical Engineering, University of Wisconsin, Madison, WI 53706 **[475]**

M.A. Gleeson, Biochemicals Division, Allelix Inc., Mississauga, Ontario, Canada L4V 1P1 **[355]**

Craig Goldensoph, Biochemistry Department, Gortner Laboratory, University of Minnesota, St. Paul, MN 55108 **[131]**

Robert Greenberg, Schering-Plough Corp., Bloomfield, New Jersey 07003 and DNAX, Palo Alto, CA 94304 **[383]**

Robert P. Gunsalus, Department of Microbiology and Molecular Biology Institute, University of California, Los Angeles, CA 90024 **[401]**

David I. Gwynne, Biochemicals Division, Allelix Inc., Mississauga, Ontario, Canada L4V 1P1 **[355]**

T. Alan Hatton, Department of Chemical Engineering, Massachusetts Institute of Technology, Cambridge, MA 02139 **[117]**

David H. Hawke, Applied Biosystems, Inc., Foster City, CA 94404 **[35]**

Cora A. Henwood, Department of Protein Chemistry, Wellcome Biotech, Beckenham, Kent BR3 3BS, England **[367]**

W. Herlihy, Repligen Corporation, Cambridge, MA 02139 **[491]**

Gary D. Hodgen, Eastern Virginia Medical School, Norfolk, VA 23507 **[225]**

Kelly J. Hoke, Eli Lilly and Company, Indianapolis, IN 46285 **[239]**

Chung-Ho Hung, Cambridge BioScience Corporation, Worcester, MA 01605 **[443]**

R. Inacker, Biopharmaceutical R&D, Smith Kline and French Labs, Swedeland, PA 19479 **[337]**

Jerry Jendrisak, Promega Corporation, Madison, WI 53711 and Department of Horticulture, University of Wisconsin, Madison, WI 53706 **[75]**

Robert Kastelein, Schering-Plough Corp., Bloomfield, New Jersey 07003 and DNAX, Palo Alto, CA 94304 **[383]**

B. Kaylos, Repligen Corporation, Cambridge, MA 02139 **[491]**

Charlotte Kensil, Cambridge BioScience Corporation, Worcester, MA 01605 **[443]**

Steven B.H. Kent, Division of Biology, California Institute of Technology, Pasadena, CA 91104 **[49]**

Mark W. Knuth, McArdle Laboratory for Cancer Research and U.W. Biotechnology Center, University of Wisconsin-Madison, Madison, WI 53706 **[279]**

Maria-Regina Kula, Institut für Enzymtechnologie der Universität Düsseldorf, in der KFA Jülich, Postfach 20 50, D-5170 Jülich, Federal Republic of Germany **[99]**

Andrew A. Kumamoto, Department of Microbiology and Molecular Biology Institute, University of California, Los Angeles, CA 90024 **[401]**

Hung V. Le, Schering-Plough Corp., Bloomfield, New Jersey 07003 and DNAX, Palo Alto, CA 94304 **[383]**

Michael Lennick, Connaught Research Institute, Willowdale, Ontario, Canada M2R 3T4 **[375]**

Joseph K.K. Li, Department of Biology, Utah State University, Logan, Utah 84322 and Becton Dickinson Research Center, Research Triangle Park, NC 27709 **[263]**

E.N. Lightfoot, Department of Chemical Engineering, University of Wisconsin, Madison, WI 53706 **[475]**

Leo S. Lin, Department of Protein Chemistry, Cetus Corporation, Emeryville, CA 94608 **[409]**

Patrizia Lorenzoni, Sclavo Research Center, Siena, Italy **[217]**

R. Love, Repligen Corporation, Cambridge, MA 02139 **[491]**

Rex Lovrien, Biochemistry Department, Gortner Laboratory, University of Minnesota, St. Paul, MN 55108 **[131]**

Peter A. Lowe, Celltech Ltd., Slough, SL1 4DY, England **[429]**

Dante J. Marciani, Cambridge BioScience Corporation, Worcester, MA 01605 **[443]**

Fiona A.O. Marston, Celltech Ltd., Slough SL1 4DY, England **[429]**

David G. Maskalick, Eli Lilly and Company, Indianapolis, IN 46285 **[239]**

Carol Mays, Schering-Plough Corp., Bloomfield, New Jersey 07003 and DNAX, Palo Alto, CA 94304 **[383]**

Thomas Mercolino, Becton Dickinson Research Center, Research Triangle Park, NC 27709 **[263]**

Leslie Meyer-Leon, Department of Biochemistry, University of Wisconsin-Madison, Madison, WI 53706 **[197]**

Sam Morris, Beckman Instruments, Columbia, MD **[207]**

Richard A. Mosher, Center for Separation Science, University of Arizona, Tucson, AZ 85721 **[315, 329]**

Cristina Mottola, Sclavo Research Center, Siena, Italy **[217]**

Tattanahalli L. Nagabhushan, Schering-Plough Corp., Bloomfield, New Jersey 07003 and DNAX, Palo Alto, CA 94304 **[383]**

Satwant Narula, Shering-Plough Corp., Bloomfield, New Jersey 07003 and DNAX, Palo Alto, CA 94304 **[383]**

Bruce Odegaard, Biochemistry Department, Gortner Laboratory, University of Minnesota, St. Paul, MN 55108 **[131]**

Bauke Oudega, Department of Molecular Microbiology, Vrije Universiteit, Amsterdam, The Netherlands **[393]**

James D. Pearson, Departments of Biotechnology Research and Diabetes, The Upjohn Company, Kalamazoo, MI 49001 **[27]**

A. Profy, Repligen Corporation, Cambridge, MA 02139 **[491]**

Reza S. Rahaman, Department of Chemical Engineering, Massachusetts Institute of Technology, Cambridge, MA 02139 **[117]**

Ursula Rdest, Institut für Genetik und Mikrobiologie, Universität Würzburg, D-8700 Würzburg, Federal Republic of Germany **[271]**

F.J.S. Reed, Sterling Organics R&D, Newcastle upon Tyne, NE3 3TT England **[247]**

Paul Reichert, Schering-Plough Corp., Bloomfield, New Jersey 07003 and DNAX, Palo Alto, CA 94304 **[383]**

Stephen K. Rhind, Celltech Ltd., Slough, SL1 4DY, England **[429]**

M. Rosenberg, Biopharmaceutical R&D, Smith Kline and French Labs, Swedeland, PA 19479 **[337]**

Robert K. Scopes, Department of Biochemistry, La Trobe University, Bundoora, Victoria, Australia 3083 **[1]**

Shi-Hsiang Shen, Connaught Research Institute, Willowdale, Ontario, Canada M2R 3T4 **[375]**

C. Cohen Silverman, Biopharmaceutical R&D, Smith Kline and French Labs, Swedeland, PA 19479 **[337]**

Michael J. Sinosich, Eastern Virginia Medical School, Norfolk, VA 23507 **[225]**

Jeffrey E. Sloan, Center for Separation Science, University of Arizona, Tucson, AZ 85721 **[329]**

Margarete Sturm, Institut für Genetik und Mikrobiologie, Universität Würzburg, D-8700 Würzburg, Federal Republic of Germany **[271]**

Richard Sugrue, Celltech Ltd., Slough, SL1 4DY, England **[429]**

Eugene Sulkowski, Department of Molecular and Cellular Biology, Roswell Park Memorial Institute, Buffalo, NY 14263 **[149, 177]**

Rosalinda Syto, Schering-Plough Corp., Bloomfield, New Jersey 07003 and DNAX, Palo Alto, CA 94304 **[383]**

Lars Thim, Novo Research Institute, Bagsvaerd, Denmark **[255]**

Wolfgang Thormann, Center for Separation Science, University of Arizona, Tucson, AZ 85721 **[329]**

Paul P. Trotta, Schering-Plough Corp., Bloomfield, New Jersey 07003 and DNAX, Palo Alto, CA 94304 **[383]**

Anita Van Kimmenade, Schering-Plough Corp., Bloomfield, New Jersey 07003 and DNAX, Palo Alto, CA 94304 **[383]**

Arnold J. van Putten, Department of Molecular Microbiology, Vrije Universiteit, Amsterdam, The Netherlands **[393]**

G. Folena Wasserman, Biopharmaceutical R&D, Smith Kline and French Labs, Swedeland, PA 19479 **[337]**

Lloyd Waxman, Department of Biological Chemistry, Merck Sharp & Dohme, West Point, PA 19486 **[459]**

Meir Wilchek, Department of Biophysics, The Weizmann Institute of Science, Rehovot, 76100 Israel **[163]**

Kenneth J. Wilson, Applied Biosystems, Inc., Foster City, CA 94404 **[35]**

Jaclyn M. Woll, Department of Chemical Engineering, Massachusetts Institute of Technology, Cambridge, MA 02139 **[117]**

David D.L. Woo, Department of Medicine, UCLA, Los Angeles, CA 90024 **[49]**

Elizabeth A. Wood, Department of Biochemistry, University of Wisconsin-Madison, Madison, WI 53706 **[197]**

Ralph Yamamoto, Department of Protein Chemistry, Cetus Corporation, Emeryville, CA 94608 **[409]**

Pau M. Yuan, Applied Biosystems, Inc., Foster City, CA 94404 **[35]**

Preface

The Cetus-UCLA Symposium on *Protein Purification: Micro to Macro* was held at Frisco, Colorado, March 29–April 4, 1987. This meeting was attended by about 239 scientists who came together with a common interest in learning more about protein purification. The meeting presentations were focused on several areas including micropurification and analysis, protecting proteins during purification and storage, precipitation and phase partitioning methods, chromatography, overproduction of proteins in bacteria and other hosts, coping with problems of insolubility and proteolysis, and scale-up considerations. This book contains a sampling of the topics presented at plenary sessions and poster sessions.

The interest in protein purification has increased dramatically as numerous enzymes of research, pharmaceutical, and industrial importance are identified. In many cases, the technology for cloning the genes for these proteins has developed more rapidly than the technology for purifying the expressed gene product. In addition, the need to produce some of these proteins at the kilogram to ton level has required a collaborative effort between biochemists and chemical engineers. This meeting was organized to bring together a wide spectrum of researchers (theoretical and practical, academic, governmental and industrial, microscale, laboratory scale and very large scale, protein biochemists, genetic engineers, and chemical engineers) to review together the state of protein purification, to share new concepts and techniques, and to pinpoint theoretical and practical problems yet to be solved.

At a time when the bulk of the advances in large scale protein purification and process design are occurring in biotechnology companies, it is crucial that the growing body of knowledge find its way to people who will teach the next generation of protein purifiers and who will be needed to fuel future innovation in separations technology. Dr. Robert Scopes, whose excellent textbook on protein purification has contributed greatly to this teaching process, was an appropriate keynote speaker.

Micropurification of proteins at the microgram level was reported using a variety of methods including HPLC, affinity chromatography, and elution from SDS polyacrylamide gels and renaturation.

Numerous examples of overproduction of cloned gene products were presented. Although overexpression in *E. coli* was the most common method, other host systems reported included yeast, filamentous fungi, mammalian cell culture, and insect cells.

Enzymes destined for pharmaceutical use pose special considerations. Products have to be consistent, well-defined, and highly purified, in some cases exceeding 99.99% purity. This requires high resolution final "polishing" steps and the need for very high sensitivity methods to detect trace impurities. An important problem was protein heterogeneity; either naturally occurring, or introduced due to production in heterologous hosts or post-harvest modification.

In designing a large-scale protein purification process new considerations arise such as viscosity, heat transfer rates, phase separations, and concentration of dilute solutions. For non-clinical commercial enzymes, cost becomes the prime design consideration. It was encouraging to see the biochemists and chemical engineers working together to try to learn and appreciate each other's perspective.

Overall, there was a sense of excitement and satisfaction generated by this meeting: excitement at the new knowledge gained, the rapid progress of the field, and the new personal connections established; satisfaction at the opportunity to participate in a meeting where the main focus was protein purification, not the protein being purified. Most of us thoroughly enjoyed being part of a meeting where protein purification emerged as a discipline rather than merely as a means to an end.

Special thanks are due Cetus Corporation for the generous Sponsorship of this meeting. I also gratefully acknowledge additional gifts from Cambridge Bioscience Corporation; Celltech Limited; Pharmacia, Inc., Biotechnology Group; Millipore Corporation; The Upjohn Company; Boehringer Mannheim GmbH; Bristol-Myers Company, Pharmaceutical Research & Development Division; DNAX Research Institute; and Interferon Sciences. I wish to thank the UCLA Symposia staff, especially Robin Yeaton, who played a crucial role in helping me organize the meeting, and Bill Coty, who skillfully saw to the details at the meeting so that I could relax and enjoy the meeting myself.

Richard Burgess

Protein Purification: Micro to Macro, pages 1–15

CLASSICAL AND MODERN TECHNIQUES IN PROTEIN PURIFICATION

Robert K. Scopes

Department of Biochemistry, La Trobe University,
Bundoora, Victoria, Australia, 3083.

Our understanding of the structure and function of proteins has progressed so rapidly over the past few decades that we tend to forget that scientists were studying them over two centuries ago. My first reference - incompletely documented for unavoidable reasons - dates from the year of the French Revolution, 198 years ago [1], and it describes the purification of coagulable substances from plants having similar properties to egg albumen, a protein already well known at that time. Of course, none of these substances were anything like as "pure" as a journal editor would like us to demonstrate these days, but they were recognised as being distinct from each other; by the early 19th century the words gelatin, albumen, fibrin, casein, gluten, gliadin, zein and legumin had been invented. The origin of the word "protein" is attributed to Berzelius in about 1838 [2] and publicised by Mulder [3,4] who was working on the concept of a "proto-radical" which was thought to be the building block from which the proteins were constructed. It was to be nearly 100 years before any further great strides in the understanding of protein structure were made. Nevertheless, proteins were not only being isolated from a wide variety of plants and animal tissues, but were already being obtained in crystalline form as early as 1859 [5]. 100 years ago egg albumen was crystallized by Hofmeister [6], these days better known for his series, but not until 1926 were crystals of a bioactive protein, the enzyme urease, obtained [7]. For the most part these early proteins were isolated from plant sources in which they existed at a high concentration, so that a relatively simple process such as acidification, or simply letting stand, might suffice to cause crystallization from the water-extract of the raw material. Genuine fractionation methods using variations of solubility in water, salt solutions and alcohol began to be used in the mid 19th century, and formed the basis for purifying proteins for the next 100 years. These we must regard as the "classical" methods, even though some of the more modern techniques, in particular column chromatography, have been in use for over 30 years. I am indepted to an article by M.V. Tracey [8] on the history of protein chemistry for these preliminary remarks.

Let us consider some of the classical techniques in protein fractionation. It is important to recognise that the modern high-tech systems are not always the best just because they are the newest, and the scientists 50 years ago were just as clever and innovative as we are today (arguably more so, as they did not have to spend so much time repairing instruments and puzzling over computer software!). They devised techniques mainly relying on solubility to fractionate protein mixtures that still form the basis of most protein purification procedures in use today. Probably the best example is the method developed for fractionation of plasma proteins during World War II, the Cohn fractionation [9], which still is the main method for the bulk production of blood proteins. One important reason for this (apart from the desire to make continued use of the investment in plant and equipment) is that the only compound added to the plasma to cause precipitation is ethanol, which is not exactly foreign to human blood, so need not be as rigorously removed before transfusion as other compounds might require. Fractionation by precipitation with organic solvents - mainly ethanol and acetone - has been widely used, especially in the first half of this century, but has recently lost favour in comparison with the use of ammonium sulphate as a precipitant. There are several reasons for this, the main one being that organic solvent precipitation must be carried out in the cold, because addition of the solvent destabilizes proteins, whereas ammonium sulphate stabilzes them. Thus with ammonium sulphate losses in bioactivity are minimised, and the process can be interrupted for long periods without fear of further losses. Ammonium sulphate is generally regarded as the best "salting-out" precipitant because of its high solubility (to 4 M), low heat of solvation, low density of solution, and cheapness. It is used in perhaps as many as 80% of all described protein isolation procedures; moreover, the final preparation is often stored as an ammonium sulphate suspension of amorphous precipitate or crystals. Of the many other precipitants that have been used, only polyethylene glycol has found a widespread use; although the precipitation method may be described as classical, this particular compound is relatively modern.

The newly-hatched scientist entering a protein laboratory equiped with the latest in column HPLC, electrophoresis apparatus etc., may well wonder why we should be bothering with all these old-fashioned techniques when all you have to do (according to the advertisements) is to put your sample on a column, press a button, and in 20 min you have a complete separation. But then he is asked to isolate the protein from 2 kg of beef liver, or 40 l of milk; the realisation comes that something else has to be done with the raw material before it is applied to the latest $600 HPLC column (maximum capacity 10 mg) - not only must the non-proteinaceous and insoluble material be got rid of, but the quantities must be reduced - a considerable amount of purification is needed before the refined processing can commence. Although there are now "low performance" column systems capable of dealing with large amounts, as well as other modern techniques such as phase partitioning [10], a preliminary "classical" process is usually used at the initial stage. By precipitating out the wanted protein, there is a good chance that much of the non-protein material: nucleic acids, soluble carbohydrates and lipids, will be left in the supernatant. This messy begining of protein purification is really as important as the final stages, when it comes to evaluating recoveries and costs.

Until relatively recently, most protein purification took place in academic research laboratories with a view simply to obtaining a sample of pure protein which could be analysed and studied structurally for the benefit of scientific knowledge. We now have different criteria being applied to purification techniques. In commercial production the protein is required in large amounts, reproducible quality and at low cost; parameters that are not uppermost in the academic researcher's mind. It is not surprising, then, that many of the recent developments have come not from university research groups (although they usually have provided the germinal ideas), but from companies whose interest is to develop products suitable for large-scale operation. And although operating costs are important for the ultimate producer, the support companies are not always enthusiastic about advocating and supplying the cheapest equipment and materials. In particular, the developments in high performance liquid chromatography and many of the new materials used in columns have come into being because of the potential of large markets for protein products in the clinical, agricultural and diagnostic fields. Superimposed on these developments has been the revolution in genetic engineering which allows so many of these proteins to be produced at reasonable cost. Indeed the developments in molecular biology are shifting the strategies of protein purification procedures from a classical fractionation of a crude extract, to a changing of the composition of the raw material to make the purification easier.

So the objectives for commercial protein fractionation can be summed up as high throughput, large-scale procedures that have a high "value-added" content - techniques nowadays described as "downstream processing".

At the other end of the scale, micro-techniques for sequencing and analysis allow useful information to be obtained from microgram amounts of pure protein, and even though these small amounts may originate from large quantities of raw material, at the later stages special techniques for handling sub-milligram quantities are necessary. It is just becoming possible to sequence small proteins after elution from a 2-dimensional electrophoretic gel, the analytical electrophoretic technique having sufficient capacity to be used preparatively. Having a sequence, an oligonucleotide probe can be constructed to find the gene, which may subsequently be expressed in a host organism or tissue culture at far higher levels than in the original material. Any further protein purification will be made not from the material in which the protein is naturally found, but from the expression system containing the cloned gene. Eventually this may be scaled up to produce kilogram amounts of protein. Yet the original purification, which needed to be done only once, had to employ micro techniques to isolate the minute amounts originally present. The title of this symposium reflects the importance of both micro- and macro-techniques in today's world of protein purification.

Against this background the routine medium-scale processes of protein isolation continue, and new techniques are introduced, tried out, and find a place amongst the plethora of methods already available. Some find widespread application, others fade out as better methods take their place.

The routine laboratory worker interested in those proteins and enzymes with relatively less sex-appeal than the upcoming products of the biotechnological revolution, will mostly require medium-scale isolation. Studies in enzymology, X-ray crystallography and n.m.r require at least milligrams, preferably 100s of milligrams of pure product. Here, as elsewhere, a combination of classical and modern techniques are involved, to reduce the raw material to a state where the more refined techniques can be applied. Low abundance proteins may be cloned and expressed at high levels, but the man-hours required to achieve this may not be justifiable to granting agencies.

As previously noted, the "modern" techniques can be dated from the introduction of column chromatography in the 1950s. Ion exchangers, which at first were very crude fibrous cellulose particles, showed the power of the technique for resolving complex protein mixtures. For the first time one of the major properties of proteins, namely net charge, was being exploited to separate them. (Electrophoresis, does the same, but despite having been used analytically since about 1900, no satisfactory preparative electrophoretic method had yet been developed). The vast majority of ion exchange steps in protein purifications have made use of DEAE-substituted matrices, i.e. anion exchangers. The main reason for this is that in general proteins are more likely to bind to anion exchangers at pHs where the proteins remain stable (mainly in the range 5 to 9), than to cation exchangers. This especially applies to plant and bacterial proteins, as their isoelectric points tend to be somewhat lower than the equivalent animal proteins; cation exchangers are only really suited for proteins with isoelectric points above 5 (Fig. 1).

Over the years column packing materials have become more refined to improve the performance of the chromatography, so that today we have high pressure, high performance systems (HPLC) on the small scale, and on the larger scale beaded particles for column packing that are far superior to the original cellulosic fibres. Column chromatography has become so universal in application that it is worth spending some time discussing the principles

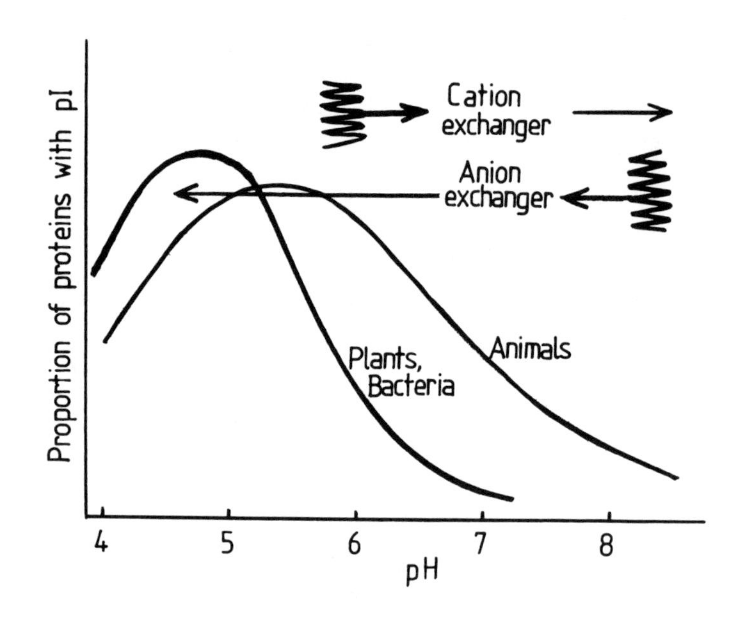

Figure 1. Approximate distribution of isoelectric points of proteins in animals, plants and bacteria. Whereas anion exchangers can be used for most proteins, cation exchangers are not generally suited for proteins with isoelectric points less than about 5.

and advantages of the different methods and materials. Chromatographic theory has been elucidated in terms of "plate heights" [11], a somewhat imaginary concept related to the minimum height on the column that a protein band can occupy. The smaller its value, the more "plates", meaning more chromatographic steps, and so higher resolution. So for optimum resolution, the plate height should be minimised. Van Deemter [12] developed an equation describing the factors determining plate height (for a very small sample), according to the flow rate of buffer through the column:

$$\text{Plate height } H = \frac{A}{v} + Bv + C$$

The parameter A in the equation relates to the diffusional spreading, which becomes significant for proteins only at very low flow rates. B is the parameter which depends mainly on the bead particle size, for it relates to the attainment of equilibrium between adsorbed and non-adsorbed protein molecules within the beads, and their transfer between the beads and the buffer flowing outside. It is also affected by the rate at which the protein molecules find binding sites within the beads. C is a column packing quality factor independent of flow rate. To minimise H, C should be small--the column packed evenly and the bead sizes have a narrow distribution. B decreases with particle size, and A depends on the molecular size (diffusion coefficient) of the solute being separated. To minimise B and C, high performance columns employ small spherical beads (5-20 μm) of very even size distribution; columns are pre-packed to maximise quality and reproducibility. But the parameter A remains user-defined; with small solutes the factor A/v may be of predominant importance in determining H, and fast flow rates are selected to minimise it. But with large solutes like proteins, A/v is relatively less important, and the factor Bv is the main determinant of H in HPLC of proteins (Fig. 2). From this we can see that whereas flow rate may not be very important in HPLC of small solutes provided it is fast enough, similarly fast flow rates for proteins result in a steep increase in plate height, thus a decrease in resolution. It is probable that the (unnecessary) stress on short separation times in protein HPLC has considerably compromised resolution in many cases. The optimum flow rate from Fig. 2, using appropriate theoretical parameters, is considerably slower than normally used, by a factor of 10 or more. This applies both to HPLC and to low pressure (LPLC) protein chromatography. In the latter case the optimum flow rates are much lower because of the larger particle size (giving a higher value for B); the theoretical optimum of 2-5 cm/h would be excessively slow and impractical on a large scale.

When considering the pressure needed to obtain optimum flow rates, theory tells us that it increases as the inverse cube of the bead diameter. Thus a reduction of average bead diameter from 80 μm to 20 μm requires a 64-fold increase in pressure to obtain an optimum flow

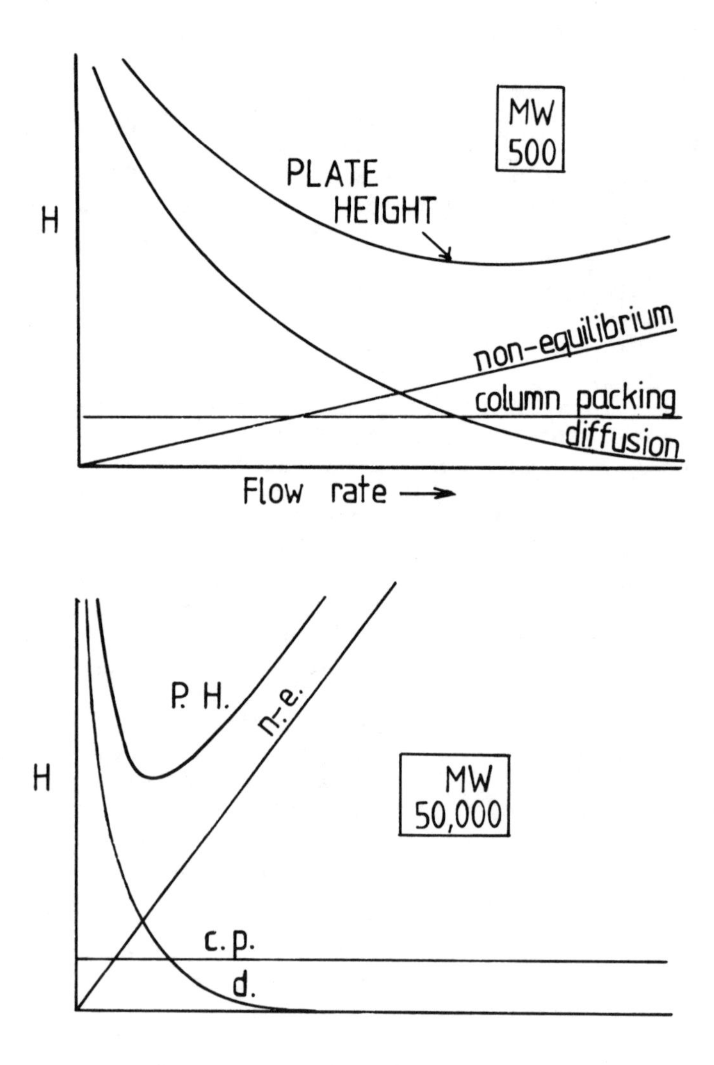

Figure 2. Effect of flow rate on Plate Height (H) for (a) small solutes and (b) large solutes. Optimum flow rate is considerably slower for large molecules because of non-attainment of equilibrium at fast flow rates.

rate (which is 4 times faster for the smaller beads). Hence the necessity of developing high pressure systems, with rigid beads that can withstand the pressure, once the bead diameter drops below about 50 μm. The costs of the equipment and packing materials also follow approximately an inverse cube law!

High performance chromatography is becoming widespread in its uses in protein purification. The chief advantages are resolution (for small loadings), reproducibility, automation and speed. But resolution is compromised at high loadings (mg rather than μg), simply because the plate height increases proportionately with the quantity of sample applied. Above a certain loading, it would be better to use a much larger "LPLC" column to achieve the capacity with no further loss in resolution. On the other hand the convenience, reproducibility and speed possible with HPLC systems makes their use desirable even if their full resolving potential is not being exploited. There is an increasing use of what might be termed "low plate number" HPLC, in which high loadings are used, but the separation of the desired component is adequate, since high resolution is not necessary in the particular application. Advantages and disadvantages of HPLC versus LPLC are listed below:

COMPARISON OF HPLC AND CONVENTIONAL "LPLC"

	HPLC	LPLC
Resolution	High	Moderate
Loading	Low	High
Speed	Fast	Slow-Moderate
Costs (equipment)	High	Low
Reproducibility	Good	Moderate
Best suited for:		
	Low MW proteins, peptides, stable proteins	High MW proteins and enzymes
Scale	Micro-Medium	Medium-Macro

The principles of adsorption in HPLC are for the most part the same as for LPLC, namely ion-exchange, hydrophobic (salt-promoted) and some special affinity techniques. But much of the earlier HPLC work employed the "reverse phase" organic solvent chromatography, adapted from the smaller-sized solute methods. This has been very successful for peptide separation and for small proteins, especially the more compact, stable, disulphide-cross-linked proteins. For larger molecules and enzymes, the stresses of organic solvent concentration usually lead to denaturation.

Gel filtration was the next major advance in protein separation [13]. Separation of protein molecules on the basis of size was a new concept, and so completely complementary to ion-exchange chromatography and the precipitation methods. It also enabled molecular sizes to be determined with the simplest of apparatus, and within a few years the analytical ultracentrifuge business was in decline. One of the delights of gel filtration is that it is the gentlest of techniques, since each molecule remains in solution all the time and the buffer composition is constant. Although the earlier materials (Sephadex) were slow flowing, later more rigid gel filtration beads have speeded up the separations, within the diffusional limitations of chromatogaphy as outlined above.

Gel filtration columns are also used in HPLC extensively, but in this case there has been no advance in resolution compared with LPLC. The reason is that the gel filtration materials developed for use under high pressure cover a wider range of protein sizes, so that the separation of two proteins of closely similar sizes although sharp, occurs over a very narrow elution volume. In the better gels in LPLC, the separation can be much greater, but the peaks broader (Fig. 3). Resolution is defined as separation divided by average peak width; in HPLC these values are smaller, but resolution is much the same.

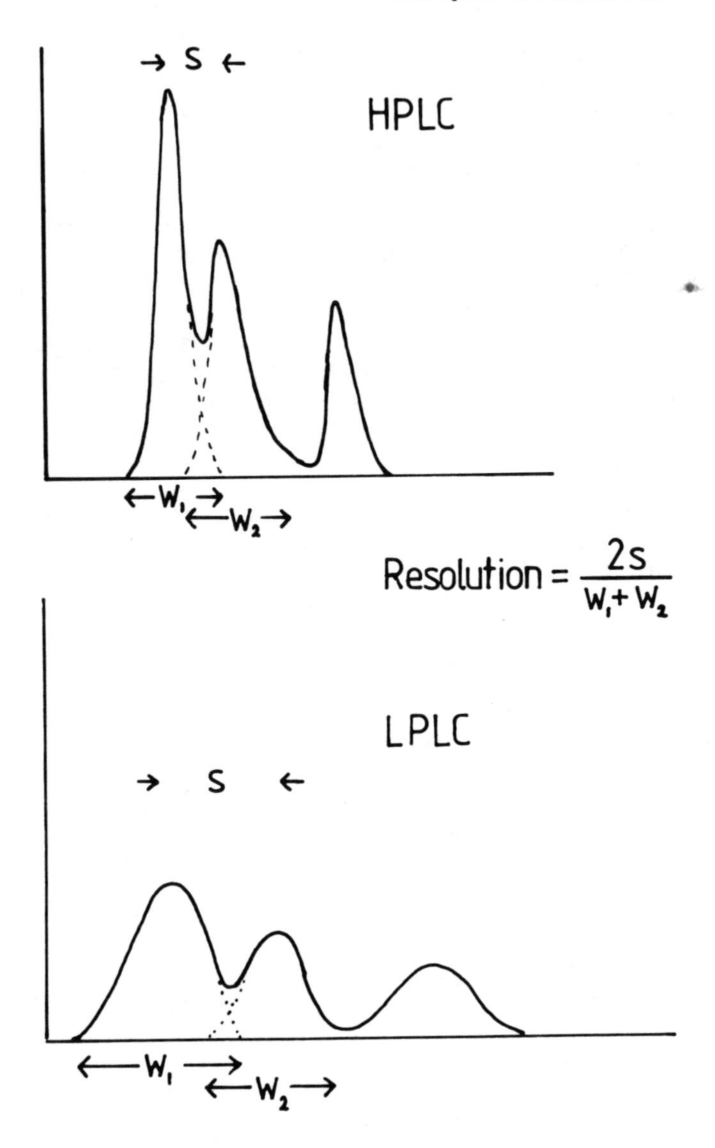

Figure 3. Comparison of the separation of three proteins on a typical HPLC gel filtration column, and on a low pressure column packed with a material that achieves greater separation, but more spreading. Resolution, defined as separation divided by average peak width, is similar in each case.

Affinity chromatography was a major development in the 1970s, and has found widespread application, especially in enzyme purification. For the first time a specific function of proteins, their ability to selectively bind to particular ligands, was exploited for their fractionation. True affinity chromatography as originally defined, describes the interaction between an immobilized ligand (which need not necessarily itself be the natural ligand) and the binding site on the protein for the natural ligand. This definition has been extended recently to include specific interactions with other parts of protein molecules. Although affinity chromatography has been highly successful in many cases, there have been drawbacks. Firstly there is the development of suitable specific adsorbents, which can occupy a lot of time, and the chemistry involved in attachment of the ligand may be complex and expensive. Secondly, the capacity of affinity adsorbents has frequently turned out to be low, with as little as 1% of the immobilized ligand being effective in binding protein. Thirdly, the potentially exquisite selectivity of an affinity adsorbent has often been elusive; non-specific binding of proteins not intended to adsorb often greatly reduces the purification factor at an affinity step. Except for very valuable products, these problems can make affinity methods uncompetitive compared with other less specific procedures. But affinity elution procedures, which do not require a high degree of selectivity at the adsorption step, can give excellent further purification; introduction of an enzyme's substrate into the elution buffer, to which the enzyme binds preferentially, can leave behind all impurities and elute the desired enzyme alone [14]. Pseudo- or empirical affinity chromatography, in which a range of random (but carefully chosen) adsorbents is screened for ability to bind the desired protein, is finding increasing use in large-scale work, mainly employing dyes as the affinity adsorbents [15,16]. The interaction of dyes with enzyme binding sites, especially with the dinucleotide fold of dehydrogenases [17,18] is well known and extensively exploited. Other affinity techniques, especially using the triazine dyes as ligands, have been developed, so that we now have affinity precipitation [19,20], affinity electrophoresis [21], and affinity phase partitioning [22,23] as useful procedures to exploit the specific properties of protein-ligand interactions.

The ultimate affinity technique, immunoadsorption,

does not (necessarily) involve the protein's natural ligand-binding site. Antibodies raised against protein molecules may interact with any part of the exposed surface, and are extremely specific for their antigen. Thus immunoadsorbents have the advantages of very high selectivity and qute high binding capacity too. Disadvantages include the time needed for development of the antibody in the first place, difficulties in eluting the antigen without inactivating it, and column stabilty problems. Nevertheless, for reasonably high value proteins (>$1000 per g), immunoadsorbents are finding an important niche in protein purification procedures, especially using monoclonal antibodies.

There are other modern techniques which can only be mentioned by name - hydrophobic adsorbents, immobilized metal affinity adsorbents, isoelectric focussing systems, ultrafiltration, selective heat denaturation, detergent extraction for membrane prtoeins, affinity tagging of recombinant proteins, and many more. Some of these we shall hear more about in this symposium, and still newer techniques will unfold as the years pass. The biotechnological revolution in protein products is poised to burst on the world as surely as Robespierre _et al_. were waiting to start another revolution two centuries ago, just at the time when Fourcroy was coagulating his plant proteins.

REFERENCES

1. Fourcroy AF (1789). Ann Chim 3:252.
2. Bezelius JJ, through Vickery HB (1950). Yale J Biol Med 22:387.
3. Mulder GJ (1838). Bull des Sciences physiques et naturelles en Nearlande 1:104.
4. Mulder GJ (1838). Natuur-en scheikundig Archief 6:87.
5. Maschke O (1859). Bot Z 17:409.
6. Hofmeister F (1889). Z Physiol Chem 14:165.
7. Sumner JB (1926). The isolation and crystallization ofthe enzyme urease. J Biol Chem 69:435-441.
8. Tracey MV (1985). "Themes in the Development of Protein Chemistry", Proc Aust Biochem Soc, Canberra.

9. Cohn EJ, Strong LE, Hughes WL, Mulford DJ, Ashworth JN, Melin M, Taylor HL (1946). Preparation and properties of serum and plasma proteins. IV. A system for the separation into fractions of the proteins and lipoprotein component of biological tissues and fluids. J Am Chem Soc 68:459-475.
10. Albertsson PÅ (1971). "Partition of Cell Particles and Macromolecules", 2nd edn. pp 24-25, Almquist & Wiksell, Stockholm; Wiley, N.Y.
11. Martin AJP, Synge RLM (1941). A new form of chromatography employing two liquid phases. 1. A theory of chromatography. Biochem J 35:1358-1368.
12. van Deemter JJ, Zudeweg FJ, Klinkenberg A (1958). Longitudinal diffusion and resistance to mass transfer as causes of nonideality in chromatography. Chem Eng Sci 5:271-289.
13. Porath J, Flodin P (1959). Gel filtration: A method for desalting and group separation. Nature 83:1657-1659.
14. Scopes RK (1977). Purification of glycolytic enzymes by using affinity elution chromatography. Biochem J 161:253-263.
15. Scawen MD, Hammond PM, Comer MJ, Atkinson T (1983). The appliation of triazine dye affinity chromatography to the large-scale purification of glycerokinase from Bacillus stearothermophilus. Anal Biochem 132:413-417.
16. Sherwood RF, Melton RG, Alwan SH, Hughes P (1985). Purification and properties of carboxypeptidase G2 from Pseudomonas sp. strain RS-16: Use of a novel triazine dye method. Eur J Biochem 148:447-453.
17. Rossman MG, Liljas A, Branden CI, Banasak LJ (1975). Evolutionary and structural relationships among dehydrogenases. "The Enzymes" (3rd edn) 11:61-102.
18. Thompson ST, Cass KH, Stellwagen E (1975). Blue dextran-Sepharose: an affinity column for the dinucleotide fold in proteins. Proc Natl Acad Sci 72:669-672.
19. Larsson P-O, Mosbach K (1979). Affinity precipitation of enzymes. FEBS Lett 98:333-338.
20. Hayet M, Vijayalakshmi MA (1986). Affinity precipitation of proteins using bis-dyes. J Chromatog 376:157-161.

21. Horesjsi V, Ticha A (1986). Quantitative and qualitative application of affinity electrophoresis for the study of protein-ligand interaction: a review. J Chromatog 376:44-67.
22. Kopperschlager G, Lorenz G, Usbech E (1983). Application of affinity partitioning in an aqueous two-phase system to the investigation of triazine dye-enzyme interaction. J Chromatog 259:97-105.
23. Johansson G, Kopperschlager G, Albertsson PÅ (1983).Affinity partitioning of phosphofructokinase from baker's yeast using polymer-bound Cibacron Blue F3GA. Eur J Biochem 131:589-594.

Protein Purification: Micro to Macro, pages 17–26

MICROPURIFICATION OF CYTOKINES

Bharat B. Aggarwal

Department of Molecular Immunology
Genentech, Inc., 460 Point San Bruno Boulevard
South San Francisco, CA 94080

ABSTRACT Lymphokines and monokines are the cytokines produced in trace amounts by activated lymphocytes and monocytes respectively. Several investigators have attempted the isolation of cytokines from normal peripheral blood leukocytes, but have achieved only minimal success due to the small amounts and the heterogeneity of the secreted molecules. Tumor Necrosis Factors (TNFs) -α and -β are two such cytokines which we have purified and characterized. Several human cell lines were screened for the production of TNFs and it was found that the promyelocytic cell line HL-60 and the B-lymphoblastoid cell line RPMI-1788 are good sources for TNF-α and TNF-β respectively. These cell lines can be grown on a large scale to provide sufficient starting material for isolation of the cytokines.

Special procedures were developed to purify these trace proteins from large volumes of cell conditioned media. This was essential since the activity of these proteins is labile. Our purification procedures included removal of cells from hundreds of liters of media by filtration through Pall Sealkleen 3 μM filter, followed by batch adsorption of the given cytokine activity to controlled pore glass beads. The binding to the beads could then be reversed by ethylene glycol. This step was followed by DEAE ion-exchange chromatography, Mono-Q and chromatofocusing fast protein liquid chromatography and reverse phase high performance liquid chromatography. Preparative denaturing and nondenaturing polyacrylamide gel electrophoresis were also employed to purify these cytokines. These procedures have provided materials of high purity, rapidly and reproducibly. The purity of cytokines was determined by amino acid sequence analysis. The physicochemical characteristics of these highly purified cytokines will be discussed.

INTRODUCTION

The activation of cells of the immune system by antigens and mitogens leads to the production of soluble mediators which are collectively called cytokines. The mediators produced specifically by lymphocytes and monocytes have been named lymphokines and monokines, respectively. These cytokines are similar to the hormones derived from the endocrine system in that they are produced in very minute amounts by cells of the immune system and usually act as mediators of immune surveillance at a distant site. Due to the trace amounts, the identification of these mediators as discreet biochemical entities has been a difficult task. It is not certain exactly how many cytokines are produced by the cells of the immune system. Originally, most of these cytokines were named on the basis of their biological activities. However, recent purification and characterization of some of these cytokines have revealed that a single cytokine could exhibit a variety of biological activities and that some of these activities are overlapping with those of other well-characterized cytokines (See review, 1).

In this article, we describe our experience with the purification of two such cytokines, one derived from monocytes, called tumor necrosis factor-α and the other derived from lymphocytes, named tumor necrosis factor-β (previously described lymphotoxin). Both of these activities have been suggested as selectively toxic towards tumor cells. These factors are produced in nanogram amounts by the cells of the immune system. Due to the limitation in obtaining enough leukocytes from human subjects, several human cell lines were screened and it was found that the promyelocytic cell line HL-60 and the lymphoblastoid cell line RPMI-1788 were optimal sources for TNF-α and TNF-β respectively. These cell lines had to be grown in large volumes to isolate enough protein for physicochemical characterization (2-4).

METHODS

Bioassay: We developed a rapid assay to monitor a large number of fractions during purification. The assay is based on the ability of both TNF-α and TNF-β to lyse actinomycin-D or mitomycin-C-treated mouse fibroblast L-929 (2). It involved plating of 30,000 cells in 0.2 ml of culture medium in each well of 96-well plate. Cells were incubated at 37°C with 0.05 ml of a serial diluted test sample in the presence of 1 μg/ml actinomycin-D. Sixteen hours later the cell lysis was monitored by crystal violet staining and absorbance was read at 540 nM using a Multiscan autoreader. One unit of activity was defined as the amount of protein required for 50 percent lysis of cells.

Cell Culture: HL-60 and RPMI-1788 were grown in RPMI-1640 medium containing 10 percent fetal bovine serum. HL-60 cells were grown in two-liter roller bottles and RPMI-1788 cells were grown in fifteen-liter spinner flasks. For production of the factors, both cell lines were grown under serum-free conditions in the presence of 20 ng/ml phorbol myristate acetate (PMA) for 48 to 72 hours.

Preparation of Cell-free Supernatants: Several liters of conditioned culture media from each cell line was pooled and cells were removed by filtration through a 3 μM filter (Pall Sealkleen) at 4°C. Cells and particulate debris could be removed from 100 liters of conditioned media in less than two hours due to the large surface area of this filter. The clear filtrate was subjected to further purification.

Controlled Pore Glass (CPG) Chromatography: Cell-free filtrates were stirred with CPG beads (120/200 mesh size) for 1 hour at 4°C. Approximately one liter of beads equilibrated in 5 mM sodium phosphate buffer pH 7.0 was used for 100 liters of filtrate. Thereafter, the beads were allowed to settle by gravity and the supernatants decanted. Under these conditions, both TNF-α and TNF-β activities (as determined by lysis of actinomycin-D-treated L-929 cells) were bound to the beads. The CPG beads were then packed in a column and washed with the equilibration buffer. The elution of activity was carried out either with 50 percent ethylene glycol (for TNF-β) or with 20 percent ethylene glycol containing 1 M NaCl (for TNF-α).

Ion Exchange Chromatography: The eluate from the CPG column was either diluted and directly applied to a pre-equilibrated DEAE ion-exchange column or dialyzed on a two-liter Amicon ultrafiltration unit and then applied to the column. Both TNF-α and TNF-β bind to DEAE-cellulose in 5 mM sodium phosphate buffer pH 7.4 (equilibration buffer) and the binding can be reversed with 50 to 100 mM NaCl in the equilibration buffer. For TNF-α and TNF-β a step-up and linear gradients were used, respectively.

Fast Protein Liquid Chromatography: Further purification of these cytokines was achieved by fast protein liquid chromatography (FPLC) using Mono-Q (ion-exchange) and Mono-P (chromatofocusing) columns. These columns were eluted either with salt gradients in the case of Mono-Q or with pH gradients for Mono-P. Because of the high resolution of the FLPC system, it provided several-fold purification for TNF-α.

Lectin Affinity Chromatography: TNF-α is a nonglycoprotein whereas TNF-β is a glycoprotein. Several lectins were tested for their binding to TNF-β. It was found that TNF-β binds to lentil lectin and the binding can be reversed by α-methyl mannoside. This resin provided several hundred-fold purification of TNF-β. Even though concanavalin-A has higher affinity than lentil lectin for galactose and mannose residues, TNF-β showed considerably higher binding to lentil compared to concanavalin-A lectin.

Preparative Polyacrylamide Gel Electrophoresis: This gel system under both denaturing and nondenaturing conditions was used to further purify TNFs. Denaturing gels have an advantage for their good resolution as compared to nondenaturing gels. The lack of good resolution with nondenaturing gels may be due to the micro-heterogeneity of the protein.

Reverse Phase High Performance Liquid Chromatography: This technique has a major disadvantage in that the solvents used usually destroy the biological activity. Therefore, this technique is useful only when one needs to isolate the protein for structure determination. We have used both C-4 and C-18 column attached to reverse phase HPLC system with propanol/acetonitrile gradients in 0.1 percent trifluoroacetic acid to obtain homogeneous proteins. The biological activity of TNF-α was partially destroyed by these solvents, whereas that of TNF-β was totally destroyed.

RESULTS AND DISCUSSION

Source. Some of the important criteria which led to the successful purification of TNF-α and TNF-β to homogeneity included: 1) the discovery of suitable cell lines (i.e., HL-60 and RPMI-1788) as a source of these cytokines; 2) the use of PMA to induce these cell lines to secrete elevated amounts of cytokines; and 3) the production of cytokines by these cell lines under serum-free conditions. In contrast to fresh human peripheral blood leukocytes (PBLs) which are routinely used as a source of these cytokines, the availability of cell lines enabled us to grow these cells in a large scale with relatively little labor and expense. Another advantage of cell lines is the relative homogeneity of the cytokine obtained in comparison to that derived from PBLs, probably because the latter usually requires pooling of cells from several different donors as well as heterogeneity of the cell population even from a single donor. Furthermore, the optimization of cell culture conditions for a given cell line with respect to the maximum production of the cytokine was found to be important.

Purification. The purification scheme used for the isolation of TNFs is shown in Figure 1. The cell conditioned media from several different batches were pooled and cell debris was cleared by filtration. This method was fast and convenient as compared to routinely used centrifugation. The clear filtrates were stirred with CPG beads. All the activity of cytokines was adsorbed to the beads and then eluted with ethylene glycol. This provided a very rapid method of

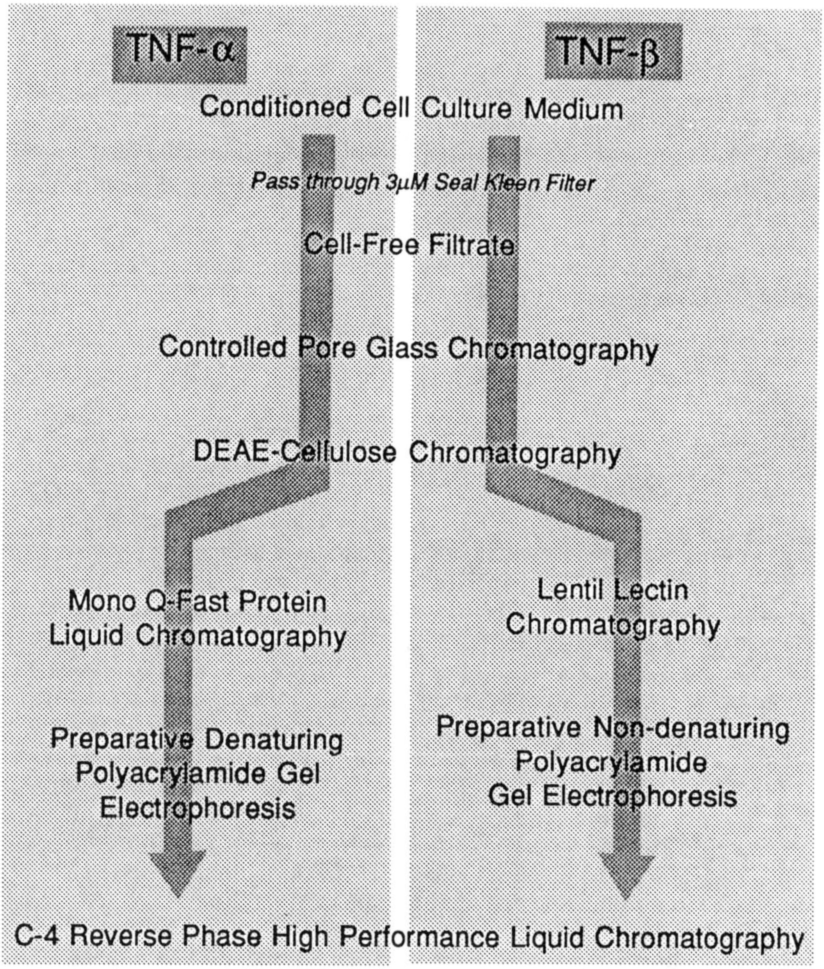

TNF-α

TNF-β

Conditioned Cell Culture Medium

Pass through 3μM Seal Kleen Filter

Cell-Free Filtrate

Controlled Pore Glass Chromatography

DEAE-Cellulose Chromatography

Mono Q-Fast Protein Liquid Chromatography

Lentil Lectin Chromatography

Preparative Denaturing Polyacrylamide Gel Electrophoresis

Preparative Non-denaturing Polyacrylamide Gel Electrophoresis

C-4 Reverse Phase High Performance Liquid Chromatography

Figure 1. Flow Sheet Diagram for the Micropurification of TNF-α and TNF-β

concentration of filtrates as well as a purification of the desired protein (Table I). Furthermore, the dialysis of the samples for column chromatography could also be avoided. The ability of both cytokines to bind to CPG and reversal of this binding by ethylene glycol suggests that these proteins are hydrophobic in nature. When the eluate from CPG was applied to DEAE-cellulose resin, the activity bound at 5 mM sodium phosphate buffer pH 7.4 and this binding could be reversed by 50–100 mM salt. TNF-α was further purified by Mono-Q (ion-exchange) fast protein liquid chromatography. Since TNF-β is a glycoprotein, it was further purified by lentil lectin affinity chromatography and preparative native polyacrylamide gel electrophoresis. Whenever necessary, reverse phase high performance liquid chromatography was used to further purify both TNF-α and TNF-β. This latter step enabled us to obtain amino acid sequence grade materials, even though most of the biological activity was destroyed. The biological activity was also found to be sensitive to sodium dodecyl sulfate.

Table I
Balance Sheet for the Purification of TNF-α and TNF-β

Purification Step	Cumulative Purification Fold	Final Recovery of Activity (percent)
TNF-α:		
Controlled pore glass chromatography	17	79
DEAE-cellulose chromatography	140	63
Mono-Q – fast protein liquid chromatography	2,240	49
Reverse-phase high performance liquid chromatography	13,638	19
TNF-β:		
Controlled pore glass chromatography	2.5	80
DEAE-cellulose chromatography	19	36
Lentil lectin chromatography	83	29
Preparative polyacrylamide gel electrophoresis	384	21

Characterization. It appears that most of the cytokines behave as oligomers under nondenaturing conditions. Both TNF-α and TNF-β were found to have molecular weights by gel filtration method 2 to 3 times higher (i.e., 45 Kd vs. 64 Kd) than that determined by SDS-PAGE analysis (i.e., 17 Kd vs. 25 Kd). This may be due to the hydrophobic nature of these proteins as also suggested both by their binding to glass beads as well as hydropathy analysis of the amino acid sequence. Furthermore, the isoelectric point of most of the cytokines of the immune system is usually in the neutral or acidic range with TNF-α and TNF-β being around pI 5-6. Both TNFs exhibit an average amino acid length for cytokines i.e., 157 and 171 amino acids for TNF-α and TNF-β respectively. TNF-α can be distinguished primarily from TNF-β by the former containing one disulfide bond, no methionine residue and biological activity being trypsin-sensitive whereas TNF-β is a glycoprotein, containing no cysteine residues and biological activity being trypsin-insensitive. The overall amino acid sequence of TNF-α is 32 percent identical and 51 percent homologous to that of TNF-β.

Both TNF-α and TNF-β share a common high affinity and low density receptors on the surface of most cells (5-7), a situation analogous to that for interferon-α and -β, or interleukin-1-α and -β (8-9). The presence of common receptors may explain the reason for a wide variety of biological activities which were originally assigned to TNF-α, now appears to be also due to TNF-β (see review 1). The ability of TNFs to selectively kill tumor cells and synergize for this function with interferons and existing chemotherapeutic agents (10,11) make these proteins the subject of intense investigation for clinical use.

ACKNOWLEDGEMENTS

I would like to thank several of my ex-colleagues for their contribution to the work described here. I would also like to thank Dr. Ramani Aiyer and the Scientific Review Board of Genentech for critically reviewing the manuscript, Ms. Socorro Cuisia for typing this paper and Ms. Carol Morita for the drawing work.

REFERENCES

1. Aiyer, R.A., and Aggarwal, B.B. (1987) Tumor necrosis factors. CRC Press (in press).

2. Aggarwal, B.B., Moffat, B., and Harkins, R.N. (1984) Human lymphotoxin: production, purification and initial characterization. J. Biol. Chem. 259:686.

3. Aggarwal, B.B., Kohr, W.J., Hass, P.E., Moffat, B., Spencer, S.A., Henzel, W.J., Bringman, T.S., Nedwin, G.E., Goeddel, D.V., and Harkins, R.N. (1985) Human tumor necrosis factor: Production, purification and characterization. J. Biol. Chem. 260:2345.

4. Aggarwal, B.B., Henzel, W.J., Moffat, B., Kohr, W.J., and Harkins, R.N. (1985) Primary structure of human lymphotoxin derived from 1788 lymphoblastoid cell line. J. Biol. Chem. 260:2334.

5. Hass, P.E., Hotchkiss, A., Mohler, M., and Aggarwal, B.B. (1985) Characterization of specific high affinity receptors for human tumor necrosis factor on mouse fibroblasts. J. Biol. Chem. 260:12214.

6. Shalaby, M.R., Palladino, M.A., Hirabayashi, S.A., Eessalu, T.E., Lewis, G.D., Shepard, H.M., Aggarwal, B.B. (1987) Receptor binding and activation of polymorphonuclear neutrophils by tumor necrosis factor-alpha. J. Leukocyte Biol. 41:196.

7. Aggarwal, B.B., Eessalu, T.E., and Hass, P.E. (1985) Characterization of receptors for human tumor necrosis factor and their regulation by gamma-interferon. Nature 318:665.

8. Branca, A.A., and Baglioni, C. (1981) Evidence that type I and II interferons have different receptors. Nature 294:768.

9. Dower, S.K., Kronheim, S.R., Hopp, T.P., Cantrell, M., Gillis, S., Deele, M., Henney, C.S., and Urdal, D.L. (1986) The cell surface receptors for IL-1α and IL-1β are identical. Nature 324:266.

10. Lee, S.H., Aggarwal, B.B., Rinderknecht, E., Assisi, F., and Chiu, H. (1984) The synergistic antiproliferative effect of γ-interferon and human lymphotoxin. J. Immunol. 133:1083.

11. Sugarman, B.J., Aggarwal, B.B., Hass, P.E., Figari, I.S., Palladino, M.A., Jr., and Shepard, H.M. (1985) Recombinant human tumor necrosis factor-α effects on proliferation of normal and transformed cells _in vitro_. Science 230:943.

Protein Purification: Micro to Macro, pages 27–34

SEPARATION, ISOLATION, AND SEQUENCING OF POLYPEPTIDES FROM RODENT PITUITARY AND WHOLE BRAIN BY TWO-DIMENSIONAL GEL ELECTROPHORESIS COUPLED WITH PRESSURE EXTRACTION.

James D. Pearson[†,*], Daryll B. DeWald[†], and Jerry R. Colca[‡]

Departments of Biotechnology Research[†] and Diabetes and Gastrointestinal Diseases Research[‡], The Upjohn Company, Kalamazoo, MI 49001.

ABSTRACT A procedure for resolving polypeptides on non-urea two-dimensional polyacrylamide gels has been coupled with pressure extraction onto a short HPLC column to allow collection for direct sequencing. The system is called "LC-gel elution." Examples are given for two-dimensional gel mapping and subsequent partial sequencing of rodent polypeptides. This technique allows for proteinaceous components from crude extract to be rapidly isolated without time consuming chromatographic procedures. Only a single, routine reversed-phase gradient is required to recover the purified polypeptide. The method is ideal for micro-scale preparations where minimal sample is available.

INTRODUCTION

We are interested in structural elucidation of a wide variety of naturally occurring polypeptide regulatory molecules in the 2kD to 30kD molecular weight range. To achieve rapid isolation of such molecules a non-urea polyacrylamide two-dimensional gel system was previously devised (1). Various methods of extracting polypeptides from this gel matrix have been tried in our laboratory, but limitations have existed in each case. Passive diffusion in ammonium bicarbonate buffer is a slow process and recoveries are commonly low. Electroelution is extremely slow, requiring

*Author to whom correspondence should be addressed.

a couple of days to prepare sample for sequencing. Our experience with the electroblotting procedure (2) has shown that small polypeptides are difficult to evenly transfer onto amine-modified glass filter paper, and readily pass through the matrix.

An alternative technique termed "LC-gel elution" has recently been reported (3) whereby polypeptide contained in a band or 2-D gel spot is excised, minced,placed in a holding cartridge, and then pressure extracted out of the gel by an HPLC pump. Recovery is fast and simple because the polypeptide is adsorbed onto a reversed-phase HPLC matrix and then desorbed by common, volatile reversed-phase solvent gradient elution. The rationale for use of ultra-short reversed-phase cartridges for increasing recovery of micro-scale samples has been recently examined (4). The utility of LC-gel elution from 2-D gels for sequence analysis of rodent brain and pituitary polypeptides is presented in this work.

METHODS

Electrophoresis

Two-dimensional electrophoresis was performed by procedures already described (1). Briefly, the SDS-PAGE system consisted of 9-26% acrylamide and 0-8% glycerol with no urea. After electrophoresis polypeptides were visualized in 0.25M KCl plus 1mM DTT at 4C for 5 min (5), or stained with Coomassie brilliant blue. The KCl method was used when a spot contained more than 10ug, otherwise Coomassie staining

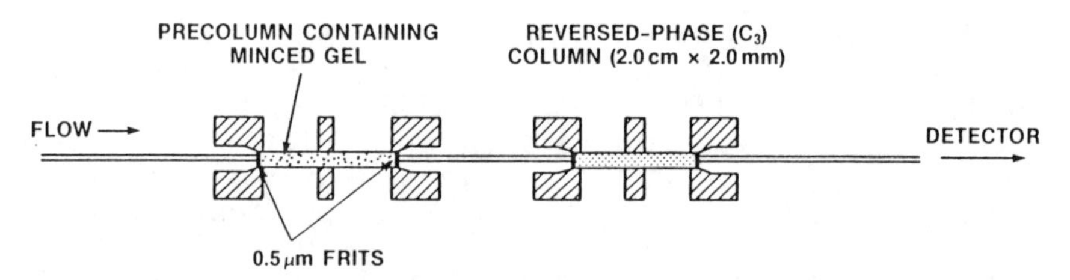

FIGURE 1. Configuration of gel precolumn chamber and HPLC cartridge for "LC-gel elution" technique. See Materials and Methods for details. Reprinted from Ref. 3 with permission.

was used to locate polypeptides. Spots were cut out of gels and kept in capped polypropylene tubes until the liquid chromatography elution step was performed.

LC-gel Elution

Protein spots in gels were sliced into cubic pieces of about 1mm, and hand packed into a precolumn chamber (2mm x 2cm) with a spatula. This chamber was then plumbed in front of the reversed-phase cartridge as illustrated in Fig. 1. 0.1% TFA was pumped through the system to remove gel contaminants, then a 0-80% acetonitrile gradient with 0.1%TFA was passed through the gel chamber and column to achieve polypeptide elution. Not shown in Fig. 1 is the contaminant containing void volume peak which elutes prior to the beginning of the gradient.

MATERIALS

Ribonuclease A, adrenocorticotropin, and myoglobin were purchased from Sigma. Fresh mouse pituitary and rat forebrain were placed in 2M acetic acid containing 10mM PMSF, homogenized, extracted with methylene chloride, and clarified by C-18 Sep-Paks prior to 2-D gel electrophoresis. HPLC was performed on a Varian 5000 connected to a Kratos model 773 detector (flow cell 2.4 ul). Sequence analysis was done on a model 470A protein sequencer equipped with an on-line model 120A PTH amino acid analyzer (Applied Biosystems). Acetonitrile and water were HPLC grade. HPLC/spectro grade trifluoroacetic acid was used (Pierce Chemical Co.). The C-3 reversed-phase support was prepared using Vydac silica by a previously described procedure (6). Empty 2mm x 2cm precolumns were purchased from Anspec (Ann Arbor, MI) and used for both the gel chamber and short HPLC cartridge. The HPLC cartridge was slurry packed at 9000 psi in 2-propanol in an upward fashion.

RESULTS AND DISCUSSION

The LC-gel elution system illustrated in Fig. 1 shows how the pressurized solvent flows from an HPLC pump to the precolumn chamber and extricates the polypeptide contained in the gel. The polypeptide then adsorbs onto the reversed-phase cartridge while small molecular weight contaminants flow through the reversed-phase bed. The adsorbed polypeptide is then eluted with a simple 0.1% TFA/acetonitrile gradient, and channeled through a flow cell for detection and subsequent collection for direct sequencing. In one example, extract from one rat forebrain was divided in half, run on two 2-D gels, spots excised, combined, and LC-gel eluted. Fig. 2

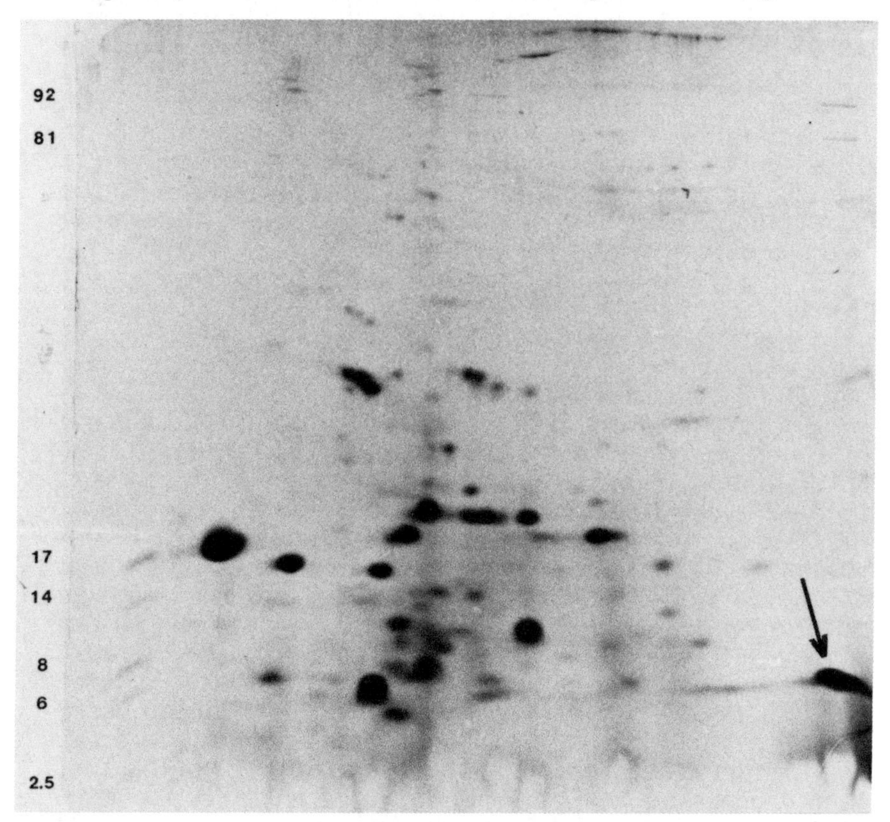

FIGURE 2. 2-D gel profile of extract from one rat forebrain. Gel conditions: isoelectric focusing mode with wide-range ampholines as previously described (1), acidic to basic (left to right); Sizing gel, 9-26% acrylamide, 0.18-1.05% Bis, 0-8% glycerol, with no urea. Gel was Coomassie stained. Arrow points to spot that was LC-gel eluted (see Methods) and identified as ubiquitin (Table 1). Molecular weight markers (MW x 10^3) indicated at left. Reprinted from Ref. 3 with permission.

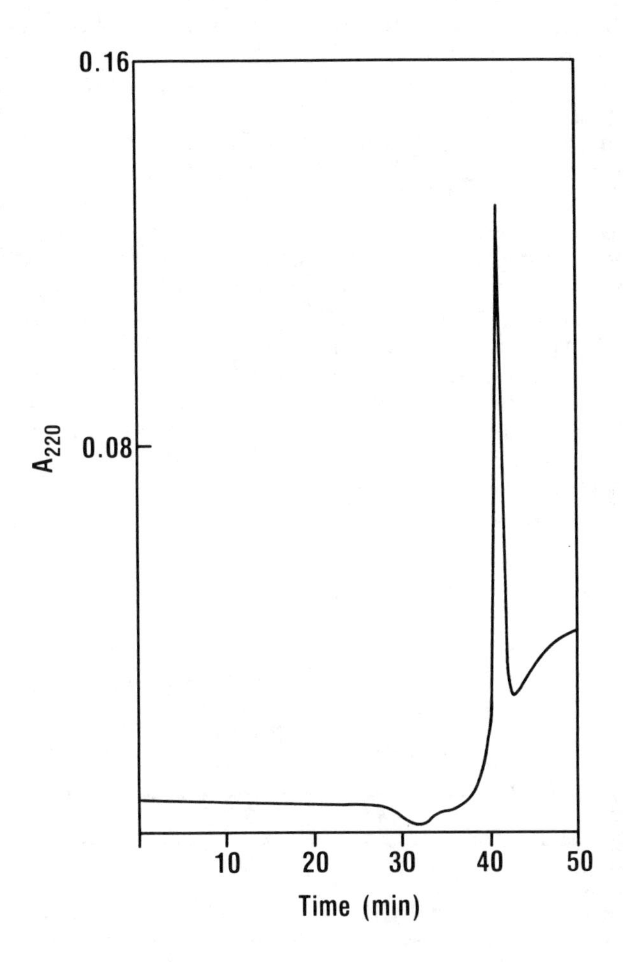

FIGURE 3. Chromatogram for LC-gel elution profile of rat brain ubiquitin. Protein gel spot was extracted directly from a 2-D gel as shown in Fig. 2 and sequenced after lyophilization to remove the volatile reversed-phase solvent components.

shows the Coomassie stained 2-D gel of the extract in which the selected spot was excised, and Fig.3 shows the LC-gel elution chromatogram. It should be noted that prior to starting the gradient (not shown) in Fig. 3., a gel contaminant containing void volume peak is first cleared by pumping 0.1%

TFA through the system until the detector gives a flat baseline, then the polypeptide is eluted in a gradient manner. The first twelve residues of N-terminal sequence analysis identified the spot in Fig. 2 as ubiquitin (Table 1). In another example, twenty mouse pituitary glands were homogenized, divided and run on four 2-D gels, spots excised, combined, and LC-gel eluted. Fig. 4 shows this 2-D gel spot that was identified as somatotropin. Table 1 summarizes N-terminal sequence data obtained for ubiquitin, somatotropin, and other polypeptides sequenced via 2-D gel isolation with subsequent LC-gel elution.

In conclusion, methods for high resolution polypeptide 2-D gel analysis and a simple polypeptide extraction technique are presented. The LC-gel elution method takes about an hour and yields purified polypeptides in a manner analogous to common TFA/acetonitrile HPLC isolations. Other than an HPLC and detector, supplies to construct the configuration shown in Fig. 1 may be purchased at minimal cost.

Table 1

PROTEIN	MOLECULAR SIZE (daltons)	AMOUNT LOADED ON GEL (nmol)	#CYCLES / #RESIDUES SEQUENCED*
ACTH	4,541	2.2	10 / 10
Ribonuclease A	12,600	0.79	10 / 10
Myoglobin	16,900	0.59	10 / 10
Mouse pituitary somatotropin	23,000	N.D.	16 / 18
Rat Brain Ubiquitin	8,500	N.D.	12 / 12

* No attempt was made to sequence more than 10 cycles of LC-gel eluted standards. Average repetitive yields were 93%. Total recovery starting from initially applying samples onto first-dimensional IEF gels to final calculation from sequence analysis yields: ACTH, 17%; ribonuclease A, 2.6%; myoglobin, 12%.

N.D. not determined

Some data in this table are reprinted from reference 3 with permission.

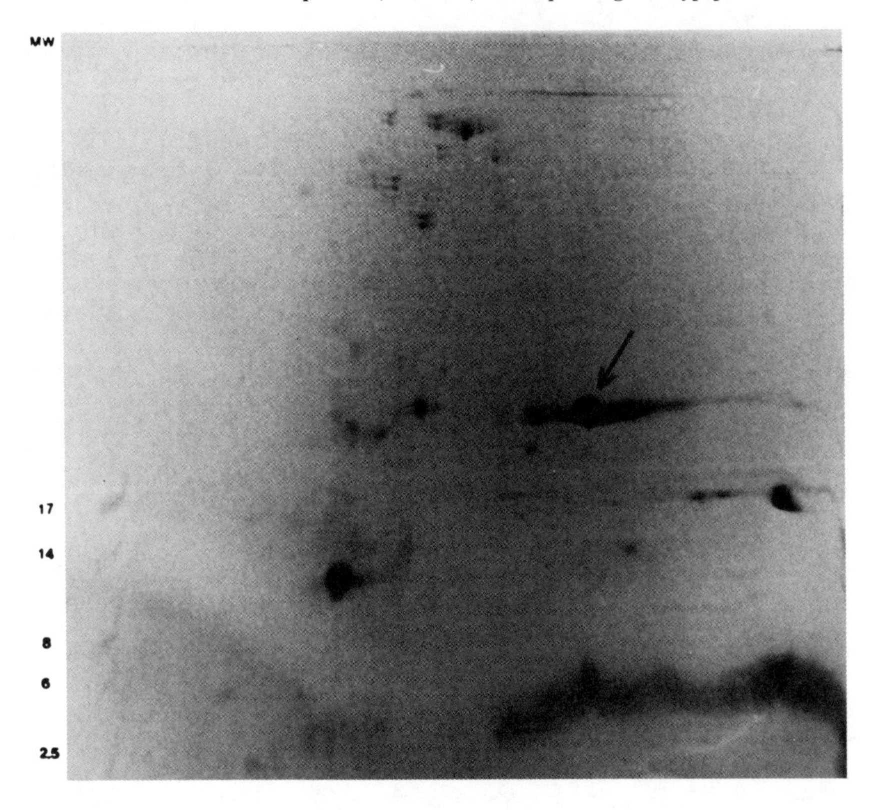

FIGURE 4. 2-D gel profile of extract from five mouse pituitary glands. Same gel conditions as in Figure 2. Arrow points to spot identified as somatotropin from partial N-terminal sequence (Table 1).

ACKNOWLEDGEMENTS

The authors thank Dr. Bob Heinrikson and Heidi Zurcher-Neely for gas-phase sequence analysis.

REFERENCES

1) DeWald DB, Adams LD, and Pearson JD (1986). Wide-Range Sodium Dodecyl Sulfate-Polyacrylamide Gel Electrophoresis; Low-Molecular Weight Polypeptides of

M_r 2,000 to M_r 200,000 Proteins; Applications in Two-dimensional Gel Electrophoresis. Anal Biochem 154:502-508.

2) Aebersold RH, Teplow DB, Hood LE, and Kent SBH (1986). Electroblotting onto Activated Glass. J Biol Chem 261:4229-4238.

3) Pearson JD, DeWald DB, Zurcher-Neely HA, Heinrikson RL, and Poorman RA (in press). Sequencing Proteins from Two-dimensional Gels Using a Liquid Pressure Extraction Technique. In Proceeding of the 6th Intl Conf on Methods in Protein Sequence Analysis.

4) Pearson JD (1986). High Performance Liquid Chromatography Column Length Designed for Sub-microgram Scale Protein Isolation. Anal Biochem 152:189-198.

5) Hager DA and Burgess RR (1980). Elution of Proteins from Sodium Dodecyl Sulfate-polyacrylamide Gels, Removal of Sodium Dodecyl Sulfate, and Renaturation of Enzymatic Activity: Results with Sigma Subunit of *Escherichia coli* RNA Polymerase, Wheat Germ DNA Topoisomerase, and Other Enzymes. Anal Biochem 109:76-86.

6) Pearson JD, Lin NT, and Regnier FE (1982). The Importance of Silica Type for Reversed-Phase Protein Separations. Anal Biochem 124:217-230.

Protein Purification: Micro to Macro, pages 35–47

INSTRUMENT AND SAMPLING OPTIMIZATION FOR MICROSAMPLE ANALYSIS

Kenneth J. Wilson, David H. Hawke, and Pau M. Yuan

Applied Biosystems, Inc., 850 Lincoln Centre Dr., Foster City, CA 94404

ABSTRACT The isolation and chemical analysis of biological samples are demanding procedures often complicated by the limited availability of starting materials. Conventional sample handling techniques frequently add problems such as poor recoveries at the ug level and extensive dilution of the desired sample. The need for concentration, dialysis or lyophilization subsequently increases the risk of chemically modifying, contaminating and/or losing the sample. Each of these methods obviously limits attainable sensitivities and the quantitative value of the data.

Narrow and microbore HPLC has proven useful in analyzing and isolating polypeptides at or below the ug level. Samples present in relatively large volumes can be conveniently concentrated and recovered in small volumes. Similarly, chemical modifications in denaturants, buffers, and/or reagents can be monitored and the final derivatives easily recovered. The utility of reduced column diameters and lengths improves overall yields, especially for more hydrophobic proteins and peptides.

INTRODUCTION

As more "interesting" proteins become scarce, protein chemists require new high-sensitivity techniques commensurate with the lower available quantities. The elaborate isolations and studies carried out on interleukins, colony stimulating factors, interferons and tumor necrosis factors, as well as many of the oncogene products, are excellent examples where only limited starting quantities were available. The methods used for isolating such substances required "optimization" to achieve both high recoveries and increased detection sensitivities. Optimization, in turn, provided the means for carrying out preparative isolations, as well as analytical characterizations, on ever decreasing amounts of starting material.

The two "basic" techniques used for microsample analysis are high-performance liquid chromatography (HPLC) and polyacrylamide gel electrophoresis (PAGE). Due to their broad dynamic ranges, ca. 10^3-10^4, both methods are useful in both analytical and preparative applications. Sample recoveries from these methods, however, differ greatly. Methods for recovering samples from gels have relied on passive elution, electoelution/dialysis or electroblotting. The latter, electroblotting (1,2), has recently attracted considerable attention because transfers are made onto solid supports, relatively easy to carryout, and yields samples amenable to subsequent chemical characterizations (amino acid analysis, primary sequencing, and peptide mapping). Samples from HPLC separations are collected directly. By optimizing the instrumentation through reduction of column ID a 10 to 15-fold improvement in detection can be achieved (3). This, in turn, allows for the tandem arrangement of the two techniques, eg. an HPLC isolation followed by PAGE, or the reverse (4).

Conventional sample handling techniques exhibit poor recoveries at the ug level and are therefore inappropriate for many applications. In addition to increasing the risk of chemically modifying, contaminating or losing the sample, these methods limit the attainable sensitivities and the quantitative value of the data. This report covers methods for concentrating, desalting and chemically modifying, ug amounts of polypeptides in solution as well as on solid-phase surfaces, all compatible with micro sample analysis.

METHODS

Chromatography was carried out with a prototype Model 130A separation system (Applied Biosystems) equipped with a 2.4 uL flow cell. All columns were from Brownlee Labs; protein standards were from Applied Biosystems.

RESULTS AND DISCUSSION

Microbore Optimization and Sample Collection

In order to carryout gradient elutions at low flow-rates optimized chromatographic hardware is essential. The solvent delivery system must be capable of virtually pulse-free pumping at low flow-rates (1-1000 uL/min) and medium pressures (10 to 5000 psi), while providing high flow-rate precision and gradient accuracy. At present only the dual syringe pump developed by Schwartz and Brownlee (5) meets these requirements. When combined with an appropriately sized dynamic mixer (200 to 400 uL volume), a detector insensitive to the refractive index changes, with low noise and drift, and tubing and connections that minimize extra-column dispersion one has a system capable of high sensitivity microbore chromatography.

Increased HPLC mass sensitivities can be achieved through reduced flow-rates and smaller column diameters (3). By appropriately adjusting these variables (Table 1) peak volumes ranging from 200 uL down to 15 uL on normal, narrowbore and microbore columns, respectively, are possible. These elution volumes are significantly smaller than the 0.5 to 2 mL ranges typical with 4.6 x 250 mm ID columns operated at 1-2 mL/min flow-rates. Samples collected in larger volumes usually require concentration by lyophilization, or evaporation, before further characterization can be performed. Sample insolubility may then be a problem, one that seriously reduces overall recovery.

TABLE 1
PEAK ELUTION VOLUMES AS A FUNCTION OF FLOWRATE AND COLUMN DIAMETER[1]

Column Size (mm)	Flow-rate (uL/min)	Peak Volume (uL)
4.6 x 100	400	210
	200	120
	100	75
2.1 x 100	200	90
	100	55
	50	35
1.0 x 100	100	35
	50	22
	20	16
1.0 x 50	100	45
	50	26
	20	12

[1]All columns were packed with Aquapore RP-300 support, a C_8 bonded phase on a 30-nm pore, 7 um support from Brownlee Labs. Peak volumes were determined by chromatographing a β-lactalbumin and carbonic anhydrase and averaging the calculated elution volumes. See reference (3) for further details.

Small elution volumes, and consequently higher sample concentrations, can be collected into micro-centrifuge tubes (vials) or onto glass fiber filters. In the latter instance liquid is soaked into the filter (held or suspended in a

1.5 mL Eppendorf tube), dried at 40-50°C, and then stored sealed at -20°C under argon. Storage of liquid samples simply requires maintenance at -20°C.

Collection by either of these methods requires that physical contact be made between the tubing (stainless steel or fused-glass capillary, 0.005 in ID) through which the effuent exits the chromatographic unit and the surface onto (or into) which it is to be collected. At flow-rates of 25 uL/min, evaporation rates of the aqueous-organic mixtures used are substantial and often equal to the chromatographic flow rate. Maintaining the samples in solution, rather than lyophilized or dried, assures higher recoveries and minimizes contamination. Keeping dried samples in an argon atmosphere and at low temperatures minimizes oxidations that can occur on glass surfaces (Trp alterations have been noted (6)).

An Isolation Example: Human Saliva Proteins

Frequently the amount of a particular sample available determines the method(s) used for its characterization. Although human saliva is usually available in mL quantities from single individuals, one requires only uL volumes to carryout both comparative mapping and isolation of specific proteins. Figure 1 illustrates the saliva protein pattern of 6 individuals. The specific, as well as relative, concentrations of many of the proteins vary greatly from person to person. All samples were identically collected (7) and only 50 uL of the acidified, filtered solution chromatographed. Considering the low sensitivity detector settings required for these separations, ie. 0.5 AUFS at 215nm, analytical comparative mapping could have been done with as little as 5 to 10uL starting volume.

Each of the numbered peaks (1 to 5) were preparatively isolated and amino terminally sequenced by a gas-phase or pulsed-liquid sequencer (automated Edman degradation). The resultant sequences, when compared to those in the Protein Data Bank (8), identified peaks 1 to

4 as the proteins statherin, lysozyme, cystatin SN and cystatin S, respectively. Peak 5 was resistant to N-terminal degradation and later identified as α-amylase from the sequences of several fragments generated by tryptic digestion and microbore peptide isolation (9).

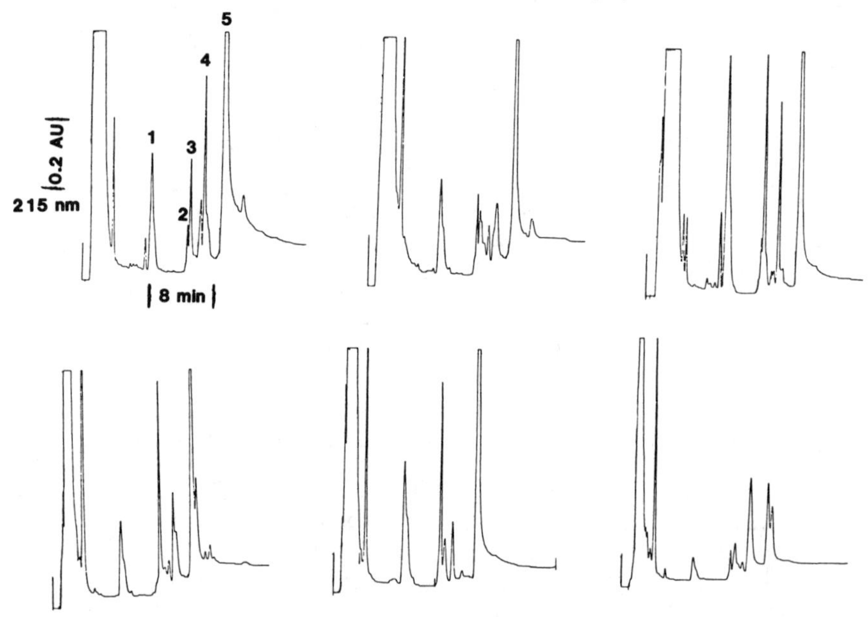

Figure 1. HPLC mapping of the hydrophobic proteins in individual human saliva samples. Injections (50 uL) were performed on acidified and filtered samples prepared according to Methods. Chromatography was carried out on an Aquapore RP-300 column (2.1 x 30 mm) at room temperature using a flow-rate of 150 uL/min. A 30 min linear gradient from 35 to 85% B used 0.1% TFA as A buffer and the combination 0.085% TFA/70% acetonitrile as B buffer. Peak indentifications (see upper, right tracing):1, Statherin;2, Lysozyme; 3, Cystatin SN; 4, Cystatin S ; 5, α-amylase.

In comparison to the original structures determined for cystatins S and SN, N-terminal extentions of 8 residues were found. These results suggest that proteolytic processing might have occurred during their respective isolations prior to the original structural determination (10,11).

The protein recoveries (Table 2) were more than sufficient to determine identity. Given that 45% of the cystatin S (110 residues) sequence was identified using only 2ug of material, working with chemically or proteolytically derived fragments from a few more ugs would complete the comparison.

TABLE 2
RECOVERIES OF HUMAN SALIVA PROTEINS FROM NARROWBORE RP-HPLC[1]

Peak #/ identity	Amount recovered (pmol)[2]	Number residues identified/ sequence
1/statherin	30.1(D)	29/30
2/lysozyme	37.5(K)	37/40
3/cystatin SN	60.9(W)	50/53
4/cystatin S	25.0(S)	51/51
5/α -amylase	?	NA[3]

[1]A 50 uL aliquot was chromatographed as described in Fig. 1 and the individual peaks collected.

[2]Amounts recovered are expressed as initial yields of the N-terminal PTH-amino acid (identified in parenthesis). In the case of α-amylase this value could not be determined since the protein was resistant to degradation.

[3]NA, not applicable since the protein identified by sequencing numerous tryptic peptides isolated by microbore mapping.

Micro-Methodology for Peptide Mapping

Carrying out mapping experiments on ug

quantities in an aliquot from a larger volume is quite different from performing the digestions on a smaller amount and having to chromatograph the entire sample. The need to chemically modify the sample prior to fragmentation complicates the undertaking even further. Reductions, alkylations, oxidations and blockage reactions all require buffer-reagent combinations, as well as the frequent use of a denaturant, that need to be removed prior to the fragmentation step.

Sample desalting and/or concentration is easily performed by simple chromatography on the appropriate support using reduced column sizes. As shown by Nice et. al. (12) and Pearson (13), samples can be recovered in high yields (>90-95 %) in minimal volumes from columns of 2mm ID and lengths of 3 cm or less. Figure 2 illustrates the purification (desalting/concentration) of pyridylethylated cystatin S. In a single step the sample was recovered in approx. 100 uL, free of contaminants from the modification reaction, and ready for further fragmentation after removal of the organic eluent.

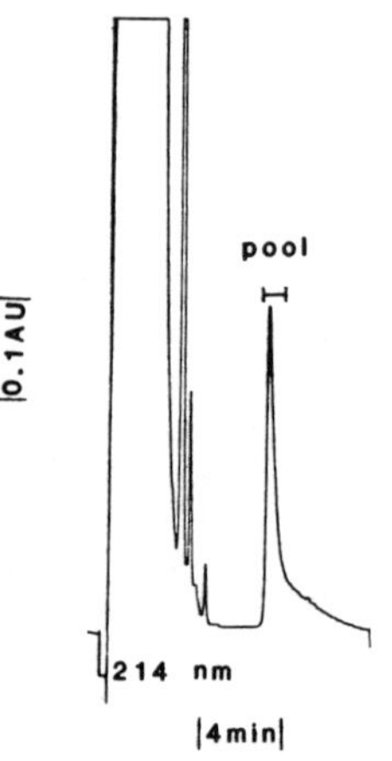

Figure 2. Sample desalting/concentration. Cystatin S (2 ug) was alkylated with 4-vinylpyridine and chromatographed on an Aquapore RP-300 column using the buffer system given in Fig. 1. The pyridylethylated derivative was injected, washed for 5 min with A buffer, and then eluted with B buffer.

In a second example, the PE-derivatives of β-lactoglobulins A and B were recovered as outlined above, carefully lyophilized and digested with trypsin prior to comparing their peptide maps by narrowbore HPLC (Figure 3). The peak designated as number 1 in the upper chromatogram contains the variant differences between the two polypeptides. The salient feature of the above series of manipulations is that all reactions were carried out in polypropylene tubes and that the drying/lyophilization step was carried out under closely monitored conditions. The bottom line is that by eliminating/minimizing handling overall results will improve.

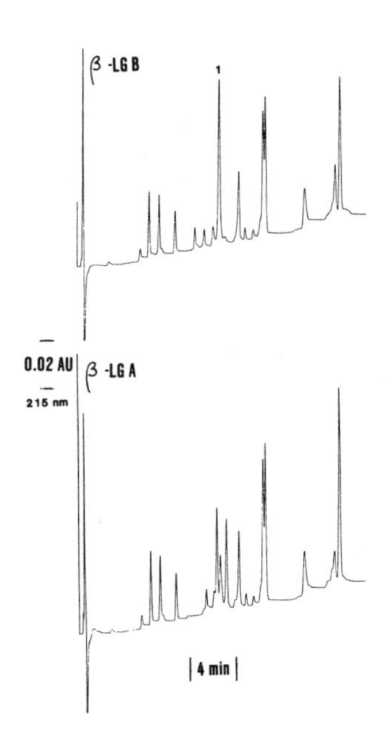

Figure 3. Enzymatic digestions of β-lactoglobulin A and B variants. S. aureus V8 protease cleavages were carried out and 100 pmol equivalents chromatographed on an Aquapore RP-300 column (2.1 x 30 mm) at 200 uL/min. A linear gradient at room temperature from 0 to 80% B buffer was used for peptide elution (see Fig. 1 for buffer compositions).

Solid-Phase Surfaces for Sample Manipulation

Table 3 indicates surfaces that are currently utilized in various sample manipulation applications. The glass fiber disc has been used for a number of years as a support medium for samples being subjected to gas-phase or pulsed liquid-phase sequencing. PVDF membranes have only recently been used for this application (14); nitrocellulose is soluble in some of the organic solvents and therefore unsuitable. Excluding untreated glass fiber filters, all of the materials are applicable to trapping electroblotted samples. Only the various glass fiber variations have been employed for direct sample collection.

TABLE 3
SOLID-PHASE SURFACES UTILIZED FOR
MICRO-MANIPULATION OF PROTEIN/PEPTIDE SAMPLES

Surface	Uses		
	sequencing	electro-blotting	collection
Glass fiber			
* untreated	Y	N	Y
* polybrene containing	Y	Y	Y
* QA- or AP-derivatized[1]	Y	Y	Y
Nitrocellulose	N	Y	?
PVDF Membranes	Y	Y	?

[1]Abbreviations: QA, quarternary amine; AP, aminopropyl.

As suggested above, results can be improved by eliminating handling steps. Some of the chemical reactions can, in fact, be carried out on the absorbed sample (Table 4). Those that have found utility are alkylation of cysteines, CNBr cleavages at methionine, and blockages of primary but not secondary amines. This latter reaction is frequently used to reduce the background produced during Edman sequencing by non-specific acid cleavage of the polypeptide chain. Obviously this is only useful when sequencing is interrupted at a proline residues and the blockage reaction performed.

TABLE 4
CHEMICAL AND ENZYMATIC PROTEIN MODIFICATIONS CARRIED OUT IN SOLID-PHASE GLASS SURFACES

Chemical	Reference
pyridylethylation-cysteinyl residues	15
cyanogen bromide cleavage at methionyl residues	16
ortho-phthalaldehyde blockage	17
Enzymatic	
trypsin	see text

Enzymatic hydrolyses can also be employed to fragment proteins on solid surfaces. Although the reaction kinetics are quite different than in solution, digestions will occur on the glass fiber support used in sequencing. Approximately 30% and 2% digestion of cyctochrome c occurs in the absence and presence of polybrene, respectively.

In summary, the methods or techniques one uses for handling ug quantities of proteins directly influence results. If "optimized" instrumentation is employed, higher detection sensitivities are achieved and one can carryout multistep isolations starting with only mg amounts of material.

REFERENCES

1. Vanderkerckhove J, Bauw G, Puype M, Van Damme J, Van Montagu M, (1985). Protein-Blotting on Polybrene-Coated Glass-Fiber Sheets. Eur J Biochem 152:9.
2. Aebersold RH, Teplow DB, Hood LE, Kent SBH (1986). Electro-blotting onto Activated Glass. J Biol Chem 261:4229.
3. Schlabach TD, Wilson KJ (1987). Microbore Flow-Rate and Protein Chromatography. J Chromatogr 385:65.
4. Yuen S, Hunkapiller MW, Wilson KJ, Yuan PM (1987). Applications of Two Dimensional Microbore LC and PAGE Electroblotting in Microsequencing Analysis. Submitted for publication.
5. Schwartz HE, Brownlee RG (1984). A Dual Syringe LC Solvent Delivery System for Use with Microbore Columns. Am Lab 16:43.
6. Yuan PM, Wilson KJ (1987). Unpublished results.
7. Hawke DH, Yuan PM, Wilson KJ, Hunkapiller MW (1987). Indentification of a Long Forum of Cystatin from Human Saliva by Rapid Microbore HPLC Mapping. Biochem Biophys Res Comm, in press.
8. Protein Identification Resource, National Biomedical Research Foundation.
9. Yuan PM, Hawke DH, unpublished results.
10. Isemura S, Saitoh E, Ito S, Isemura M, and Sanada K (1984). Cystatin S: A Cysteine Proteinase Inhibitor of Human Saliva, J Biochem (Japan) 96:1311.
11. Isemura S, Saitoh E, and Sanada K (1986). Characterization of a New Cysteine Proteinase Inhibitor of Human Saliva, Cystatin SN, which is Immunologically Related to Cystatin S. FEBS Letters 198:145.
12. Nice EC, Grego B, Simpson RJ (1985). Application of Short Microbore HPLC "Guard" Columns for the Preparation of Samples for Protein Microsequencing. Biochem Intl 11:187.

13. Pearson JD (1986). High-Performance Liquid Chromatography Column Length Designed for Submicrogram Scale Protein Isolation. Anal Biochem 152:189.
14. Matsudaira P (1987). Sequence From Picomole Quantitites of Protein Electroblotted Onto Polyvinylidene Difluoride Membranes. Submitted for Publication.
15. Andrews PC, Dixon JE (1987). A Procedure for In Situ Alkylation of Cystine Residues on Glass Fiber to Protein Microsequence Analysis. Anal Biochem 161:524.
16. Simpson RJ, Nice EC (1984). In Situ Cyanogen Bromide Cleavage of N-Terminally Blocked Proteins in a Gas-Phase Sequencer. Biochem Intl 8:787.
17. Bond MW (1985). Detailed Protocols for Extended Sequence Determinations: A Method for Recombinant Proteins. Paper 109, Sym Am Protein Chem, San Diego, CA.

Protein Purification: Micro to Macro, pages 49–73

PURIFICATION OF SYNTHETIC PROTEINS

David D.L. Woo
and Steven B.H. Kent*

Department of Medicine, UCLA
Los Angeles, California 90024
and
*Division of Biology,
California Institute of Technology,
Pasadena, California 91104

ABSTRACT Total chemical synthesis of small proteins (up to 20,000 daltons) and their analogs is a powerful and practical research approach to structure-function analysis of proteins. On rare occasions, synthetic proteins will spontaneously assume their active forms after release from the solid-phase synthesis resin. More often, they are inactive immediately after synthesis and initial isolation. They require deliberate refolding to produce the active forms. Refolding methods used will vary depending on preferences. We have learned and applied modern principles of refolding, criteria for purity and methods of analysis from our recent success in obtaining sufficient active synthetic human transforming growth factor-alpha (TGF-α) for structure-activity analysis.

Human TGF-α was synthesized on a fully automated peptide synthesizer (Applied Biosystem, 430A) reprogrammed to use a double coupling protocol. Preformed symmetrical anhydride of protected amino acids (t-Boc AA's) in dimethylformamide was used in the first coupling, followed by in situ activation of t-Boc AA's in dichloromethane during the second coupling step. Using the technique of quantitative Edman degradation (preview sequencing), the yield of full length target sequence was 84.5%, representing an average coupling yield of 99.65% per residue. The peptide was then cleaved and deprotected. This sample, when analyzed on HPLC (Vydac C4 column), showed a single major peak, representing 44% of the total product. The major peak was isolated using preparative HPLC. The purified peptide had no biological activity at this stage and was refolded under controlled oxidative conditions in guanidinium hydrochloride. After refolding, several forms of TGF-α were apparent. They were separated from each other by preparative HPLC and their activities measured. One peak was found to have activities indistinguishable from isolated natural murine EGF in receptor binding, mitogenic and soft agar colony formation assays. The purified product focussed as a single band at pI = 6.2 on Immobiline gels, and displayed a MW = 5,546.2 (The. = 5546.3) by mass spectrometry.

INTRODUCTION

A fundamental goal of protein biochemistry is to understand, at the molecular level, how the activity of a protein is reflected in its three-dimensional arrangement of amino acids. A related problem is to understand how the information for the activity of a protein and its three-dimensional structure are encoded by its linear amino acid sequence. Studies designed to gain insight into these problems traditionally rely on detailed analysis of structure-function relations of model proteins such as hemoglobin and its numerous naturally occurring mutants. With the advent of molecular cloning, in-vitro mutagenesis and high level protein expression systems, combinations of these techniques have been employed to experimentally produce targeted and site-saturated analogs for detailed structure-function studies in model protein systems where mutants and analogs are difficult to identify or do not occur naturally at great frequencies. Significant progress has been achieved in several systems including subtilisin and dihydrofolate reductase. Undoubtedly, these techniques will find applications in many other systems in the near future.

Detailed structure-function analysis of small biologically active peptide and their analogs have been extensively studied using methods of chemical peptide synthesis in the past fifteen years. Studies on the encaphalin, the bradykinin and the angiotensin families of peptides are some examples of many outstanding analyses. In these studies, literally thousands of peptide analogs with various amino acid sequences replaced by chosen amino acid substitutes are produced by total chemical synthesis. The desired analogs are then purified and their properties carefully analyzed and compared to the native peptide.

By far the techniques of solid-phase peptide synthesis have been used most extensively in these studies. Modification, refinements and optimization on the solid-phase synthesis protocols have been accumulating over the years. However it is not until very recently that these accumulated improvements have been integrated and fully implemented in a reliable and versatile commercial instrument that automates the chain assembly process of solid-phase peptide synthesis (1,2) making chemical synthesis a practical technique to study structure-function analyses of not just peptides but also small proteins.

In the following, we will briefly describe the principle of solid-phase peptide synthesis, list the key features of the synthetic protocol currently adopted in our laboratory for chemical synthesis of proteins and peptides, discuss strategies for purification of synthetic proteins and illustrate the process with the synthesis and purification of active human transforming growth factor-alpha (TGF-α).

SOLID PHASE SYNTHESIS OF PEPTIDES AND PROTEINS

Synthesis of peptides on a solid phase polymeric support was originally conceived by

Merrifield more than twenty years ago (3). The underlying strategy employed in solid phase peptide synthesis, outlined in Figure 1, has remained essentially unchanged as the techniques are practiced today with improved and optimized reaction conditions.

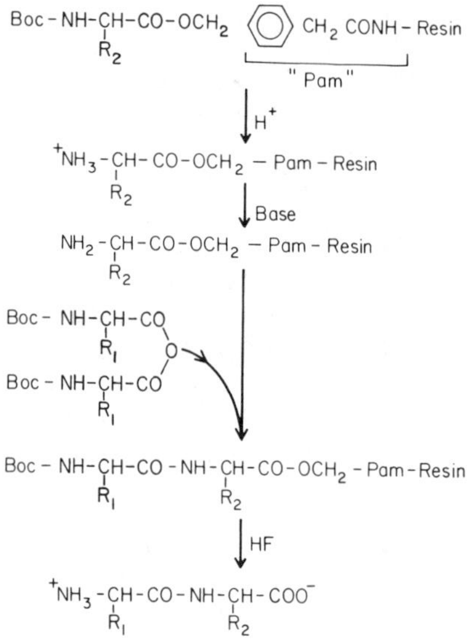

Boc-NH-CH(R_2)-CO-OCH_2-C_6H_4-CH_2 CONH-Resin ("Pam")

↓ H^+

$^+NH_3$-CH(R_2)-CO-OCH_2 – Pam – Resin

↓ Base

NH_2-CH(R_2)-CO-OCH_2 – Pam – Resin

+ (Boc-NH-CH(R_1)-CO)$_2$O

↓

Boc-NH-CH(R_1)-CO-NH-CH(R_2)-CO-OCH_2-Pam-Resin

↓ HF

$^+NH_3$-CH(R_1)-CO-NH-CH(R_2)-COO^-

FIGURE 1. Schematic of steps in solid-phase peptide synthesis.

The procedures consist of a stepwise "chain assembly" followed by "deprotection" of the assembled chain. The "chain assembly" stage of solid phase peptide synthesis begins with the C-terminal of a peptide and proceeds toward the N-terminal by stepwise addition of respective amino acids. The reaction begins with an insoluble beaded polymeric resin onto which the carboxyl-terminal amino acid, protected as an α-tertiary butyloxycarbonyl (t-Boc) derivative, is covalently linked. This α-amino protecting group is removed by treatment with trifluoroacetic acid in dichloromethane. The resulting α-amine salt is neutralized with diisopropyl ethyl amine, a tertiarly amine, to provide a free amine to react with the activated carboxyl group of an incoming Boc-amino acid, resulting in a dipeptide on the solid segment. These steps are repeated for each of the succeeding amino acids until the desired protected peptide chain is formed. The peptide is then cleaved from the solid support and the side chain protecting groups removed (deprotected) by treatment with strong anhydrous hydrofluoric acid, HF.

Recently, substantial advances in both the chemistry of chain assembly on solid supports and removal of protecting groups have been achieved. Aspects of these improvements have been reviewed in detail (4).

KEY FEATURES OF SYNTHETIC CYCLES USED FOR THE SYNTHESIS OF TGF-α

Because of the stepwise nature of the synthesis, it is essential that a near-quantitative yield be achieved at each step. All chemical reactions must be driven to completion and any side reactions reduced to extremely low levels.

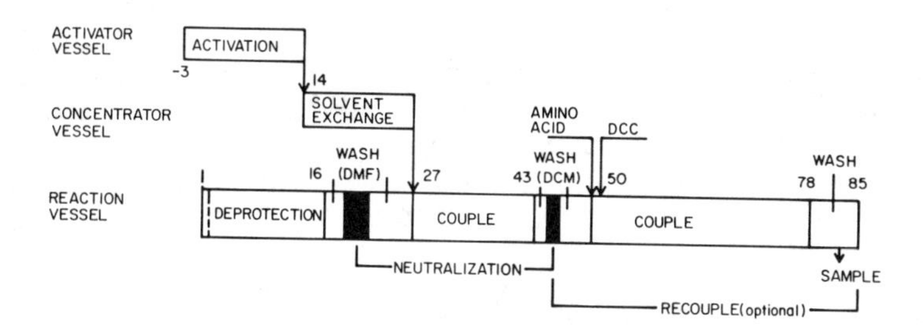

FIGURE 2. Graphic outline of the automated chain assembly cycles use for TGF-α synthesis.

The synthetic cycle involved in the addition of each amino acid is developed at Caltech by modifying and optimizing every step of the automatic chain assembly program supplied with the commercial peptide synthesizer and is shown in figure 2.

Deprotection Cycle.

The N-Boc group was removed by two treatments with trifluoroacetic acid in dichloromethane. Because of the rapid rate of removal of the N-Boc group under these pseudo-first order acidolysis conditions (estimated half-time <15 seconds), the first treatment removes more than 90% of the Boc group and the reactive tert.butyl trifluoracetate formed is rapidly removed from the peptide chain. The second acid treatment ensures complete removal of the N-Boc group. The TFA-dichloromethane solution used in the deprotection is a powerful solvating agent for protected peptide chains because of its strong interactions with peptide bonds. For this reason there is no evidence of sequence-dependent problems in removal of the N-Boc groups during assembly of the peptide chain.

Coupling Cycle.

Sequence dependent problems can also arise during the amino acid coupling step in the assembly of protected peptide chains by stepwise solid phase synthesis. These difficulties have been shown to be due to the tendency of the protected peptide chain to undergo intermolecular aggregation and precipitate. It is now known that protected peptides which are covalently attached to a swollen resin support are more highly solvated and available for reaction than the corresponding peptides in free solution. To reinforce this favorable influence of the resin on the properties of the protected peptide intermediates in the synthesis, it is essential to maximize the solvation of the protected peptide chain at every step of the synthetic cycle. Thus, the neutralization and subsequent washes, and the coupling reaction, were carried out in DMF, an excellent solvent for protected peptides. This has been shown to minimize sequence-dependent coupling problems and give a generally applicable chain assembly chemistry.

Based on this enhanced solubility of resin-bound peptides, the chain assembly was carried out at a substitution of 1 millimole peptide per gram polystyrene resin. This allowed the use of very high concentrations of activated amino acid and growing peptide chain, with only a two-fold excess of activated amino acid. Coupling was carried out using the symmetric anhydrides of the Boc-amino acids. These highly reactive species were preformed in dichloromethane, and the solvent exchanged to DMF. The DMF solution of the Boc-amino acid symmetric anhydride was then added to the neutralized peptide-resin in DMF. Under these conditions, peptide bond formation was very rapid, with typical yields of >99% using coupling times of only 15 minutes.

Because of the 50-residue length of the polypeptide chain, it was essential to achieve near quantitative coupling yield for each amino acid addition. If this is not achieved, the product will be contaminated with significant amounts of peptides with an internal amino acid missing. In many instances, the properties of these deletion peptides will be extremely similar to those of the target peptide, so that the complete purification of the product will be difficult or impossible.

The protocol shown in figure 2 uses a second coupling involving "in situ" activation of the amino acid in the presence of the peptide-resin. In previous work we have found an average of 0.3% increased yield per amino acid by use of this second coupling, from an average coupling yield of 99.3% to a yield of 99.6% per step. For the assembly of the TGFα protected peptide chain, this would result in an increase from 73% to 84% yield of the target sequence and, more importantly, an almost two-fold reduction in the level of deletion peptides (those missing an internal amino acid) from 27% to 16% of the crude product.

Several chemical side reactions that affected the assembly of peptides by stepwise solid phase synthesis in the past were also eliminated by the use of a derivatized resin support that was stable to the conditions of chain assembly and free of extraneous functional groups that cause side reactions. In this way we have avoided chain termination due to trifluoroacetylation, that was caused by the presence of hyoxymethyl groups on the resin or by instability of the peptide-resin linkage. We also minimized the formation of deletion peptides by elimination of the aldehyde groups on the resin that prevented complete coupling by forming Schiff's bases with the alpha-amino moiety of the growing peptide chain. Use of Boc-amino acids free of contamination eliminated another source for the formation of shortened peptide chains during the synthesis. The cumulative effect of these improvements in the chemistry is a reduction in total side reactions from an average of 3-5% per step to less than 0.2% per step. Use of the preformed symmetric anhydride and rapid couplings also minimized the exposure of the dipeptidyl-resin to the free carboxyl group of the amino acid and avoided any detectable loss of peptide from the resin due to diketopiperazine formation, a side reaction which can assume serious proportions.

This optimized chemistry was carried out on a versatile fully automated peptide/protein synthesizer, expressly designed to accommodate the necessary protocols. Thus, the chain assembly was completed in only 75 hours, without operator interruption, and yielded more than 3 grams of protected peptide from which over a gram of crude product was obtained. Automation of the chemistry of chain assembly is essential for the synthesis of the long peptide chains that fold to form proteins for reasons of speed, but even more importantly for reproducibility. This reproducibility is of critical importance if chemical synthesis of long peptide chains is to be used for studies of structure-function relationships in proteins.

STRATEGY FOR PURIFYING SYNTHETIC PROTEIN

By applying the optimized automated chain assembly and improved peptide deprotection procedures outlined above, the crude cleaved product should contain the desired molecule with the correct primary structure represented as the major product (usually greater than 50% by mass). Obtaining maximum synthetic efficiencies in every step is extremely important because contaminating impurities will consist of a complex mixture of terminated peptides, deletion peptides and incompletely deprotected peptides. These side products of synthesis, although low in amounts percentage-wise, will have very similar physical sizes, biological and chemical properties to the desired peptide and it will be difficult to separate them from the desired main product. Thus highly specific separation procedures based on biological or chemical properties such as affinity chromatography using specific antibodies will not work well to purify the desired products. Separations based on size differences such as gel filtration and polyacrylamide gel electrophoresis are also ineffective because of their low resolving power.

The greatest difference between synthetic peptides and synthetic impurities are to be found amongst differences in their physical properties such as charge and hydrophobicity. Two high resolution separation procedures based on differences in molecular charges and hydrophobicities are isoelectric focussing using immobilized pH gradients and reverse-phase HPLC. Both of these techniques have extremely high resolving power and can be scaled up to purify 100 mg aliquots of crude peptides without much technical difficulty. On some occasions, application of one of the two procedures will produce the desired product in homogeneous form, simply by isolating the major peak or band. However, sequential application of HPLC followed by IEF will allow purification of the desired product to homogeneity in most instances. The use of a simple HPLC step to purify chemically synthesized human transforming growth factor-alpha (TGF-α) to homogeneity and the application of IEF to prove its homogeneity is described below.

HUMAN TRANSFORMING GROWTH FACTOR-α

Natural human TGF-α is very difficult to obtain in amounts sufficient for biochemical and biophysical studies. The isolation 1.5 ug of human TGF-α required the use of 136 liters of conditioned media as starting material (5). Purified TGFα is a single chain polypeptide that migrates at an apparent molecular weight of 6,400. It competes with ^{125}I-EGF for binding to EGF receptor (6,7) and enables normal anchorage-dependent rat kidney cells to grow into colonies in soft agar in the presence of 10% calf serum (8).

The complete amino acid sequence of rat TGF-α and the N-terminal amino acid sequence of human TGF-α have been determined by protein microsequencing (9). The complete amino acid sequence of human TGF-α was deduced from sequencing cDNA and genomic clones (10). Human and rat TGF-α amino acid sequences are 92% homologous to each other, differing in only four out of 50 amino acid residues. Three of these four differences are conservative substitutions. However, TGF-αs have only 40% sequence identities with either human or mouse EGF (Fig 3).

```
HTGF   VVSHFNDCPDSHTQFCFH GTCRFLVQEDKPACVCHSGYVGARCEHADLLA
RTGF   ......K.......Y... .........E............V.........
HEGF     NSDSE..L..DGY.L.D.V.MYIEAL..Y..N.VV..I.E..QYR..KWWELR
MEGF     NSYPG..S.YDGY.LNG.V.MHIESL.SYT.N.VI..S.D..QTR..RWWELR
```

FIGURE 3. Amino acid sequences of TGFs and EGFs.

Most of the sequence homologies are accounted for by six conserved cysteines distributed throughout the molecule and their adjacent prolines and glycines. This low degree of homology is of interest since all known biological activities of EGF are shared by TGF-α, however the converse is not true.

A systematic structural-activity study of TGF-α in relation to EGF may reveal fundamental principles of protein structure and activities in relation to their primary sequences. We report here the synthesis, refolding, purification and biological characterization of human TGF-α and the determination of disulfide bonding pattern in this molecule.

EXPERIMENTAL PROCEDURES

Quantitative Ninhydrin Assay.

Resin samples (approx. 10mg, 2micromol each) was automatically taken by the instrument at the end of each cycle. The efficiency of the coupling reactions were determined using the quantitative ninhydrin reaction for the colorimetric determination of unreacted alpha-amino groups (11). Resin sampling protocol was developed to avoid artifactually-high ninhydrin values associated with resins left for prolonged standing (hours or more). His-containing peptides gave rise to slowly increasing values for residual free amino groups. This was overcome by adding a small amount of DIEA to the test tube into which the resin sample was deposited in DMF. In this solvent, the peptide-resin sample sank and was thus kept in basic solution. Provided resin samples were not allowed to stand for excessive amounts of time (days), this protocol minimized the problem giving typically an order of magnitude reduction in the artifact.

Quantitative Edman Degradation.

Quantitative Edman degradation of the protected peptide-resin was used to assess the accuracy and overall efficiency of chain assembly (12). The expected amino acid PTH derivative was monitored at each cycle of the degradation to confirm the desired target sequence of amino acids had been synthesized. The exact quantitation of PTH-amino acids in successive Edman degradation cycles, after correction for background due to random chain scission during the sequencing experiment, is a quantitative measure of overall yield of the target protected peptide chain (including both deprotection and coupling).

Mass Spectroscopy.

Synthetic TGF-α was analyzed by plasma-desorption time-of-flight mass spectrometry. The instruments used have resolution of $\pm$ 0.2 A.M.U. One nanomole of the refolded, purified synthetic material was deposited on a nitrocellulose membrane and inserted into the source of the instrument. Spectra was then accumulated to determine the absolute molecular mass of the synthetic product.

Mitogenic Assay.

NRK fibroblasts trypsinized into single-cell suspension were seeded into 96-well microtiter plates in Dulbecco's modified Eagles' medium (DMEM) supplemented with 10% calf serum (Gibco) at a density of 8 x 10^3 cells/well. Cells were incubated at 37°C in a humidified chamber under 5% CO_2 in air. On the following day, culture medium were replaced with a 50% DMEM, 50% HAM's medium containing insulin (2μg/ml), transferrin (5μg/ml), fatty acid-free BSA (1mg/ml) and linoleic acid (5μg/ml). Cells were incubated for three more days before samples of TGF-α of EGF were added. Twenty hours after mitogen addition ^{3}H-thymidine incorporated were determined by scintillation counting of precipitated radioactivity solubilized with 0.5 M NaOH.

Soft Agar Colony Formation Assay.

The soft agar colony formation activity of TGF-α and EGF were assayed essentially as described (13). NRK fibroblast were trypsinized to obtain a single cell suspension. 2 x 10^3 cells were resuspended in 0.5% agar contain DMEM supplemented with 10% calf serum and varying concentrations of human TGF-α or EGF. 0.5 ml of each test suspension was overlaid in duplicates, onto one of a 6 wells Costar cell culture clusters that are pre-coated with 1.0 ml of 0.5% agar in DMEM and the exact concentration of growth factor being tested. The cells were incubated at 37°C in moisture saturated atmosphere of 5% CO_2 in air. Colonies formed were visually scored after seven days of incubation, using a Nokon inverted microscope. Eight fields at 10X magnification were scored for each concentration and averaged to obtain the data shown. Colonies containing 50 cells or more were scored as positive.

EGF Receptor Binding Assay.

Murine EGF were purified and iodinated by the chloramin-T method as described (3). The specific activity was between 1.2-1.5 x 10^6 cpm/pmol. Human EGF receptor was purified to homogeneity from Triton X-100 extracts of human epidermoid carcinoma (A431) cells by affinity chromatography on Fractogel TSK-immobilized ricin-binding subunit, followed by EGF-Fractogel affinity chromatography (14). The purified receptor migrated as a single 160 kDa band in SDS-PAGE, binds EGF with a K_d of 65.8 ± 3.5 mM and phosphorylate up to 430 nmol of a tyrosine protein kinase substrate peptide per mg of receptor per min.

The binding of TGF-α to EGF receptor was assayed indirectly by comparing its ability to compete with ^{125}I-EGF in binding to EGF receptor with that of native EGF. Two microliters of TGF-α or EGF at 10X the test concentration were added to eighteen microliters of a solution containing 17 μg/ml of EGF receptor and 100 mM ^{125}I-EGF in a 50 mol Hepes pH = 8.0 buffer containing 0.1% Triton X-100 (Binding buffer) and vortexed in a 1.5 ml polypropylene test tube at 26°C. After a 30 min incubation, the assay was terminated by adding 0.3 ml bovine IgG (1 mg/ml) and 0.3 ml of 25% polyethylene glycol (PEG-8000), both in binding buffer, in rapid succession. Each tube was capped and mixed vigorously to precipitate macromolecules. Precipitates were collected by centrifugation in an Eppendorf microfuge for 5 minutes. Supernatants were aspirated and pellets were washed once by vortex mixing with 0.5 ml of 8% PEG in 50 nM Hepes pH = 8.0 and recovered by a 1 minute spin in an Eppendorf microfuge. After removal of the supernatant, pelletes associated ^{125}I-EGF was determined by gamma counting.

Refolding of Human TGF-α.

The crude peptide (100mg) was treated with 10 mls of 2M 2-mercaptoethanol in 6M guanidine-HCl, 50 mM tris acetate at pH =8.5 at 37°C for two hours to remove residue DNP-groups. The excess reagents and the DNP released were separated from the peptides by gel-filtration chromatography on Sephadex G-25 equilibrated with 1M guanidine-HCl in 50 mM tris acetate pH = 8.5 (Standard refolding buffer). The reduced and deprotected peptide was then diluted to 100μg/ml with degassed standard refolding buffer. Refolding was carried out for 24 hrs at 25°C. The oxidation of free thio into disulfide during refolding was mediated by atmospheric oxygen introduced into the buffer by rapidly stirring the sample in a 2 liter flask with a magnetic stirrer such that a smooth vortex almost reaching the bottom of the flask was formed. No visible precipitate was formed during this process. At the end of the refolding period the pH of the sample was slowly adjusted to 4.0 with glacial acetic acid to prevent further thio-disulfide exchange. About 40% of the peptide precipitated out of solution during this procedure. These were removed by centrifugation and saved for further refolding. The soluble refolded products were concentrated into 20 mls under nitrogen on YM-2 membranes using an Amicon ultrafiltration apparatus. The concentrated refolded peptide was then dialyzed extensively against 1M acetic acid at 4°C (4 changes of 5 liters at 8 hr intervals). Five milligrams of peptides precipitated during guanidine HCl removal by dialysis, and were separated by centrifugation. The remaining soluble refolded peptides were then purified by reverse-phase HPLC on a 10z 250 mm Vydec C-4 column. The gradient used was from 0 to 60% acetonitrile in 0.1% trifluoroacetic acid over 4 hours at a flow rate of 3.0 ml per minute. Six ml fraction were collected in tared tubes. Solvents were removed by lyophilization in a Speed-Vac concentrator. Dried samples were redissolved in 0.5% acetic acid to 1 mM peptide. TGF-α activities were monitored by EGF receptor binding assay.

RESULTS

Chemical Synthesis of Human TGF-α.

The entire automatic synthesis ran uninterrupted for 72 hours yielding 3.276 gm of total resin.

The stepwise coupling yield monitored by quantitative ninhydrin assay at the end of each double coupling cycle is listed in Table 1. The calculated yield of the full-length target sequence based on these values is 67.2%. The magnitude of resin ninhydrin background could be estimated (11) from the ninhydrin value obtained for the residue after proline cycle 20 and cycle 41 to be 0.73% and 1.06%, respectively. Therefore, the actual yield of full-length protected peptide on the peptide was higher because the values used for the calculation were not corrected for increasing ninhydrin background contributed by the resin.

The yield of the target sequence on the resin based on quantitative preview Edman sequencing (12) of the resin-bound protected peptide through leucine residue 48 was found to be 84.5%, corresponding to an average repetitive coupling yield of 99.65% per residue. Preview sequencing also established that correct sequence of amino acids have been assembled on the resin.

Following DNP removal by thiolysis and removal of N-protecting group, 374 mg of crude cleaved and deprotected peptides were obtained from 600 mg of protected TGF-α-resin after deprotection and removal of the peptide by the low-high HF cleavage (15). Analytical HPLC of this material on a C4 reverse-phase column revealed one major peptide peak representing 44% of the total products, figure 4 (top).

The remainder represents a mixture of closely related products consisting primarily of peptides with a single amino acid missing and incompletely deprotected DNP-containing peptides. Human TGF-α contains five histidines in its amino acid sequence. The residual DNP protecting group on histidine used in human TGF-α synthesis is not removed by HF treatment and is not quantitatively removed by the initial thiophenol treatments used. This is evident from the characteristic yellowish color of fractions in the trailing peaks. The major peptide peak is colorless and showed no absorbance at 410 mm.

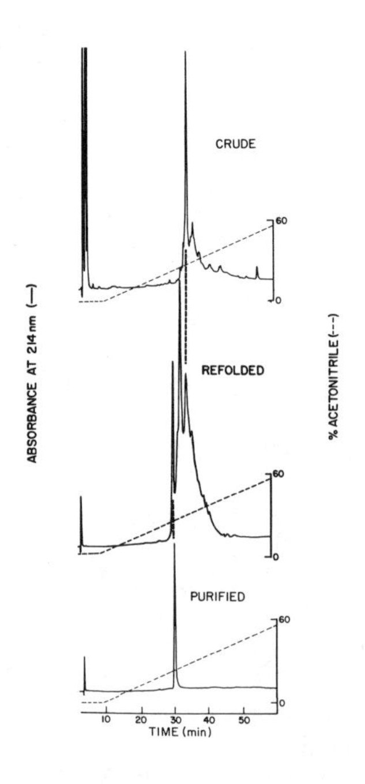

FIGURE 4. Analytical C-4 RP-HPLC of crude, refolded and purified synthetic human TGF-α.

Folding the Polypeptide Chain and Formation of Disulfide Bonds.

Virtually no EGF receptor binding activity was detected at this stage. We therefore proceeded to compare several refolding conditions for their efficiency to generate TGF-α activity by monitoring the specific ability of the crude refolded product to compete with ^{125}I-EGF binding to purified human EGF receptor. The complexity of the crude refolded TGF-α was monitored by C4-reverse phase HPLC.

Overnight air oxidation of 100 μg/ml of crude cleaved TGF-α at 25°C in 50 mM Tris, pH 8.5, int he presence of 1 M guanidium HCl was judged to be the refolding condition of choice. This procedure is simple and the yield of EGF receptor binding activity under these conditions was comparable to more complex systems tested. The addition of a 10:1 molar ratio of reduced to oxidized glutardione at a final concentration of 5 mM as a thio-disulfide exchange catalyst did not significantly improve the yield. The omission of guanidium chloride in the refolding buffer led to the formation of a complex mixture of folding intermediates detected by HPLC. These intermediates displayed specific EGF receptor binding activity ranging from 0-100% of native EGF (data not shown).

An analytical HPLC profile of the crude TGF-α refolded under conditions described above is shown in Figure 4 (middle). Only the earliest eluting of the four major peaks detected possessed ^{125}I-EGF competing activity.

This activity peak was isolated from the rest of the sample by semipreparative C4-reverse phase HPLC. Figure 4 (bottom) illustrates the HPLC purity of this purified TGF-α. Approximately 10 mg of this purified TGF-α can be obtained after refolding and purification of 100 mg of the crude cleaved TGF-α.

Physical and Chemical Characterization of Human TGF-α.

The HPLC purified synthetic human TGF-α were hydrolyzed with 6N HCl and the amino acid composition of the hydrolysate determined by amino acid analysis. The results are in close agreement with the values calculated from the known sequence of human TGF-α. The HPLC purified TGF-α was found to have isoelectric point of 6.2 by analytical electrofocusing on Immobiline gels. Only one focused band at pI = 6.2 could be detected after staining a nitrocellulose blot of the gel with rollodial carbon (ink) (Figure 5).

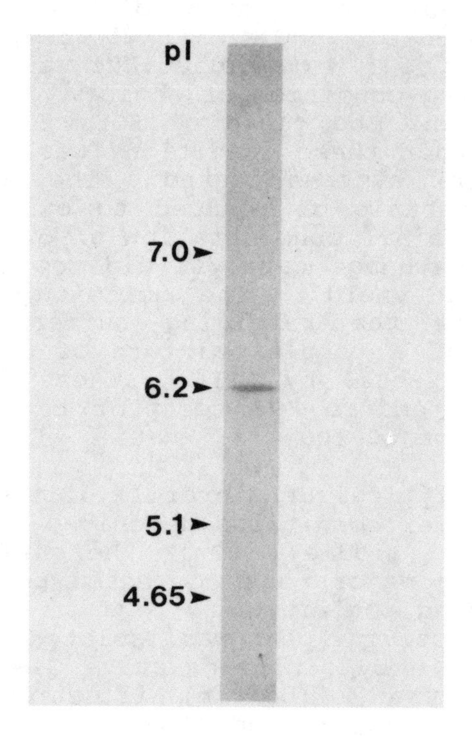

FIGURE 5. Isoelectric focussing of synthetic human TGF-α.

The HPLC purified TGF-α was then analyzed on a plasma desorption mass spectrometer for determination of its absolute mass. The mass spectrum of TGF-α is shown in Figure 6. The $(M+H)^{+}$ and $(M+2H)^{2+}$ molecular ions are detected as clean, sharp peaks, indicating no other peptides with closely related mass are present.

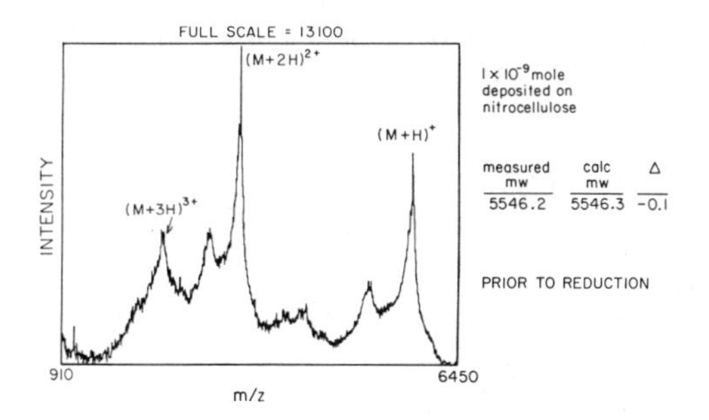

FIGURE 6. Mass spectrometric analysis of synthetic human TGF-α.

The measured molecular masses of $(M+H)^+$ and $(M+2H)^+$ are within 0.2 and 0.4 atomic mass unit of the calculated molecular weight of a peptide having the human TGF-α sequence. Since the mass spectrometric measurements were carried out on samples that had been refolded under oxidative conditions, the absence of detectable signal with molecular mass greater than 5,546 in the spectrum also indicates no dimer or other higher order covalent complex were present in the sample. Taken together, the above analytical data led us to conclude that active peptides with human TGF-α sequence have been synthesized and purified to homogeneity.

Biological and Biochemical Characterization of Synthetic Human TGF-α.

The biological and biochemical properties of the homogenous active synthetic human TGF-α were characterized and its specific activity compared to purified native mouse EGF in three assays. The ability of synthetic human TGF-α to compete with EGF in binding to EGF receptor purified to homogeneity from human A431 cells (14) is illustrated in Figure 7. The interaction of EGF to solubilized EGF receptor has an intrinsic dissociation constant of 65.8 ± 3.5 mM at 25°C (14). Equal molar amounts of TGF-α and EGF displace identical amounts of ^{125}I-EGF bound to EGF receptor throughout the concentrations tested.

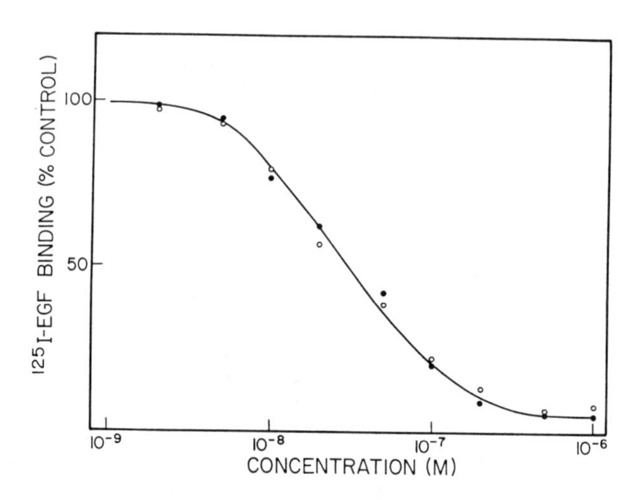

FIGURE 7. Competition with ^{125}I-EGF in binding to EGF receptor (○, murine EGF; •, synthetic human TGF-α).

These results indicated that human TGF-α is equipotent to EGF in its ability to bind to EGF receptor. Synthetic human TGF-α also stimulated the tyrosine kinase activity of purified human EGF receptor (data not shown). These two activities have been observed for purified natural TGF-α acting on A431 cytoplasmic membrane fractions and were the basis for the widely accepted view that TGF-α exerts its biological effect via interactions with cell surface EGF receptors. Our data using purified EGF receptor and synthetic TGF-α definitely demonstrated the in vitro abilities of TGF-α to bind directly to and consequently activate EGF receptor in the absence of interfering or accessory biological molecules. However, the existence of an TGF-α receptor independent of EGF receptor although unproven cannot be formally ruled out.

The mitogenic properties of purified synthetic human TGF-α on quiescent NRK cells was compared with that of native EGF in Figure 8. Half maximal mitogenic response to TGFα by NRK cells in a defined medium containing TGF-α insulin, transferrin, linoleic acid, as measured by ^{3}H-thymidine incorporation, occurs at 4 mM TGF-α.

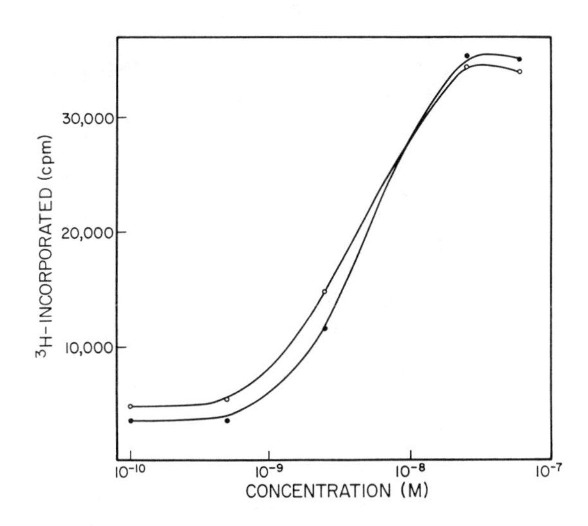

FIGURE 8. Mitogenic response by quiesent NRK cells. (O, murine EGF; •, synthetic human TGF-α).

Half maximal mitogenic response of these cells to EGF under identical conditions was obtained at 3.3 mM. Thus, within experimental error, purified synthetic human TGF-α is as active as native EGF in stimulating mitogenic response in quiescent NRK fibroblasts.

The dose response of NRK cells in soft agar colony formation to varying concentrations of purified synthetic human TGFα in the presence of a fixed concentration of TGF-β, supplied in the form of 10% calf serum, is illustrated in Figure 9.

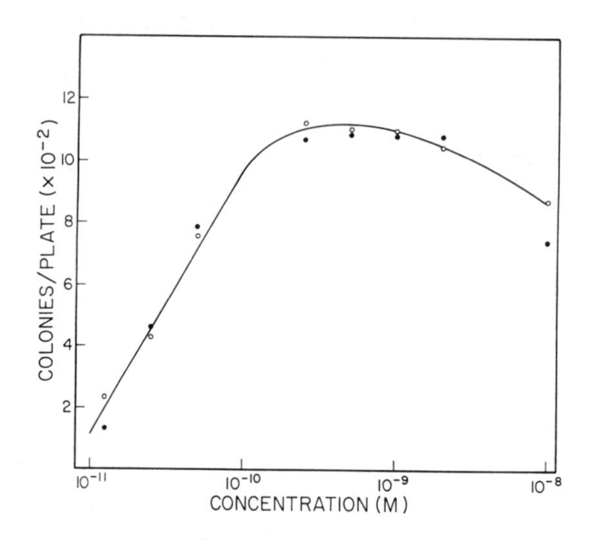

FIGURE 9. Dose response of colony formation in soft agar (O, murine EGF; •, synthetic human TGF-α.

After seven days of incubation in soft agar medium containing only 10% calf serum and no TGF-α, cells did not grow and remained as a suspension of viable single cell suspension. A large number of aggressively growing colonies (Figure 10) can be detected in all samples containing TGF-α at concentrations in excess of 100 pM. The half maximal soft agar colony formation response to synthetic human TGF-α was estimated to be 30 pM. EGF is as potent as our synthetic human TGF-α in this assay. Also noted was a toxicity or inhibitory effect towards the number and the size of soft agar colonies formed at TGF-α or EGF concentration above 1 mM. This effect is more pronounced with TGF-α than with EGF.

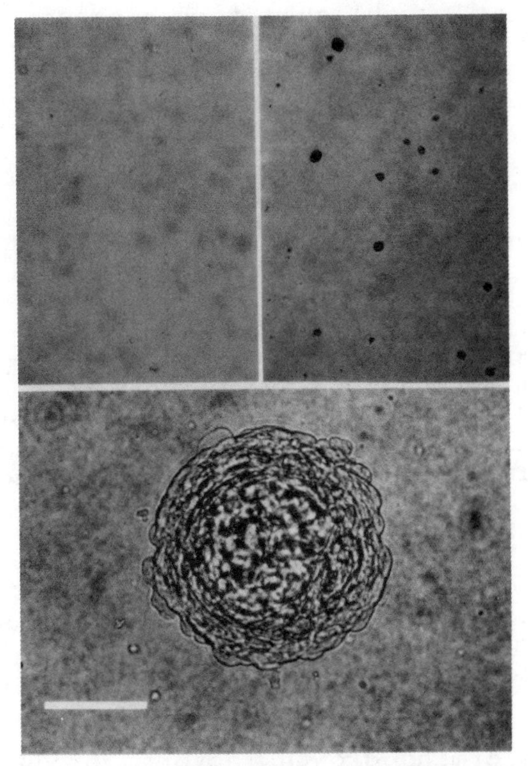

FIGURE 10. Colonies of NRK cells in soft agar grown with (left) and without (right) TGF-α. A large colony of cells is shown in the bottom panel.

We have therefore demonstrated that, chemically homogenous synthetic human TGF-α, with specific biochemical and biological activities indistinguishable from EGF purified to homogeneity from mouse submaxillary glands can be readily prepared with the use of automated solid-phase peptide synthesis instruments. A variety of full-length analogs of TGF-α is currently being synthesized and purified for structure-function analysis using the methods developed.

In conclusion, we wish to emphasize that all elements critical to the successful application of solid-phase protein synthesis to structure-function analysis described here are now readily available as established technologies. They are complimentary and in some instances preferable to molecular biological methods in the study of protein structure and functions.

REFERENCES

1. Kent SBH, Hood LE, Beilan H, Meister S, Geiser T (1985). In Ragnarsson U (ed): "Peptides 1984," Stockholm: Almquist and Wiksell Intl, pp 185-188.
2. Kent SBH, Hood LE, Geiser T, Beilan H, Bridgham J, Marriot M, Meister S (1985). A novel approach to automated peptide synthesis based on new insights into solid phase chemistry. In Izumiya N (ed): "Peptide Chemistry," Osaka: Protein Research Foundation, pp 217-222.
3. Merrifield
4. Kent SBH, Clark-Lewis I (1985). Modern methods for the chemical synthesis of biologically active peptides. In Alitalo K, Partanen P, Vaheri A (eds): "Synthetic Peptides in Biology and Medicine," Amsterdam, 1985.
5. Marquardt H, Todaro GJ (1982). Human transforming growth factor: production by a melanoma cell line, purification, and initial characterization. J Biol Chem 257:5220-5225.
6. Todaro GJ, Fryling C, De Larco JE (1980). Transforming growth factors produced by certain human tumours: Polypeptides that interact with epidermal growth factor receptors. Proc Natl Acad Sci USA 77, 5258-5262.
7. Massague J (1983). Epidermal growth factor-like transforming growth factor. II. Interactions with epidermal growth factor receptors in human placenta membranes and A431 cells. J Biol Chem 258, 13614-13620.
8. Massague JJ (1983). Epidermal growth factor-like transforming growth factor. I. Isolation, chemical characterization and potentiation by other transforming growth factors from feline sarcoma virus-transformed rat cells. J Biol Chem 258, 13606-13613.

9. Marquardt H, Hunkapillar MW, Hood LE, Twardzik DR, De Larco JE, Stephenson JR, Todaro GJ (1983). Transforming growth factors produced by retrovirus-transformed rodent fibroblasts and human melanoma cells: Amino acid sequence homology with epidermal growth factor. Proc Natl Acad Sci USA 80, 4684-4688.
10. Derynck R, Roberts AB, Winkler ME, Chen EY, Goeddel DV (1984). Human transforming growth factor-α: Precursor structure and expression in _E. coli._ Cell 38, 287-297.
11. Sarin VK, Kent SB, Tam JP, Merrifield RB (1981). Quantitative monitoring of solid phase peptide synthesis by the ninhydrin reaction. Anal Biochem 117:147-157.
12. Clark-Lewis I, Kent SBH (In press). Chemical synthesis, purification and characterization of peptides and proteins. In Kerlavage AR (ed): "Receptor Biochemistry and Methodology".
13. Todaro GJ, De Larco JE, Cohen S (1976). Transformation by murine and feline sarcoma viruses specifically blocks binding of epidermal growth factor to cells. Nature 264, 26-31.
14. Woo DDL. UCLA Ph.D. dissertation, 1984.
15. Tam JP, Health WF, Merrifield RB (1983). SN2 deprotection of synthetic peptides with a low concentration of HF in dimethyl sulfide: evidence and application in peptide synthesis. J Am Chem Soc 105:6455-6461.

Protein Purification: Micro to Macro, pages 75–97

THE USE OF POLYETHYLENEIMINE IN PROTEIN PURIFICATION

Jerry Jendrisak

Promega Corporation, Madison, WI 53711 and
Department of Horticulture, University of Wisconsin
Madison, WI 53706

ABSTRACT The use of polyethyleneimine (or PEI) as a reagent for protein purification is discussed. PEI, which has the structure: $H_2N.X(C_2H_4NH)C_2H_4NH_2$, is a positively charged molecule in solutions at neutral pH values (the pKa value of the imino groups is 10-11). This property makes PEI a very effective agent for the precipitation of nucleic acids in crude cellular extracts (presumably due to cooperative charge neutralization). PEI has been exploited as a selective protein fractionation agent. PEI is capable of precipitating a subset of cellular proteins along with nucleic acids under low ionic strength conditions. Proteins are extracted (eluted) and separated from the PEI-nucleic acid-protein complex (precipitate) with buffers of higher ionic strength followed by centrifugation. The selectivity of purification is optimized by judicious choice of buffer salt concentrations during the precipitation and subsequent elution steps. The PEI fractionation step can complement other fractional precipitation methods (where proteins are precipitated due to other chemical and physical properties) to afford efficient, inexpensive purification prior to the use of column chromatographic steps.

INTRODUCTION

Protein precipitation remains a valuable step for the purification of most enzymes. The use of water soluble, charged polymers as precipitating agents has been suggested but has received relatively little attention in contrast to the use of other agents, despite potential unique benefits. The reader is referred to an excellent review by Bell et al. (1) which summarizes protein precipitation by various agents.

In this report I will discuss the use of the charged polymer polyethyleneimine or PEI as a reagent for protein purification. PEI forms ionic complexes with macromolecules containing acidic domains (nucleic acids and some proteins) resulting in their precipitation. Precipitation behavior is affected by salt concentration, pH, and the concentratin of precipitable components in the extract. Some of these parameters affecting precipitation as well as methods for recovering protein from PEI precipitates are illustrated in the following experiments with crude extracts prepared from wheat germ. Optimized conditions for the use of PEI in the purification of several enzymes are also summarized in this report to illustrate the variety of conditions required for optimal selectivity with PEI. Finally a reference list of enzymes purified with a PEI step is presented as an aid in locating reports of specific interest.

MATERIALS AND METHODS

Crude extract preparations. These were done at 0-4° C. 20 g quantities of wheat germ were homogenized for 30 seconds in 100 ml of TEB (0.05 M Tris-HCl, pH 7.5, 0.1 mM EDTA, and 15 mM 2-mercaptoethanol) and the indicated concentrations of NaCl with the use of a Polytron™ homogenizer run at half speed. Homogenates were centrifuged at 10,000 g for 10 minutes to remove gross debris and the resulting supernatant solutions are referred to as crude extracts.

PEI reagent preparation. 100 ml of 50% (w/v) PEI, which is sometimes sold under the trade name Polymin P, (BRL; mol. wt. range 30-40,000) was dissolved in 800 ml of deionized water. Concentrated HCl (approx. 40-45 ml) was used to adjust the pH to 7.5. After diluting to 1000 ml, the neutralized solution was filtered and stored refrigerated. This 5% (w/v) stock is the working PEI reagent. This reagent appears to be stable indefinitely.

Protein determinations. These were done with the Bradford reagent (2) (Bio-Rad) which was not interfered with by PEI in the concentrations used in this study.

Nucleic acid estimations. These were done by scanning sample dilutions in the ultraviolet region and observing relative absorbance values at 260 and 280 nm (3). Dilutions were made in TEB containing 0.2 M NaCl.

RESULTS AND DISCUSSION

Precipitation of macromolecules with PEI.

To illustrate some of the parameters that affect fractionation of proteins and nucleic acids by PEI, the experiments described in MATERIALS AND METHODS and in the legends to Figures 1-5 were carried out. In Figure 1 is shown the effect of salt concentration on the precipitation of protein by PEI from crude wheat germ extracts. It can be seen that the amount of protein precipitated depended upon both the salt concentration in the extract and the amount of PEI added. At 0 M NaCl, about half of the protein in a crude extract was precipitated by PEI at a final concentration of 0.4%. Little or no additional protein was precipitated at higher concentrations of PEI at this salt concentration. When extracts were prepared at higher salt concentrations, less protein was precipitated at the various PEI final concentrations such that in extracts prepared at 0.5M NaCl, less than 10% of the protein was precipitated at a final concentration of 0.4% PEI. Similar behavior is observed for precipitation of protein from crude extracts of E. coli by PEI (data not shown).

Since PEI precipitates protein by directly interacting or essentially titrating anionic charged groups, it was expected that the amount of PEI required to precipitate a given percentage of total protein depended upon the concentration of precipitable components in a crude extract. This is indicated in Figure 2. If a crude extract was diluted 10-fold, it required one tenth as much PEI to precipitate the same percentage of total protein. This is in contrast to precipitation behavior with agents such as ammonium sulfate where more, not less of this precipitation agent is required to precipitate a given percentage of the total protein from a more dilute solution.

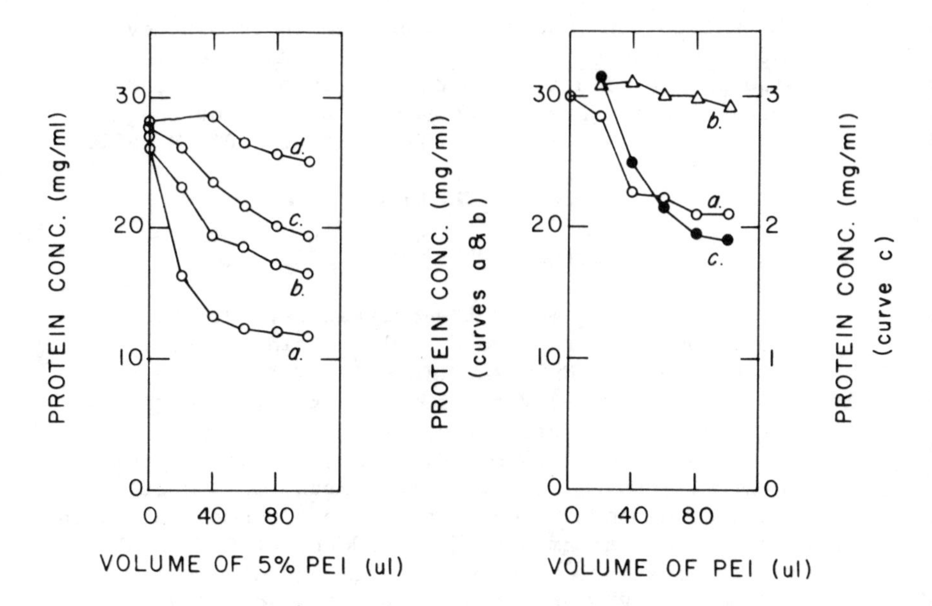

FIGURE 1. The effect of salt and PEI concentrations on the precipitation of protein. Wheat germ homogenates were prepared in TEB containing different salt concentrations: a) 0.0, b) 0.1, c) 0.2, and d) 0.5 M NaCl. 1 ml samples of the resulting crude extracts were treated with the indicated volumes of 5% PEI. After mixing and centrifugation (1 minute at top speed in a microfuge), the resulting PEI supernatant solutions were assayed for protein.

FIGURE 2. The effect of protein concentration on the precipitation of protein with PEI. A wheat germ crude extract was prepared in TEB containing 0.1 M NaCl. 1 ml samples were treated with the indicated volumes a) 5% or b) 0.5% PEI and protein concentrations determined in the resulting PEI supernatant solutions. In curve c, the crude extract was diluted 10-fold with TEB containing 0.1 M NaCl prior to treating with the indicated volumes of 0.5% PEI and supernatant protein concentrations were determined and plotted against a 10-fold expanded scale.

Salt concentration also effected the precipitability of nucleic acids from crude extracts by PEI. This was revealed by examination of adsorption spectra in Figure 3. When a wheat germ extract was prepared in 0 M NaCl, addition of twenty microliters of 5% PEI was sufficient to precipitate nearly all of the nucleic acids from the crude extract as seen from the drop in adsorbance value at 260 nm and the shift in peak adsorbance to 280 nm. When extracts were prepared in at higher salt concentrations, more PEI was required. Thus extracts prepared 0.5 M NaCl required three times as much PEI (sixty microliters) to completely precipitate the nucleic acids.

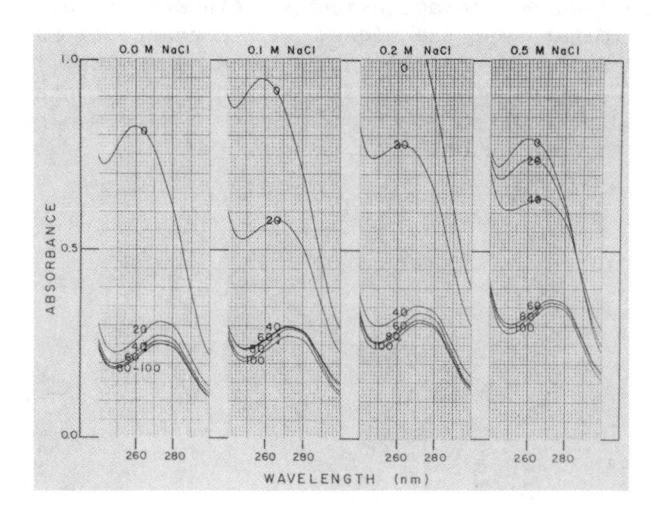

FIGURE 3. The effect of salt and PEI concentrations on the precipitation of nucleic acids as determined by ultraviolet absorption spectrophotometry. The PEI supernatant solutions used to construct the PEI precipitation curves in Figure 1 were scanned for absorbance versus wavelength in the ultraviolet region. Numerical values on the scans indicate the volumes (in microliters) of 5% PEI used to treat 1 ml samples of crude extracts prepared in TEB plus the indicated different salt concentrations.

Recovery of Proteins from a PEI Precipitate.

Proteins that are precipitated along with nucleic acids may be recovered or solubilized from a PEI precipitate simply by extracting the pelleted proteins with buffers containing higher salt concentrations. This step is conveniently done with a controlled speed motor driven homogenizer such as a Polytron (TM). Extractions may be performed in the same centrifuge bottles or tubes with no transfer of pellet. Extraction of protein from a PEI precipitate of a wheat germ crude extract prepared in 0.1 M NaCl as a function of various salt concentrations is shown in Figure 4. Increasing protein was extracted from the precipitates at the higher salt concentrations. As shown in Figure 5, nucleic acids were not solubilized from PEI precipitates even when extraction buffer contained 1 M NaCl, indicating the strong complexes are formed between PEI and nucleic acids.

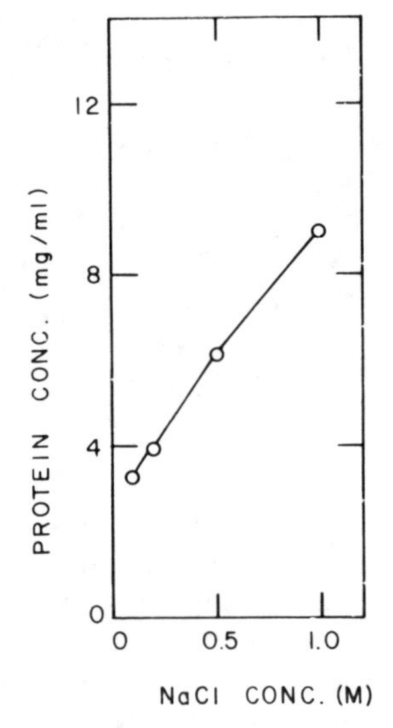

FIGURE 4. The effect of salt concentration on the recovery of protein from PEI precipitates. A wheat germ crude extract was prepared in TEB plus 0.1 M NaCl and was treated with one tenth volume of 5% PEI. The mixture was divided into 10 ml aliquots which were centrifuged at 10,000 g for 10 minutes. After removing the supernatant solutions, the precipitates were individually suspended in 5 ml of TEB containing the indicated concentrations of

NaCl. After thorough suspension (using a Polytron homogenizer), the mixtures were centrifuged (10,000 g for 10 minutes) and protein concentrations of the supernatant solutions were determined.

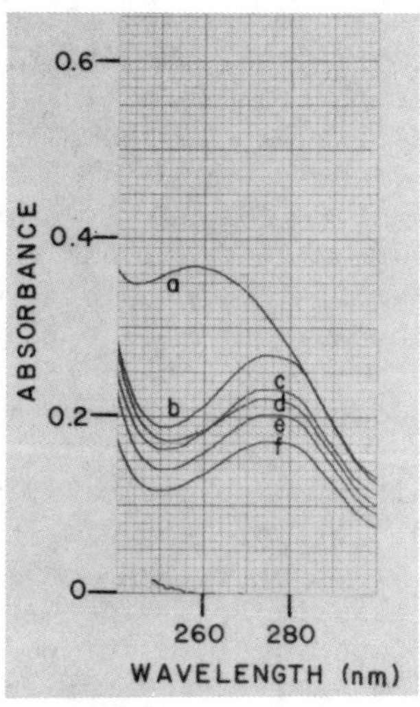

FIGURE 5. The disposition of nucleic acids after PEI fractionation. Ultraviolet absorption spectra of the wheat germ crude extract prepared 0.1 M NaCl (a), the first PEI supernatant solution (b), and the 0.1 M (c); 0.2 M (d); 0.5 M(e); and 1.0 M(f) NaCl extracts of the PEI precipitates described in the legend to Figure 4 are shown.

Examples of Ionic Conditions Used in the PEI Fractionation of Specific Proteins.

A summary of optimized conditions used for the PEI fractionation of several proteins is shown in Table 1. For the purification of *E. coli* DNA dependent RNA polymerase, the enzyme was precipitated with PEI from crude extracts prepared in 0.2 M NaCl. Approximately three-fourths of the protein was not precipitated and was removed in the supernatant solution after centrifugation. The PEI pellet was extracted with buffer containing 0.5 M NaCl to remove more contaminating protein from the still precipitated RNA polymerase after centrifugation. RNA polymerase was extracted from 0.5 M NaCl-washed PEI precipitate with buffer containing 1 M NaCl. The yield of RNA polymerase after PEI fractionation was almost 90% with almost a 30-fold purification. After concentration of proteins by ammonium sulfate precipitation (a common step after PEI fractionation), final purification was achieved by affinity chromatography on DNA cellulose followed by a sizing step. More details may be found in reference 4. A critical step in PEI fractionation is to add sufficient PEI to the crude extract to completely precipitate RNA polymerase. This occurred at final PEI concentrations slightly above that required to precipitate all of the nucleic acids from the extract. The optimal amount of PEI to add was determined by construction of PEI precipitation curves such as those described in Figure 1, by monitoring the ultraviolet absorption spectra of PEI supernatant solutions, and by monitoring specifically the fractionation behavior of the protein of interest with enzyme assays, polyacrylamide gels, or other means.

As can be seen in Table 1, a wide variety of salt conditions have been found optimal for purification of a variety of enzymes with PEI. Characterization of optimal conditions for the use of PEI is similar to finding optimal conditions for purifying an enzyme on an ion exchange column, such as DEAE-cellulose. Ionic conditions for precipitation, precipitate washing, and eluting a protein of interest from a PEI precipitate can be carried out on a small scale before applying conditions to a preparative scale purification. We have found optimized conditions to be extremely reproducible and readily scaled up (4, 5). A major factor in reproducibility of course is in the preparation of the crude extract, to ensure that reproducible lysis released expected concentrations of proteins and nucleic acids into the extract. Suboptimal lysis will result in less PEI required for complete precipitation of nucleic acid and a protein of interest.

Wheat germ RNA polymerase II was purified over thirty-fold and with good yield with a PEI step. Optimal fractionation conditions are summarized in Table 1. Many other RNA polymerases of both bacterial and eukaryotic origin have been purified with a PEI step with this general protocol (see Table 2).

In the next set of examples summarized in Table 1, four enzymes induced in E. coli after phage T4 infection were purified with one batch of infected cells.

Polynucleotide kinase was not precipitated with PEI even when extracts were prepared in 0 M NaCl. It was precipitated from the PEI supernatant solutions with ammonium sulfate and further purified on a series of columns. The other three enzymes: DNA polymerase, DNA ligase and RNA ligase were precipitated. DNA ligase was extracted from the PEI precipitate with buffer containing 0.15 M NaCl, the RNA ligase and DNA polymerase remained precipitated. These two enzymes were subsequently solubilized with buffer containing 1.0 M NaCl. The solubilized enzymes were concentrated by ammonium sulfate precipitation (which also removes residual PEI). Final purifications were achieved with a battery of column chromatographic steps. Conditions for precipitation and elution of these enzymes with PEI as well as fold purification and yield are also summarized in Table 1 and more details are found in reference 6.

In the last examples presented in Table 1, the purification of the restriction enzyme Bgl II with the use of a PEI step is summarized. PEI precipitated Bgl II at low salt concentrations (0 M NaCl) but in extracts prepared in 0.1 M NaCl, the enzyme remained in the supernatant after PEI treatment. The Bgl II in the PEI precipitate done with 0 M NaCl was recovered by extracting the precipitate with buffer containing 0.1 M NaCl. After either PEI treatment protocol, sufficiently pure Bgl II for recombinant DNA experiments could be obtained after subsequent chromatography steps. See reference 7 for more details.

The above examples all dealt with enzymes involved in nucleic acid metabolism. The majority of enzymes purified with a PEI step have been enzymes in this category. A list of references for enzymes purified with PEI is presented in Table 2. In some of the above cases, the enzymes of interest were not precipitated with PEI (even under relatively low salt concentrations), but a bulk of contaminating material which did precipitate was removed after centrifugation. PEI can be used in the place of protamine sulfate or streptomycin sulfate for removal of nucleic acids, nucleoproteins and microparticulates from crude extracts before proceeding onto other steps such as ammonium sulfate precipitation and column chromatography.

As shown above, PEI is a useful reagent for protein purification. The mechanism of precipitation, which differs from salt or solvent precipitation, occurs by the process known as floculation (1). Floculation is precipitation as the result of the formation of charge neutralization complexes (analogous to isoelectric precipitation) and the formation of cross bridges between the complexes by the precipitating agent. Nucleic acids which form strong precipitation complexes and probably contribute to extensive cross bridge formation, act as carrier during protein precipitation and ballast during centrifugation. Formation of precipitates is rapid and moderate centrifugation speeds are required to pellet the generally dense precipitates from a non-viscous supernatant solution. In most cases very clear supernatant fractions are obtained after PEI fractionation due to the precipitation of other cellular microparticulate debris.

Acidic polyelectrolytes such as polyacrylic acid and polyphosphates have also been suggested for use in protein purification by fractional precipitation. In general precipitations with these agents must be carried out at low pH values in order to impart sufficient positive charge on the protein to allow for complex formation and precipitation (1). Several extracellular, secreted proteins such as plasma proteins, lysozyme, and casein and gelatin, (8,9,10, respectively) have been precipitated with these agents. In many cases intracellular enzymes are inactivated at low pH values resulting in poor recoveries. Also, these agents are generally not useful for treating crude cellular extracts since these agents do not form complexes with nucleic acids and most microparticulates. Protein solutions remain viscous and turbid after treatment which complicates downstream processing. PEI fractionation, in contrast, is generally done under slightly alkaline pH conditions where most intracellular proteins are stable. The low cost and low amounts of agent required when coupled with high recoveries and in many cases rapid removal of large quantities of contaminating protein and all nucleic acids, make PEI fractionation a valuable and highly recommended protein fractionation agent.

TABLE 1
EXAMPLES OF THE SALT CONCENTRATIONS (IONIC STRENGTH CONDITIONS) USED FOR PEI PRECIPITATION AND ELUTION OF ENZYMES FROM THE PEI PRECIPITATE

Enzyme or Protein	Source	Salt Conc. in Crude Extract	P or S	Salt Conc. in Wash of PEI Ppt.	Salt Conc. to elute Enzyme from PEI Ppt.	Ref. no.	Fold Puri.	Yield (%)
RNA Polymerase	<u>E. coli</u>	0.233M NaCl	P	0.50M NaCl	1.00M NaCl	4	29	89
RNA Polymerase II	Wheat Germ	0.075M $(NH_4)_2SO_4$	P	0.075M $(NH_4)_2SO_4$	0.20M $(NH_4)_2SO_4$	5	31	110
T4 Polynucleotide Kinase	T4-Infected <u>E. coli</u>	0.0M Salt	S	N.A.	N.A.	6	5.0	77
T4 DNA Ligase	T4-Infected <u>E. coli</u>	0.0M Salt	P	Omitted	0.15M NaCl	6	3.3	49
T4 DNA Polymerase	T4-Infected <u>E. coli</u>	0.0M Salt	P	0.15M NaCl	1.00M NaCl	6	4.3	170
T4 RNA Ligase	T4-Infected <u>E. coli</u>	0.0M Salt	P	0.15M NaCl	1.00M NaCl	6	53	216
Bgl II	<u>Bacillus globigii</u>	0.0M Salt	P	Omitted	0.10M NaCl	7	N.G.	N.G.
Bgl II	<u>Bacillus globigii</u>	0.10M NaCl	S	N.A.	N.A.	7	N.G.	N.G.

Abbreviations: P: enzyme precipitated after PEI addition; S: enzyme soluble after PEI addition
N.A: not applicable; N.G: not given

TABLE 2
A REFERENCE LIST OF SOME ENZYMES AND PROTEINS PURIFIED WITH A PEI FRACTIONATION STEP

Enzyme/Protein	Reference	Organism/Tissue
1. Acetyl CoA carboxylase	11	plant sp.
2. Cytokinin oxidase	12	*Phaseolus vulgaris* (bean) callus
3. Dam methylase	13	*E. coli*
4. DNA ligase	6	*E. coli*, phage T4-infected
	14	*E. coli*, lambda lysogen of cloned *E. coli* (NAD-dependent) gene
	15	human placenta
5. DNA polymerase	16,17	*E. coli*, DNA pol I gene cloned and expressed
	18	*E. coli*, DNA pol I gene segment for Klenow fragment cloned and expressed
	6	*E. coli*, phage T4-infected, T4 DNA pol
	19	*E. coli*, DNA pol III
	20	cauliflower
	21	*Drosophila melanogaster*
6. Erf protein	22	*E. coli*, phage P22-infected
7. Fatty acid synthetase	23	yeast
8. Glutamine synthetase	24,25	*E. coli*
9. Gro E protein	25	*E. coli*
10. Growth hormone (human)	26	*E. coli*, gene cloned into an expression vector

TABLE 2 (Continued)

Enzyme/Protein	Reference	Organism/Tissue
11. Guanyl transferase	27	Saccharomyces cerevisiae (yeast)
12. Lambda repressor	28,29	E. coli, phage lambda-infected
13. Lex A repressor	30	E. coli
14. M U-5 methyltransferase	31	E. coli
15. Ntr A gene product	32	E. coli
16. Phytochrome	33-35	Avena sp. (oat) coleoptile
17. Polynucleotide kinase	6	E. coli, phage T4-infected
18. Pyruvate oxidase	36	E. coli
19. Q gene product	37,38	E. coli, phage lambda infected
20. Rec A protein	39	E. coli
	40	Salmonella typhimurium
21. Receptor, 1,25-dihydroxy vitamin D	41	chicken intestines
22. Restriction endonucleases		
Bgl I, II	7	Bacillus globigii
Eco RI	42,43	E. coli RY13
Mla I	44	Mastigocladus laminosus
Rru I, II	45	Rhodospirillum rubrum
Sal I	46	Streptomyces albus
Sma I	46	Serracia marescens
Xam I	46	Xanthomonas amaranthicola
23. Reverse transcriptase	47	E. coli, MoMuLV gene cloned and expressed
24. Rho (transcription factor)	48-50	E. coli
	51	Erwinia caritovora

TABLE 2 (Continued)

	Enzyme/Protein	Reference	Organism/Tissue
25.	RNA ligase	6,52	E. coli, phage T4-infected
		53	wheat germ
26.	RNA polymerase		
	Prokaryotic	4,55-70	various bacterial species
	Eukaryotic		
	Mammalian	71	calf thymus
	Plant	5,72-79	various monocot and dicot species
	Others	80-99	yeast, protozoans, invertebrates, etc.
27.	D-Serine dehydratase	100	E. coli
28.	Single-stranded DNA binding protein	101	E. coli, cloned and overexpressed
29.	Ter components	102	E. coli, phage P2-infected
30.	Topoisomerase I	103	avian erythrocytes
		104	wheat germ
	Topoisomerase II (gyrase)	105-108	E. coli
		109	yeast
31.	Trp repressor	110	E. coli
32.	Uricase	111	Neurospora crassa

REFERENCES

1. Bell DJ, Hoare M, Dunnill P (1982). The formation of protein precipitates and their centrifugal recovery. In Fiechter A (ed): "Advances in Biochemical Engineering/Biotechnology", New York: Springer-Verlag, p 1.
2. Bradford MM (1976). A rapid and sensitive method for the quantitation of microgram quantities of protein utilizing the principle of protein-dye binding. Anal Biochem 72: 248.
3. Warburg O, Christian W (1942). Isolierung und kristallisation des garungsferments enolase. Biochem Zeitschrift 310: 384.
4. Burgess RR, Jendrisak JJ (1975). A procedure for the rapid, large-scale purification of Escherichia coli DNA-dependent RNA polymerase involving Polymin P precipitation and DNA-cellulose chromatography. Biochemistry 14: 4634.
5. Jendrisak JJ, Burgess RR (1975). A new method for the large-scale purification of wheat germ DNA-dependent RNA polymerase II. Biochemistry 14: 4639.
6. Dolganov GM, Chestukhin AV, Shemyakin MF (1981). A new procedure for the simultaneous large-scale purification of bacteriophage T4 induced polynucleotide kinase, DNA ligase, RNA ligase and DNA polymerase. Eur J Biochem 114: 247.
7. Bickle TA, Pirrotta V, Imber R (1977). Simple, general procedure for purifying restriction endonucleases. Nucl Acids Res 4: 2561.
8. Pennell RB (1960). Fractionation and isolation of purified components by precipitation methods, In Putnam F W (ed): "The Plasma Proteins" Vol. 1 New York: Academic Press, p. 9.
9. Sternberg M, Herschberger D (1974). Separation of proteins with poly-acrylic acids. Biochim Biophys Acta 342: 195.
10. Ishii S, Hirata A, Watanabe T (1979). Japan Soc Food Sci Technol. 26: 279.
11. Eastwell KC, Stumpf PK (1983). Regulation of plant acetyl Coa carboxylase by adenylate nucleotides. Plant Physiol 72: 50.

12. Chatfield JM, Armstrong DJ (1986). Regulation of cytokinin oxidase activity in callus tissues of Phaseolus vulgaris- cv Great Northern. Plant Physiol 80: 493.
13. Geier GE, Modrich P (1979). Recognition sequence of the DAM methylase of Escherichia coli K12 and mode of cleavage of Dpn I endonuclease. J Biol Chem 254: 1408.
14. Panasenko SM, Alazard RJ, Lehman IR (1978). Simple, 3-step procedure for large-scale purification of DNA ligase from a hybrid lambda lysogen constructed in vitro. J Biol Chem 253: 4590.
15. Bhat R, Grossman L (1986). Purification and properties of 2 DNA ligases from human placenta. Arch Biochem Biophys 244: 801.
16. Rhodes G, Jentsch KD, Jovin TM (1979). Simple and rapid purification method for Escherichia coli DNA polymerase I. J. Biol Chem 254: 7465.
17. Kelley WS, Stump KH (1979). Rapid procedure for isolation of large quantities of Escherichia coli DNA polymerase I utilizing a lambda pol A transducing phage. J Biol Chem 254: 3206.
18. Joyce CM, Grindley NDF (1983). Construction of a plasmid that overproduces the large proteolytic fragment (Klenow fragment) of DNA polymerase I of Escherichia coli. Proc Natl Acad Sci USA 80: 1830.
19. Kobayashi Y, Kuratomi K (1982). Purification involving polymin P fractionation of Escherichia coli DNA polymerase III which shows a high sedimentation constant. FEBS Lett 138: 211.
20. Yamaguchi M, Chou MY, Matsumoto H, Fukasawa H (1979). Partial purification and characterization of DNA polymerases from the cauliflower inflorescence. Japan J Genet 54: 97.
21. Banks, GR, Boezi JA, Lehman IR (1979). High Molecular weight DNA polymerase from Drosophila melanogaster embryos. Purification, Structure, and Partial Characterization. J Biol Chem 254: 9886.
22. Poteete AR, Fenton AC (1983). DNA binding properties of the erf protein of bacteriophage P22. J Mol Biol 163: 257.
23. Werkmeister K, Johnston RB, Schweizer E (1981). Complementation in vitro between purified mutant fatty acid synthetase complexes of yeast. Eur J Biochem 116: 303.
24. Burton ZF, Sutherland T, Eisenberg D (1981). Refinements in the rapid isolation of glutamine

synthetase from Escherichia coli. Arch Biochem Biophys 211: 507.
25. Burton ZA, Eisenberg D (1980). A procedure for rapid isolation of both gro E Protein and glutamine synthetase from Escherichia coli. Arch Biochem Biophys 205: 478.
26. Prender G, Stebbing N (1981). Purified human growth hormone from Escherichia coli is biologically active. Nature 293: 408.
27. Itoh N, Yamada H, Kaziro Y, Mizumoto K (1987). mRNA guanyl transferase from Saccharomyces cerevisiae. Large scale purification, subunit functions and subcellular localization. J Biol Chem 262: 1989.
28. Craig NL, Roberts JW (1980). Escherichia coli rec A protein-directed cleavage of phage lambda repressor. J Biol Chem 256: 8039.
29. Sauer RT, Anderegg R (1978). Primary structure of lambda repressor. Biochemistry 17: 1092.
30. Schnarr M, Pouyet J, Grangerschnarr M, Daune M (1985). Large scale purification, oligomerization equilibria, and specific interaction of the lex A repressor of Escherichia coli. Biochemistry 24: 2812.
31. Greenberg R, Dudock B (1980). Isolation and characterization of M U-5-methyltransferase from Escherichia coli. J Biol Chem 255: 8296.
32. Hirschman J, Wong PK, Sei K, Keener J, Kustu S (1985). Products of nitrogen regulatory genes and ntr C of enteric bacteria activate gln A transcription in vitro: Evidence that the ntr A product is a sigma factor. Proc Natl Acad Sci USA 82: 7525.
33. Bolton GW (1979). Phytochrome: Aspects of its protein and photochemical properties. Ph.D. Dissertation, University of Minnesota.
34. Tokuhisa JG, Daniels SM, Quail PH (1985). Phytochrome in green tissue. Spectral and immunochemical evidence for 2 distinct molecular species of phytochrome in light grown Avena sativa l. Planta 164: 321.
35. Vierstra RD, Quail PH (1983). Purification and initial characterization of 124 kilodalton phytochrome form Avena. Biochemistry 22: 2498.
36. Recny MA, Hager LP (1982). Reconstitution of native Escherichia coli pyruvate oxidase from apoenzyme monomers and FAD. J Biol Chem 257: 12878.
37. Schechtman MG, Alegre JN, Roberts JW (1980). Assay and characterization of late gene regulators of

bacteriophage phi 82 and bacteriophage lambda. J Mol Biol 142: 269.
38. Grayhack EF, Roberts JW (1983). Purification of the bacteriophage lambda late gene regulator encoded by gene Q. J Biol Chem 258: 9192.
39. Kuramitsu S, Hamaguchi K, Ogawa T, Ogawa H (1981). A large scale preparation and some physiochemical properties of rec A protein. J Biochem 90: 1033.
40. Pierre A, Paoletti C (1983). Purification and characterization of rec A protein from _Salmonella typhimurium_. J Biol Chem 258: 2870.
41. Pike JW, Haussler MR (1979). Purification of chicken intestinal receptor for 1,25 dihydroxyvitamin D. Proc Natl Acad Sci USA 76: 5484.
42. Bingham AHA, Sharman AF, Atkinson T (1977). Purification of restriction endonuclease EcoRI by precipitation involving polyethyleneimine. FEBS lett 76: 250.
43. Sumegi J, Breedveld D, Hossenlopp P, Chambon P (1977). A rapid procedure for purification of EcoRI endonuclease. Biochem Biophys Res Commun 76: 78.
44. Duyvesteyn M, DeWaard A (1980). A new sequence-specific endonuclease from a thermophilic cyanobacterium, _Mastigocladus laminosus_. FEBS Lett 111: 423.
45. Duyvesteyn MCG, DeWaard A, VanOrmondt H (1980). Two sequence-specific deoxyribonucleases from _Rhodospirillum rubrum_. FEBS Lett 117: 241.
46. Carlson K, Nicolaisen B (1979). Cleavage map of bacteriophage T4 cytosine-containing DNA by sequence-specific endonucleases Sal I and Kpn I. J Virol 31: 112.
47. Kotewicz ML, Dalessio JM, Driftmier KM, Blodgett KP, Gerard GF (1985). Cloning and overexpression of moloney murine leukemia virus reverse transcriptase in _Escherichia coli_. Gene 35: 249.
48. Sharp JA, Galloway JL, Platt T (1983). A kinetic mechanism for the poly (C)-dependent ATPase of the _Escherichia coli_ transcription termination protein Rho. J Biol Chem 258: 3482.
49. Finger LR, Richardson JP (1981). Procedure for purification of _Escherichia coli_ ribonucleic acid synthesis termination protein R. Biochemistry 20: 1640.
50. Mott JE, Grant RA, Ho Y-S, Platt T (1985). Maximizing gene expression from plasmid vectors containing the lambda pL promoter: Strategies for

overproducing transcription termination factor rho. Proc Natl Acad Sci USA 82: 88.
51. Nwankwo DO, Guterman SK (1985). Purification of RNA polymerase and transcription termination factor rho from Erwinia carotovora. Eur J Biochem 146: 383.
52. Mosemanmccoy MI, Lubben TH, Gumport RI (1979). Purification of nuclease-free T4 RNA ligase. Biochim Biophys Acta 562: 149.
53. Gegenheimer P, Gabius HJ, Peebles CL, Abelson J (1983). An RNA ligase from wheat germ which participates in transfer RNA splicing in vitro. J Biol Chem 258: 8365.
54. Jaehning JA, Wiggs JL, Chamberlin MJ (1979). Altered promoter selection by a novel form of Bacillus subtilis RNA Polymerase. Proc Natl Acad Sci USA 76: 5470.
55. Heintz N, Shub DA (1982). Transcriptional regulation of bacteriophage SP01 protein synthesis in vitro and in vitro. J Virol 42: 951.
56. Zillig W, Zechel K, Walbwachs H-J (1970). A new method of large scale preparation of highly purified DNA-dependent RNA polymerase from E. coli. Hoppe-Seyler's Z. Physiol Chem 351: 221.
57. Solaiman D, Wu FYH (1985). Preparation and characterization of various Escherichia coli RNA polymerase containing one or two Intrinsic metal ions. Biochemistry 24: 5077.
58. Westin G, Djurhuus R, Skreslett U (1982). In vitro transcription of phage T4 late genes on purified DNA by partially purified RNA polymerase from T4-infected Escherichia coli B cells. Biochim Biophys Acta 699: 28.
59. Shanblatt SH, Nakada D (1982). Escherichia coli mutant which restricts bateriophage T7 has an altered RNA polymerase. J Virol 42: 1123.
60. Niles EG, Conlon SW, Summers WC (1974). Purification an physical characterization of T7 RNA polymerase from T7-infected Escherichia coli B. Biochemistry 13: 3904.
61. Murray Cl, Rabinowitz JC (1981). RNA polymerase from Clostridium acidi-urici. characterization of a naturally occurring rifampicin resistant bacterial enzyme. J Biol Chem 256: 5153.
62. Rudd KE, Zusman DR (1982). RNA polymerase of Myxococcus xanthus. Purification and selective transcription in vitro with bacteriophage templates. J Bacteriol 151: 89.

63. Gragerov AI, Chenchik AA, Aivasashvilli VA, Beabealashvilli RS, Nikiforov BG (1984). Escherichia coli and Pseudomonas putida RNA polymerases display identical contacts with promoters. Mol Gen Genet 195: 511.
64. Lotz W, Fees H, Wohlleben W, Burkardt HJ (1981). Isolation and characterization of the DNA-dependent polymerase of Rhizobium leguminosarium. J Gen Microbiol 125: 301.
65. Boyd DH, Porter LM, Young BS, Wright A (1979). In vitro detection of defects in temperature sensitive RNA Polymerase from mutants of Salmonella typhimurium. Mol Gen Genet 173: 279.
66. Allan B, Greenberg EP, Kropinski A (1986). DNA-dependent RNA polymerase from Spirochaeta aurantia. FEMS Microbiol Lett 35: 205.
67. Jones GH (1978). Sensitivity of transcription by purified Streptomyces antibioticus RNA polymerase to actinomycin. Biochem Biophys Res Commun 84: 962.
68. Jones GH (1979). Purification of RNA polymerase from actinomucin producing and nonproducing cells of Streptomyces Antibioticus. Arch Biochem Biophys 198: 195.
69. Buttner MJ, Brown NL (1985). RNA polymerase DNA interactions in Streptomyces. In vitro studies of a Streptomyces lividans plasmid promoter with Streptomyces coelicolor RNA polymerase. J Mol Biol 185: 177.
70. Meyer J, Gautschi J, Staljammarcarlemalm M, Storl J, Klaus S (1983). DNA of the streptomyces phage SH10. Binding sites for Streptomyces hygroscopicus RNA polymerase and in vitro transcription map. Gene 23: 25.
71. Hodo HG, Blatti SP (1977). Purification using polyethylenimine precipitation and low molecular weight subunit analysis of calf thymus and wheat germ DNA-dependent RNA polymerase II. Biochemistry 16: 2334.
72. Goto H, Sasaki Y, Kamikubo T (1978). Large scale purification and subunit structure of DNA-dependent RNA polymerase II from cauliflower inflorescence. Biochim Biophys Acta 511: 195.
73. Sasaki Y, Ishiye M, Goto H, Kamikubo T (1979). DNA-dependent RNA polymerase III from cauliflower. Characterization and template specificity. Biochim Biophys Acta 517: 205.

74. Jendrisak J, Guilfoyle TJ (1978). Immunological properties and alpha amanitin sensitivities of class II enzymes from higher plants. Biochemistry 17: 1322.
75. Sasaki Y, Ishiye M, Goto H, Kamikubo T (J1979). Purification and subunit structure of RNA polymerase II from the pea. Biochim Biophys Acta 564: 437.
76. Lerbs S, Briat JF, Mache R (1983). Chloroplast RNA polymerase from spinach. Purification and DNA binding proteins. Plant Mol Biol 2: 67.
77. Guilfoyle TJ, Key JL (1976). Subunit structures of soluble and chromatin-bound RNA polymerase II from soybean. Biochem Biophys Res Commun 74: 308.
78. Guilfoyle TJ, Jendrisak J (1978). Plant DNA-dependent RNA polymerases. Subunit structures and enzymatic properties of class 2 enzymes from quiescent and proliferating tissues. Biochemistry 17: 1860.
79. Jendrisak J (1981). Purification and subunit structure of DNA-dependent RNA polymerase III from wheat germ. Plant Physiol 67: 438.
80. Spindler SR, Duester GL, Dalessio JM, Paule MR (1978). Rapid and facile procedure for preparation of RNA polymerase I from *Acanthamoeba castellanii*. Purification and subunit structure. J Biol Chem 253: 4669.
81. Dalessio JM, Spindler ST, Paule MR (1979). DNA-dependent RNA polymerase II from *Acanthamoeba castellanii*. Large scale preparation and subunit composition. J. Biol Chem 254: 4085.
82. Spindler SR, Dalessio JM, Duester GL, Paule MR (1978). DNA dependent RNA polymerase III from *Acanthamoeba castellanii*. Rapid procedure for large scale preparation of homogeneous enzyme. J Biol Chem 253: 4669.
83. Bettiol MF, Irivin RT, Horgen PA (1985). Immunological analysis of selected eukaryotic RNA polymerases II. Canad J Biochem Cell Biol 63: 1217.
84. Vaisius AC, Horgen PA (1979). Purification and characterization of RNA polymerase II resistant to alpha amanitin from the mushroom *Agaricus bisporus*. Biochemistry 18: 795.
85. Stunnenberg HG, Wennekes LMJ, Vandenbroek HWJ (1979). RNA polymerase from the fungus *Aspergillus nidulans*. Large scale purification of DNA-dependent RNA polymerase I (or polymerase A). Eur J Biochem 98: 107.
86. Stunnenberg HG, Wennekes LMJ, Spierings T, Vandenbroek H W U (1981). An alpha amanitin resistant DNA-dependent RNA polymerase II from the fungus *Aspergillus nidulans*. Eur J Biochem 117: 121.

87. Sanford T, Prenger JP, Golomb M (1985). Purification and immunological analysis of RNA polymerase II from Caenorhabditis elegans. J Biol Chem 260: 8064.
88. Greenleaf AL, Borsett LM, Jiamachello PF, Coulter DE (1979). Alpha amanitin resistant D. melanogaster with an altered RNA polymerase II. Cell 18: 613.
89. Coulter DE, Greenleaf AL (1982). Properties of mutationally altered RNA polymerase II of Drosophila. J Biol Chem 257: 1945.
90. Armaleo D, Gross SR (1985). Purification of the 3 nuclear RNA polymerases from Neurospors crassa. J Biol Chem 260: 16169.
91. Smith SS, Braun R (1978). New method for purification of RNA polymerase II (or B) from lower eukaryote Physarum polycephalum. Presence of subforms. Eur J Biochem 82: 309.
92. Barreau C, Begueret J (1982). DNA-dependent RNA polymerase III from the fungus Podospora comata. purification, subunit structure, and comparison with the homologous enzyme of a related species. Eur J Biochem 129: 423.
93. Valenzuela P, Bell GI, Weinberg F, Rutter WJ (1976). Yeast DNA dependent RNA polymerase 1, polymerase 2, and polymerase 3. Existence of subunits common to 3 enzymes. Biochem Biophys Res Commun 71: 1319.
94. Bell GI, Valenzuela P, Rutter WJ (1977). Phosphorylation of yeast DNA-dependent RNA polymerases in vivo and in vitro. Isolation of enzymes and identification of phosphorylated subunits. J Biol Chem 252: 3082.
95. Bitter GA (1983). Purification of DNA-dependent RNA polymerase II from Saccharomyces cerevisiae. Anal Biochem 128: 294.
96. Winkley CS, Keller MJ, Jaehning JA (1985). A multicomponent mitrochondrial RNA polymerase from Saccharomyces cerevisie. J Biol Chem 260: 14214.
97. Kelly JL, Lehman IR (1986). Yeast mitochondrial RNA polymerase. Purification and properties of the catalytic subunits. J Biol Chem 261: 10340.
98. Sittman DB, Stafford DW (1983). Purification and characterization of the RNA polymerases of the sea urchin, Lytechinus variegatus. Prepar Biochem 13: 21.
99. Engelke DR, Shastry BS, Roeder RB (1983). Multiple forms of DNA dependent RNA polymerases in Xenopus laevis. Rapid purification and structural and immunological properties. J Biol Chem 258: 1921.

100. Schiltz E, Schnackerz KD (1976). Sequence studies on D-serine dehydratase of Escherichia coli. Primary structure of tryptic phosphopyridoxyl peptide and of N-terminus. Eur J Biochem 71: 109.
101. Lohman TM, Green JM, Beyer RS (1986). Large scale overproduction and rapid purification of the Escherichia coli ssb gene product. Expression of the ssb gene under lambda pL control. Biochemistry 25: 21.
102. Bowden DW, Modrich P (1985). In vitro maturation of circular bacteriophage P2 DNA. Purification of ter components and characterization of the reaction. J Biol Chem 260: 6999.
103. Trask DK, Muller MT (1983). Biochemical characterization of topoisomerase I purified from avian erythrocytes. Nucl Acids Res 11: 2779.
104. Dynan WS, Jendrisak JJ, Hager DA, Burgess RR (1981). Purification and characterization of wheat germ DNA topoisomerase I (nicking-closing enzyme). J Biol Chem 256: 5860.
105. Liu LF, Wang JC (1978). DNA-DNA gyrase complex. Wrapping of DNA duplex outside enzyme. Cell 15: 979.
106. Gellert M, Mizuuchi K, Odea MH, Nash HA (1976). DNA gyrase enzyme that introduces superhelical turns into DNA. Proc Natl Acad Sci USA 73: 3872.
107. Sugino A, Cozzarelli NR (1980). The intrinsic ATPase of DNA gyrase. J Biol Chem 255: 6299.
108. Klevan L, Wang JC (1980). Deoxyribonucleic acid gyrase-deoxyribonucleic acid complex containing 140 Base Pairs of deoxyribonucleic acid and an a_2b_2 protein core. Biochemistry 19: 5229.
109. Goto T, Wang JC (1984). Yeast DNA topoisomerase II is encoded by a single copy, essential gene. Cell 36: 1073.
110. Tsapakos MJ, Haydock PV, Hermodson M, Somerville RL (1985). Ligand-mediated conformational changes in trp repressor protein of Escherichia coli probed through limited proteolysis and the use of specific antibodies. J Biol Chem 260: 16383.
111. Wang LWC, Marzluf GA (1980). Purification and characterization of uricase, a nitrogen-regulated enzyme, from Neurospora crassa. Arch Biochem Biophys 201: 185.

Protein Purification: Micro to Macro, pages 99–115

USE OF PHASE PARTITIONING TO SCALE-UP PROTEIN PURIFICATION

Maria-Regina Kula

Institut für Enzymtechnologie
der Universität Düsseldorf, in der KFA Jülich
Postfach 20 50, D-5170 Jülich

Application of aqueous two-phase systems is discussed for the extraction of intracellular enzymes from microbial cell homogenates and protein purification. The method exploits the behavior of hydrophilic polymers in solution and provides a fast and gentle separation technique well suited for large scale operation.

INTRODUCTION

The surprising observation that two liquid phases consisting mainly of water may exist in equilibrium has been reported already 1896 by the dutch microbiologist Beijerinck (1), who tried to mix aqueous solutions of soluble starch with agar or gelatine. About 60 years later Per Ake Albertsson found aqueous two-phase systems in certain mixtures of polyethylen glycol (PEG) and potassium phosphate or PEG and dextran. He established a large number of phase diagrams of various aqueous phase systems and described their basic physicochemical properties (2). In fig. 1 the phase diagram of a PEG/salt system is given. Albertsson also recognized the potential of these systems to achieve separation of macromolecules and cellorganelles as well as intact cells by partitioning under conditions which preserve biological activity (2).

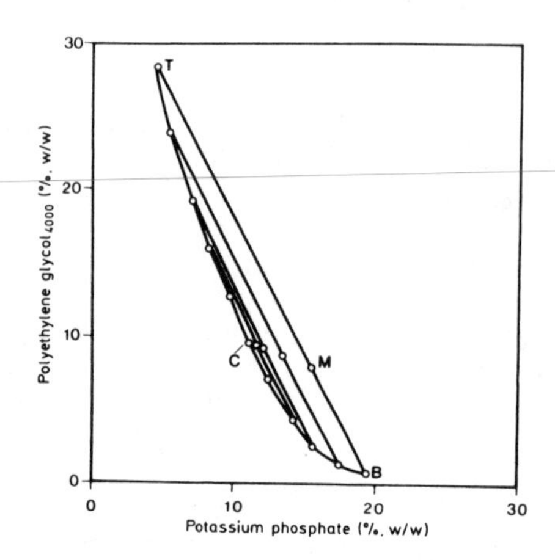

FIGURE 1. Simplified Phase diagram

Phase diagram of the system PEG 4000-potassium phosphate at 20°C and pH 7.0 (data replotted from P.A.Albertsson). Below the binodal homogeneous mixtures are obtained. If M presents the composition of the total system a top phase with composition given by T (rich in PEG) and a bottom phase with composition given by B (rich in salt) are formed. Lines connecting phases, which are in equilibrium are called tielines. Any pair of PEG-salt concentrations represented by the line T over M to B will give rise to an identical phase system however of different volume ratio.

A partition coefficient is defined by:

$$K_i = C_{iT}/C_{iB} = A_{it}/A_{iB} \qquad 1$$

where C_{iT} and A_{iT} denote the concentration or activity of the compound of interest in the top phase and C_{iB} and A_{iB} the concentration or activity in the bottom phase. The partition coefficient of a given protein is a constant and independent of the concentration provided the molecular species are identical in both phases. The partition coefficient is also independent of the presence of other proteins as long as complex formation does not occur. The numerical value of K will depend on the difference in the chemical potential of the protein in the two phases and on the surface area, which interacts with the phase forming polymers and the salts. In principle, the partition coefficient of a protein as well as phase diagrams may be calculated using thermodynamic theories (3,4,5). The Bronstedt equation provides a qualitative description of the process:

$$\ln K = \frac{\lambda \cdot M}{R \cdot T} \qquad 2$$

where R is the gas constant, T the absolute temperature, M the molecular weight and λ a coefficient reflecting the free energy of interactions in the system. For preparative purposes the exponential relation of K with the molecular weight is important. Even for very small differences in λ large values of M will result in a strongly onesided partition. For this reason and considering the favourable environment of aqueous phase systems for proteins we investigated in the last ten years the extraction of intracellular enzymes from microbial cell homogenates and developed it for large scale and industrial use (6-9).

The efficient clarification of microbial cell homogenates by conventional mechanical separation techniques is a very difficult task in large scale (10), because centrifugation as well as filtration depend strongly on the particle size. Considering the small size of microorganisms the harvest of intact bacteria (size ~ 1 μm) is already difficult and requires rather low feedrates. Desintegration of cells produces even smaller cell wall fragments with a wide distribution in size extending below 0.1 μm into the colloidal range. At the same time the viscosity will be slightly increased. Both parameters severely limit the performance of mechanical separation techniques. Partition however is a thermodynamically controlled process and as pointed out above, for very large particles onesided distribution is to be expected. Therefore conditions have to be established to partition cell debris and the desired protein into opposite phases. Separation of the two loaded, immissible, liquid phases should be a much faster process than the solid-liquid separation otherwise necessary. Process design and scale up of protein extractions in aqueous phase systems are discussed below in some detail.

PROCESS DESIGN

In the last 10 years numerous examples have been published demonstrating that indeed cell debris and desired proteins can be partitioned into opposite phases (7-9). The following operational parameters can be employed to achieve this goal:

- Choice of hydrophilic polymer (s) to generate the phase system
- average molecular weight of the polymer (s)
- length of the tieline - complex function of the concentration of the phase forming components
- pH
- choice of salt generating or added to an aqueous phase system
- ionic strength
- temperature

Presently it is much easier, to determine a partition coefficient experimentally and arrive at a useful system by trial and error or by statistic design methods, than to calculate the resulting partition coefficient in various aqueous phase systems from molecular properties of the components involved. With further development of theory, refinement of mathematical models and a more extensive data basis computer aided design of protein extraction appears feasable in years to come. The present experimental approach is limited mainly by the time necessary to quantify the compound of interest. If a fast analytical method is available an experienced person will find a useful system in a few days, often within some hours. Strategies are discussed in references 2,3 and 9.

The most useful variables in our hands for PEG-salt systems are the average molecular weight of PEG, the length of the tieline (see legend to fig. 1) and the pH, while for PEG-dextran systems beside the molecular weight of the polymers the phase potential (2,3) and the addition of salts are important. PEG is commercially available as bulk chemical in various molecular weight fractions. Crude dextran of high molecular weight (5-50 Mio dalton) and several molecular weight fractions purified after limited acid hydrolysis are also commercial products.

If the molecular weight of a phase forming polymer is lowered Flory-Huggins theory predicts, that the partition of a third component increases into the phase dominated by that polymer, conversely partition will be lower in that phase dominated by the polymer with increased molecular weight. In PEG-salt as well as in PEG-dextran systems the PEG rich phase exhibits lower density and therefore forms the top phase. Lowering the molecular weight of PEG should lead to an increase in the partition coefficient of the protein. This has been experimentally verified for numerous proteins, however exceptions from the rule are known (11). Using higher molecular weight fractions of dextran will also increase the partition coefficient, but the effect is less pronounced than altering the molecular weight of PEG (6).

The length of the tieline is a measure of the relative difference in the composition of the two phases in equilibrium and will determine in a complex way the free energy of interaction of the protein with those phases and therefore the chemical potential of the protein. Increasing the length of the tieline in general leads to a higher partition coefficient of protein but it requires also higher concentration of phase forming polymers and carries therefore higher costs.

Because of the high water content of the phases amino acid side chains in proteins and salts are dissociated depending on the pH of the system. The pH can be used to modulate the hydrophobic/hydrophilic balance on protein surfaces changing thereby the free energy of interaction. The whole range of protein stability with regard to pH can be explored to achieve the desired partition. In PEG-dextran systems the pH in addition may influence the phase potential (3,12) eg in the presence of HPO_4^{--} ions proteins carrying excess negative charges will be preferentially exported to the top phase (2,6).

A combination of the parameters discussed above will provide a useful system for extraction of proteins. In general partition coefficients of proteins do not depend strongly on temperature. Most experiments and certainly all large scale applications will be carried out at room temperature

TABLE 1: EXTRACTIVE SEPARATION OF INTRACELLULAR ENZYMES FROM MICROBIAL HOMOGENATES

Enzyme	Organism	Kind of phase system	Cell concentration (%)	Yield (%)	Purification factor
Glucose-6-phosphate DH	Saccharomyces cerevisiae	PEG/salt	30	91	1.8
Alcohol DH		PEG/salt	30	96	2.5
Hexokinase		PEG/salt	30	92	1.6
Formate DH	Candida boidinii	PEG/crude dextran	20	91	n.d.
Formate DH		PEG/salt	33	90	2.0
Isopropanol DH		PEG/salt	20	98	2.6
Fumarase	Escherichia coli	PEG/salt	25	93	3.4
Aspartase		PEG/salt	25	96	6.6
Penicillin acylase		PEG/salt	20	90	8.2
D-Lactate DH	Lactobacillus confusus	PEG/salt	20	95	1.5
L-2-hydroxy isocaproate DH	Lactobacillus confusus	PEG/salt	20	93	17
D-2-hydroxy isocaproate DH	Lactobacillus casei	PEG/salt	20	95	4.9
Glucose-6-phosphate DH	Leuconostoc species	PEG/salt	35	94	1.3
Glucose isomerase	Streptomyces species	PEG/salt	20	86	2.5
Pullulanase	Klebsiella pneumoniae	PEG/dextran	25	91	2.0
Leucine DH	Bacillus spaericus	PEG/crude dextran	20	98	2.4
Leucine DH	Bacillus cereus	PEG/salt	20	98	1.3
Fumarase	Brevibacterium ammoniagenes	PEG/salt	20	83	7.5

DH = dehydrogenase, u.d. not determined

to save energy and inconvenience. Since phase forming chemicals stabilize protein structure this can be done without jeopardizing activity yield. Table 1 shows data for the extraction of various enzymes from cell homogenates of yeast species as well as Gram positive and Gram negative bacteria. Three points are noteworthy:

1. High activity yields ($\gg$ 90 %) are obtained for a single step extraction.
2. Besides cell debris also accompanying proteins are removed resulting in a significant purification of the enzyme of interest.
3. Rather high loads of biomass (20-35 % wet cells) can be processed.

The influence of high biomass concentration on a carrier aqueous phase system is discussed in reference 6 and 8 and will not be considered here. The ability to handle rather high biomass concentrations is of particular importance for scale up and the economy of the process. In fig. 2 the results of a case study are shown analyzing the clarification of cell homogenates by alternate technology.
Cost has been analyzed according to equation:

$$\text{Production cost} = \frac{1{,}13M + 2{,}6L + 1{,}13E + 0{,}13I}{\text{yield}} \quad 3$$

where M denotes material cost including raw material, L labour expenses, E energy cost and I investment cost. Application of equation 3 assumes that in a first approximation biochemical processing has the same costs structure as chemical processing (13). The high space time yield of protein extraction in aqueous phase systems is particulary evident. The absolute values for cost are a matter of assumptions and change with time and location, however the relation between extraction in aqueous phase systems and centrifugation, drum filtration and cross-flow filtration should be much less effected. Figure 2 shows that application of aqueous phase systems is also economically feasable, especially if one considers that the clarified extract obtained by extraction has a much higher quality than that obtained by a conventional solid-liquid separation technique. The protein of interest will be enriched (see table 1) and in addition other cell constituents such as high molecular weight

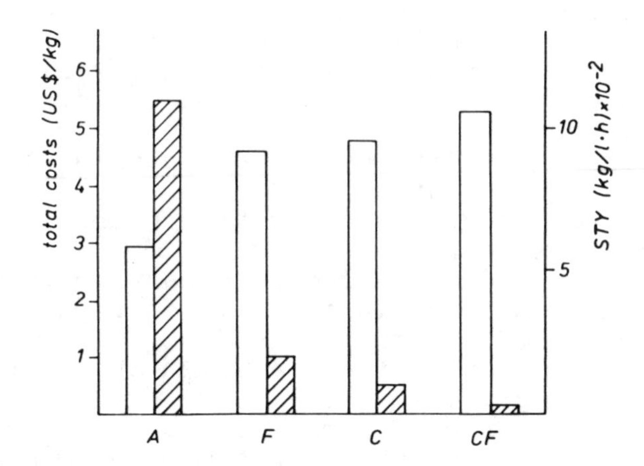

FIGURE 2. Clarification of cell homogenate

Comparison of cost and performance for alternate methods to remove cell debris from the homogenate of 100 kg Saccharomyces cerevisiae (data taken from Kroner et al, reference 13).

A = Extraction in aqueous phases
F = Drum filtration (1.5 m^2) using filter aid
C = Centrifugation, intermittent disk stack separator Σ 7000 m^2
CF = Crossflow filtration, hollow fiber device 20 m^2, average enzyme retention 0,7
open bars = cost
hatched bars = space time yield

nucleic acids, which interfere with later chromatographic separation steps, are removed too by partition (14).

Extraction may also be utilized as a fast and gentle concentration step, removing the major portion of water from eg animal cell culture supernatants containing valuble proteins and effecting a considerable enrichment at the same time (15).

Further purification of a protein can be achieved by treatment of the primary top phase with salt, forming a secondary PEG-salt system. Changing the length of the tieline and if necessary the pH reextraction of the protein of interest into the salt rich phase is possible in good yield. With

TABLE 2: ENZYMES ISOLATED BY SEVERAL SUBSEQUENT SINGLE-STAGE EXTRACTION STEPS

Enzyme	Organism	No. of extraction steps	Overall purification factor	Overall yield (%)
Fumarase	Saccharomyces cerevisiae	2	13	77
Formate dehydrogenase	Candida boidinii	3	4.2	78
Aspartase	Escherichia coli	3	18	82
Penicillin acylase	Escherichia coli	2	10	78
L-2-Hydroxyisocaproate dehydrogenase	Lactobacillus confusus	2	24	80
D-Lactate dehydrogenase	Lactobacillus confusus	2	1.9	91
D-2-Hydroxyisocaproate dehydrogenase	Lactobacillus casei	2	7	85
D-2-Hydroxyisocaproate dehydrogenase	Leuconostoc oenus	2	3.7	77
Glucose-6-phosphate dehydrogenase	Leuconostoc species	2	5	80
Leucine dehydrogenase	Bacillus sphaericus	2	3.1	87
Leucine dehydrogenase	Bacillus cereus	2	2.4	89
Glucose dehydrogenase	Bacillus species	3	33	83
Fumarase	Brevibacterium ammoniagenes	2	22	75
Aspartate ß-decarboxylase	Pseudomonas dacunhae	3	6	78

such a process design most of the PEG introduced as a phase forming polymer may be recovered in the top phase of the secondary phase system and possibly recycled (16). The product containing salt rich phase is usually concentrated and diafiltrated by an ultrafiltration step. Table 2 summarizes a number of enzymes purified by two or three consecutive single stage extraction steps, yielding products with very good performance as industrial catalysts. If homogeneous, highly purified proteins are to be produced high performance chromatographic separation steps can follow without complications from the preceeding partition steps.

APPROACH TO EQUILIBRIUM

For large scale application the time needed to reach equilibrium ist very important. Partition involves mass transport by diffusion. Since macromolecules and even particles are involved it may be argued, that the rate of mass transport will be quite low and equilibrium reached only slowly. A careful experimental study (17) in agitated vessels however demonstrated that equilibrium is reached very fast. Results are shown in table 3.

TABLE 3: APPROACH TO EQUILIBRIUM
PEG/potassium phosphate systems containing 20 % Saccharomyces cerevisiae homogenate

scale liter	volume ratio -	stirrer speed rpm	power input Wm^{-3}	time to equilibrium s^{-1}
21	4.8	200	16	<140
160	4.8	300	175	<80
2.8	3.1	500	61	<50

In the first experiment solid PEG 1500 was added to yeast homogenate together with appropiate amounts of a concentrated salt solution. This way dissolution of PEG, phase formation and partition are superimposed. The first sample was removed after 2 minutes and the phases separated within 20 seconds. The partition coefficient of α-glucosidase was determined and found to be equal to equilibrium values. In the second experiment concentrated PEG solution, yeast homogenate and salt solution were mixed together. In the first sample taken after 1 minute again equilibrium values were found. Similar results were obtained extracting enzymes from Brevibacterium ammoniagenes and E. coli. To differentiate phase formation from partition a PEG-salt system containing 20 % yeast homogenate was preformed and leucine dehydrogenase added as a tracer either to the dispersion or to the bottom phase of the system. In both cases samples were taken after mixing for 30 seconds. Under the conditions employed leucine dehydrogenase has a partition coefficient of 12, so that considerable mass transport to the top phase must have occured to reach the equilibrium observed.

Table 3 also demonstrates that power consumption for mixing is very low in aqueous phase systems, reflecting the low interfacial tension which causes the formation of very small droplets and creates a large mass transfer area and short distances for diffusion. The mass exchange is further aided by the rather fast coalescense and redispersion of the bottom phase. These results demonstrate that the time needed to approach equilibrium of partition will not limit performance in large scale. In practice either agitated vessels or static mixers may be employed.

PHASE SEPARATION

The separation of highly loaded primary aqueous phase systems under unit gravity is a rather slow process. 15 - 20 hours are required to reach a stationary interface (8). Under these conditions 10 - 15 % of product may be lost by entrainment of very small droplets in the bottom phase, which do not rise anymore in the highly viscous phase. However separation in a centrifugal separator is very fast as expected. Table 4 summarizes performance of various small separators employed for this purpose. Maximal feed rates of 110-120 l/h are reported for conventional liquid-liquid purifier bowls (9) resulting in residence times of seconds. If the bottom phase contains crude dextran or a large amount of solids the viscosity of the separated phase becomes very, high leading to flow resistance in the bowl and eventually to clogging of the disk stack. In such cases a nozzle separator may be used with advantage for phase separation (18,19). An optimal feed rate has to be selected where the entrainment in the top phase can still be tolerated and the interface line is established within the bowl, minimizing loss of product containing top phase through the nozzles.

TABLE 4: SEPARATION OF BIOMASS LOADED AQUEOUS TWO PHASE SYSTEMS IN DISK STACK SEPARATORS

separator	bowl	phase system	organism	feedrate liter/h	residence time/sec
SAOH 205[a]	LL	PEG/salt	B.ammoniagenes (20 %)	110	8
SAOH 205[a]	LL	PEG/salt	E.coli (25 %)	66	13
Gyrotester B[b]	LL	PEG/crude dextran	K.pneumonia (20 %)	48	30
Gyrotester B[b]	N	PEG/crude dextran	C.boidinii (20 %)	96	15
YEB 1334[b]	N	PEG/salt	C.boidinii (20 %)	180	16

a Westfalia Separator AG, D-4740 Oelde (FRG)
b Alfa Laval, Tumba, Sweden
LL liquid-liquid purifier bowl, N = nozzle bowl

Separators are operated with the smallest orfices available and reducing the number of nozzles as much as possible. Under such circumstances mechanical losses can be kept ≤ 2 %, which is certainly less than occuring in the interststital spaces of rather loosely packed pellets obtained by medium speed centrifugation.

Due to the rapid establishment of equilibrium and the efficient phase separation in centrifugal separators performance in large scale can be very well predicted from experiments in 10 ml scale as shown in table 5 and make scale-up of extraction processes fairly easy and straight forward.

TABLE 5: SCALE UP OF ENZYME EXTRACTION FROM CELL HOMOGENATES

enzyme	cells	phase system	Yield in 10 ml scale %	Yield in process scale %
formate DH	Candida boidinii	PEG-salt	95	94 (250 liter)
formate DH	Candida boidinii	PEG-crude dextran	99	97 (170 liter)
Leucine DH	Bacillus cereus	PEG-salt	99	98 (128 liter)
Leucine DH	Bacillus sphaericus	PEG-crude dextran	99	97 (160 liter)

After removal of solids secondary phase systems may be separated at unit gravity in a settling tank within 30-90 minutes, in a suitable flow device within a few minutes or at increased gravitational forces even within a few seconds (8). Since mixing and phase separation can easily be carried out continuously, extractions offer the potential of continuous processing.Two and three stage crosscurrent extractions have been described (8,20). With the smallest commercially available centrifual separators intracellular enzymes from 500-1000 kg of wet biomass are extracted within 24 hours. Based on online determination of enzyme activity, protein concentration and flow the process can be controlled by a personal computer. Due to the extremly high capacity of the method data for long term operation ($>$ 20 h) are still missing. Wether operating in batch or continuously there are no apparent technical difficulties in the realization of simple single stage extraction processes.

Multistage operations have been tested and appeared technically feasable, provided special attention is paid to the low interfacial tension and the inherent risk of flooding (7,21). However the data base in this area is still very small and performance is difficult to judge.

AFFINITY PARTITION

The selectivity of partition can be improved if a biospecific ligand is confined to one phase by covalent attachment to a phaseforming polymer and conditions are selected, which allow complex formation in that phase. The basic principle is very similar to affinity chromatography. Certain advantages of affinity partitioning are expected due to the almost complete absence of mass transport resistance in aqueous phase systems, the possibility to operate at higher ligand density in liquid phase and the applicability of solids containing process streams. However ligand recovery is more difficult to achieve in solution than with solid matrices. In both instances repeated use of the affinity ligand appears necessary for economic reasons.

The requirement to promote complex formation excludes PEG-salt systems from primary affinity extractions. Phase formation requires rather high salt concentrations (see fig. 1), this in turn will lead to a dissociation of the complex. Therefore affinity extraction requires PEG-dextran systems and low or moderate ionic strength. The affinity ligand should be bound to PEG so that the complex can be removed with the top phase. In the next step the complex may be dissociated by the addition of salt. This leads to the partition of the desired protein to the bottom phase, while collecting the ligand in the top phase at the same time. We developed recently a large scale process for the extraction of formate dehydrogenase from Candida boidinii according to the strategy outlined above (22). A triazine dye was chosen as ligand for economic reasons. From several available dyes Procion Red was selected because of the low dissociation constant of the enzyme dye complex. It can be shown that in case of general ligands the selectivity of extraction will depend on the strength of the interaction between protein and ligand. The enzyme with the lowest dissociation will be saturated first and taken up by the top phase. 1 mM ligand concentration was sufficient for the extraction of formate dehydrogenase. In table 7 scale up data are presented including recycling of ligand containing top phase.

TABLE 6: AFFINITY PARTITION OF FORMATE DEHYDROGENASE

Scale kg	Yield %	Specific Activity U/mg	Number of experiments	Recycle of top phase
0.005	72 ± 3	5.9 ± 0.06	5	4
5.00	78 ± 2	5.0 ± 0.05	5	4
220.0	74	3.5	1	-

The low standard deviation in the specific activity and yield in the first series of experiments demonstrates that even such highly complex systems will give reproducible results. The drop observed in the specific activity in the large scale experiment has to be attributed to another batch of cells grown under different conditions and containing only about half of the formate dehydrogenase at the start of the experiment. The yield in this case however was as expected, demonstrating again the comparatively easy scale up. A first analysis indicated that affinity partitioning with recycling is comparable in cost to the physical extraction while yielding a product of higher specific activity (22).

CONCLUSIONS

Formation of aqueous two phase systems is based on physico-chemical properties of polymers in solution. Proteins can be partitioned between such phases and easily isolated from cell homogenates converting a very difficult mechanical separation step for the clarification of crude extracts into a thermodynamically controled process. At the same time removal of nucleic acids is achieved and considerable purification of the desired product. Scale up appears easy and straight forward. Necessary chromatographic steps will be improved since less proteins have to be handled and viscous byproducts of fermentation are already separated. Extraction technology is characterized by high space time yield and high recovery. The development of enzyme technology on one hand and increased production of mamalian proteins by recombinant DNA technology should lead to an increased number of applications in the future.

ACKNOWLEDGEMENT

The work described has been carried out in Braunschweig while I was affiliated with the GBF. The contributions of my collaborators A.Cordes, H.Hustedt, K.H.Kroner, U.Menge and H.Schütte are gratefully acknowledged.

References:

1. Beijerinck MW (1896). Über eine Eigentümlichkeit der löslichen Stärke. Zentralbl Bakteriol Abt 2, 2: 697-699.
2. Albertsson PA (1960)."Partition of cell particles and macromolecules", New York: J. Wiley, 2nd edition 1971, 3rd edition 1986.
3. Walter H, Brooks D, Fisher D (eds) (1985): "Partitioning in Aqueous Two-Phase Systems: Theory, Methods, Uses and Applications to Biotechnology", Orlando: Academic Press.
4. King R, Blanch HW, Prausnitz JM (1986). Thermodynamics of aqueous polymer-polymer two-phase systems. ACS-Meeting Anaheim
5. Baskir JN, Halton TA, Suter UW (1986). Thermodynamics of biomaterial partition in two-phase aqueous polymer systems. ACS-Meeting New York
6. Kula MR (1979). Extraction and Purification of Enzymes Using Aqueous Two-Phase Systems. Appl Biochem Bioeng 2: 71-95
7. Kula MR, Kroner KH, Hustedt H (1982): Purification of enzymes by liquid-liquid extraction, Adv Biochem Eng 24: 73-118.
8. Kula MR (1985) Liquid-liquid extraction of biopolymers, in: Humphrey A, Cooney CL (eds.) "Comprehensive Biotechnology", Vol 2, pp 451-471. New York: Pergamon Press
9. Hustedt H, Kroner KH, Kula MR (1985). Applications of phase partitioning in biotechnology. in ref. 3, pp 529-587.
10. Naeher G, Thum W (1974). Production of enzymes for research and clinical use. in: Spencer B (ed)

"Industrial Aspects of Biochemistry", Vol 1, pp 47-64. New York: Elsevier
11. Hummel W, Schütte H, Kula MR (1984). New enzymes for the synthesis of chiral compounds. Enzyme Engineering 7, Ann NY Acad Sci 434: 87-90.
12. Johansson G (1974). Effect of salts on the partition of proteins in aqueous polymeric two-phase systems. Acta Chem Scand B 28: 873-884.
13. Kroner KH, Hustedt H, Kula MR (1984). Extractive enzyme recovery: economic considerations. Proc Biochemistry 19: 170-179.
14. Veide A, Smeds AL, Enfors SO (1983). A process for large scale isolation of ß-galactosidase from E. coli in aqueous two-phase systems. Biotechnol Bioeng 25: 1789-1800.
15. Menge U, Fraune E, Lehmann J, Kula MR. Purification of proteins from cell culture supernatants. Proceedings ESACT-Meeting, Baden bei Wien 1985, in press
16. Hustedt H (1986). Extractive enzyme recovery with simple recycling of phase forming chemicals. Biotechnol Lett 8: 791-796.
17. Fauquex PF, Hustedt H, Kula MR (1985). Phase Equilibration in agitated vessels during extractive enzyme recovery. J Chem Tech Biotechnol 35: 51-59.
18. Kroner KH, Schütte H, Stach W, Kula MR (1982). Scale-up of formate dehydrogenase isolation by partition. J Chem Tech Biotechnol 32: 130-137.
19. Kroner KH, Hustedt H, Kula MR (1982). Evaluation of crude dextran as phase forming polymer for the extraction of enzymes in aqueous two-phase systems in large scale. Biotechnol Bioeng 24: 1015-1045.
20. Hustedt H, Kroner KH, Kula MR (1984). Continuous enzyme purification by crosscurrent extraction. Proceedings 3rd Eur Congr Biotechnology Vol 1: pp 597-605.
21. Schütte H, Kroner KH, Hummel W, Kula MR (1983). Recent developments in separation and purification of biomolecules. Ann NY Acad Sci 413: 270-282.
22. Cordes A, Kula MR (1986). Process design for large-scale purification of formate dehydrogenase from Candida boidinii by affinity partition. J Chromatography 376: 375-384.

Protein Purification: Micro to Macro, pages 117–130

PROTEIN SEPARATIONS USING REVERSED MICELLES[1]

Jaclyn M. Woll, Anne S. Dillon,
Reza S. Rahaman and T. Alan Hatton

Department of Chemical Engineering
Massachusetts Institute of Technology
Cambridge, MA 02139

ABSTRACT Reversed micelles, surfactant aggregates in organic solvents, can be used to extract proteins selectively from aqueous solutions. Selectivity of the extraction can be controlled through changes in pH, ionic strength, surfactant concentration and surfactant type. Partition coefficients for the proteins studied are found to be independent of protein concentration, and of the presence of other proteins in solution. Affinity separations can be achieved through the addition of a second surfactant having an affinity ligand as its polar headgroup. Extraction of an extracellular alkaline protease from a clarified fermentation broth resulted in a doubling of the specific activity of the recovered protein mass.

INTRODUCTION

Liquid-liquid extraction provides an attractive alternative to conventional procedures for the recovery and purification of proteins and other biopolymers, since it combines moderate to high selectivities with high volumetric capacities and the promise of truly continuous operations. For this technology to be useful on a large scale, it is necessary to identify inexpensive solvents having the desired

---

[1]This work was supported by the NSF Biotechnology Process Engineering Center at M.I.T. and by a grant from Alfa Laval.

selectivity characteristics, but which do not damage the labile bioproducts. A class of solvents which satisfies these requirements is that of reversed micelle-containing organic phases.

Reversed micelles are surfactant aggregates in organic solvents capable of solubilising significant quantities of polar compounds, including proteins, in the bulk organic phase. As shown in Figure 1, these micelles consist of a polar core of water and other solubilised species, stabilised by a surfactant layer in which the polar headgroups point inwards, while the hydrophobic tails extend into the surrounding organic phase. The proteins solubilised within this polar core are protected from the harsh organic environment by the water shell around the protein. The protein can be induced to move from a bulk aqueous phase into the micelle-containing organic solvent, and *vice-versa*, by manipulation of the pH, ionic strength and surfactant concentration [1-9]. It has been shown that protein mixtures can be readily resolved using this technique [5], and that the extraction/stripping cycle can be operated on a continuous basis [8].

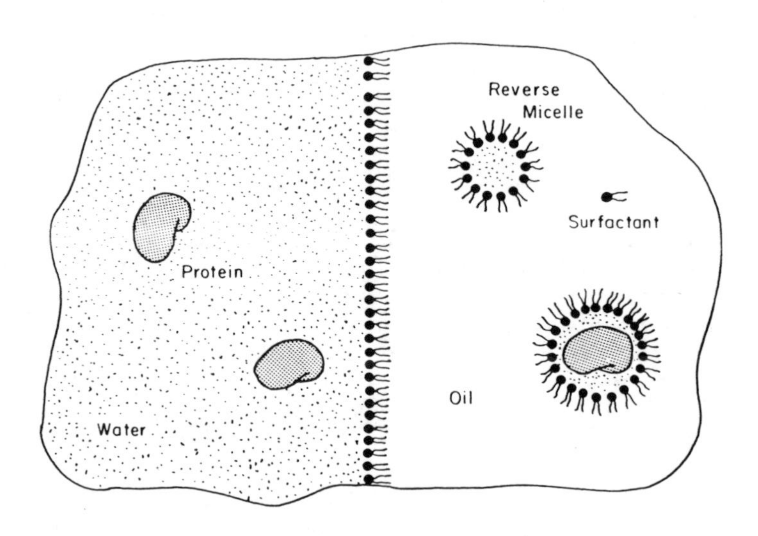

FIGURE 1. Protein solubilisation in reversed micelles.

In this paper, we demonstrate the dependence of protein partitioning between the phases on the protein and surfactant concentrations, the relative volumes of the phases being contacted, and the presence of biologically active surfactants. Preliminary results on the extraction of an extracellular alkaline protease from a clarified fermentation medium are also presented to show that this system can be used under realistic operating conditions.

EXPERIMENTAL

All solubilisation experiments were carried out by intimately contacting an aqueous protein solution with a surfactant-laden organic phase under intense agitation conditions for a time sufficient to attain phase equilibrium. This was followed by centrifugation to ensure complete phase separation. The organic solvent used was iso-octane, and the anionic surfactant employed in these studies was Aerosol-OT, or AOT (sodium di-2-ethyl-hexyl sulfosuccinate). Unless otherwise stated, the standard experiment consisted of contacting equal volumes of 50mM AOT in isooctane with an aqueous 1 mg/ml protein, 0.1 M KCl solution. Except for the studies with a real fermentation broth, the protein solutions were buffered with low concentrations of potassium phosphate to ensure constant pH conditions during the extraction process. Initial and final protein concentrations were determined by UV/visible spectroscopy, reversed phase HPLC, and/or Lowry assay.

SOLUBILISATION STUDIES

The transfer of proteins from an aqueous solution to a reversed micellar organic phase appears to be dominated by electrostatic interactions between the charged protein and the surfactant head layer forming the walls of the micellar polar core. Only when the two are of opposite polarity, i.e., when the electrostatic interactions are favourable, can solubilisation be expected. Thus, with the anionic surfactant AOT, solubilisation is observed only for pH's below the isoelectric point, pI, of the protein. This is evident from the results of Goklen and Hatton [5] shown in Figure 2, where the percent of protein transferred from an unbuffered aqueous solution to an equal volume of solvent is shown as a function of the solution pH for the small, equi-sized proteins cytochrome-c, ribonuclease-a and lysozyme. The point at which each solubilisation curve

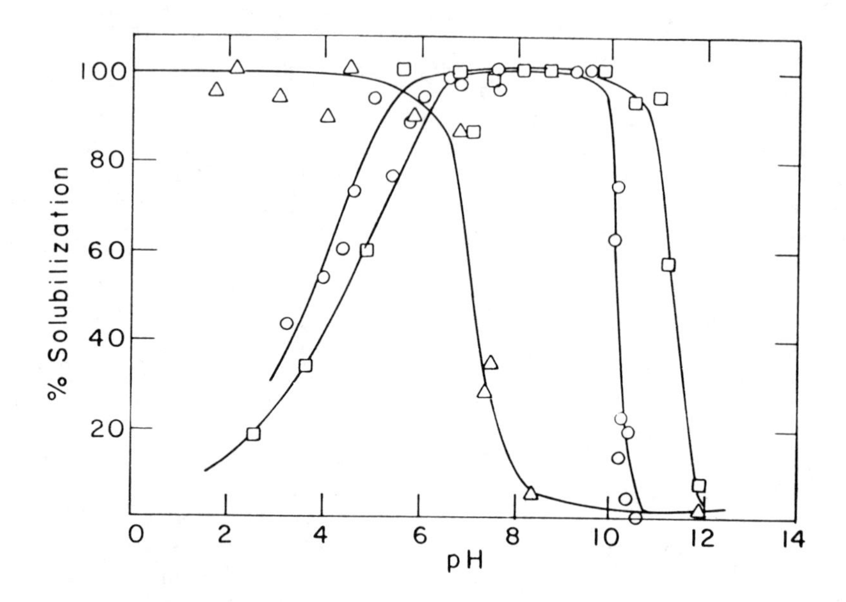

FIGURE 2. Effect of pH on solubilisation of cytochrome-c, (○), lysozyme (□), and ribonuclease-a (△). (From Ref. 5, with permission)

changes abruptly, from total transfer to no transfer of protein to the organic phase, corresponds in each case to the pI of the particular protein studied. (The fall-off at lower pH's was attributed to protein denaturation, which led to precipitate formation at the interface).

A more detailed study of the solubilisation of proteins over the narrow pH region below the pI where the transfer is not quantitative has been undertaken to explore more closely the effect of protein and surfactant concentrations on the partitioning behaviour. The results are shown in Figures 3, 4 and 5.

In Figure 3, the percent ribonuclease-a transferred is seen to be independent of the initial protein concentration. Thus, over this wide concentration range, the partition coefficient (the ratio of the concentration in the organic phase to that in the aqueous phase) is dependent only on pH and not on protein concentration. Similar results have been observed for the larger protein, concanavalin-a, which exists as a dimer of molecular weight 55,000 under these pH conditions. The partitioning behaviour of each of these

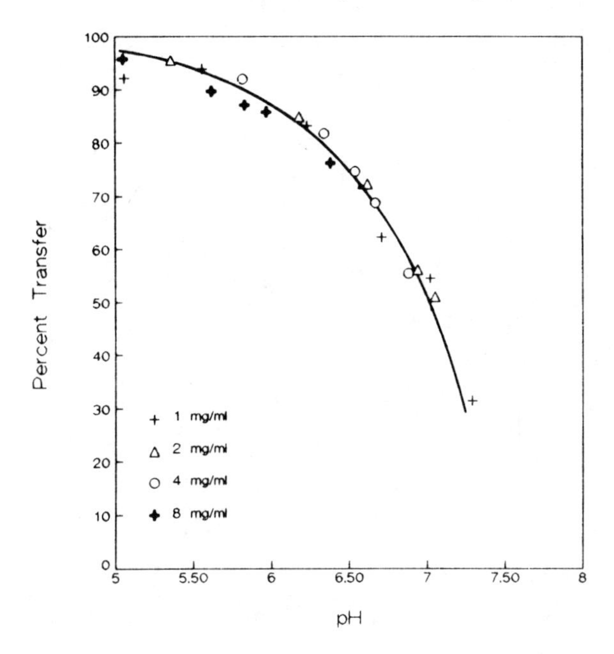

FIGURE 3. Effect of protein concentration and pH on the transfer of ribonuclease-a to a reversed micellar solution.

proteins was unaffected by the presence of the other in solution, indicating no inter-protein interactions. Only at the highest concanavlin-a loading was there seen to be some effect of the protein concentration on the partition coefficient, and this has been attributed to the limited capacity of the organic phase for the larger protein. Earlier studies have pointed to there being a size exclusion phenomenon operative in the solubilisation process, owing to the requirement for the formation of a thermodynamically less favourable larger micelle to accommodate the larger proteins [4,6].

The effect of solvent to feed ratio on the extraction of ribonuclease-a from the aqueous phase is shown in Figure 4, where now the final organic phase concentration is shown as a function of pH. The initial protein concentration was 1 mg/ml. Also shown are the curves calculated using the partition coefficients obtained from the data in Figure 3 (and shown in Figure 5 as the 50 mM AOT curve). The agreement is good, but more important is the fact that the extraction process can result in a significant increase in

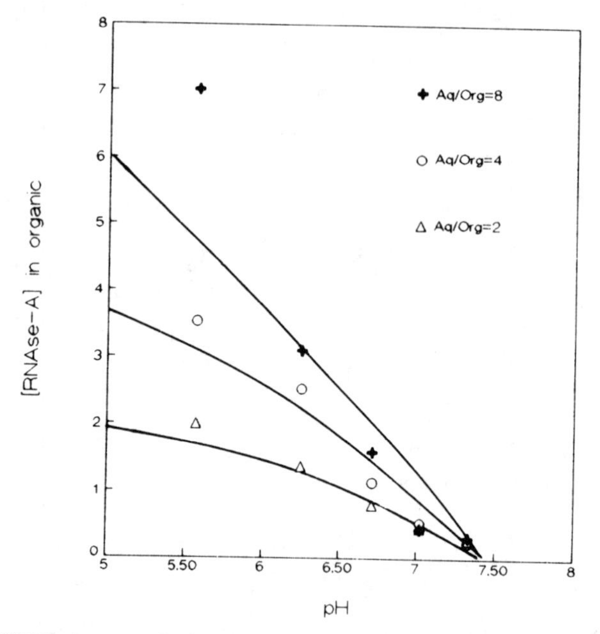

FIGURE 4. The concentration of ribonuclease-a in the organic phase as determined by the solvent-to-feed ratio.

protein concentration. At the lower pH's, where transfer is essentially complete, the concentration factor is just the phase volume ratio used in the extraction. In this case, a maximum of an 8-fold increase was obtained, but there appears to be no limitation on the concentration effect, up to the saturation limit of the solvent phase. If, in back transferring the protein to a second aqueous phase, similar concentration factors can be attained, it is evident that this technique can have strong concentrating powers indeed.

The capacity of the organic phase for the protein is limited by the total concentration of reversed micelles that can form within this phase to accommodate the solute. Thus, we anticipate that the higher the surfactant concentration, the greater will be the degree of protein solubilisation, for any given pH. This is seen to be the case in Figure 5, where partition coefficients for ribonuclease-a are shown as a function of pH for different surfactant concentrations. The net effect is for the solubilisation curve to become steeper near the pI of the protein, and for the region over which only partial transfer is obtained to become narrower.

Similar results have been observed previously for the proteins chymotrypsin, chymotrypsinogen and trypsin [4,6].

The dependence of solubilisation on surfactant concentration can be used to change the selectivity of the extraction, defined as the ratio of the partition coefficients for the proteins under consideration. In Figure 6, the selectivity of the process for ribonuclease-a over concanavalin-A is shown as a function of the solution pH for two different AOT concentrations. It is clear that increasing the surfactant loading shifts the selectivity curve to higher pH values, and increases the peak value of this selectivity. Additional sensitivity can be induced by manipulating the ionic strength of the protein solution, and it is evident that reversed micellar extraction of proteins can have a number of degrees of freedom in the choice of operating conditions to maximise separation and purification of bioproducts.

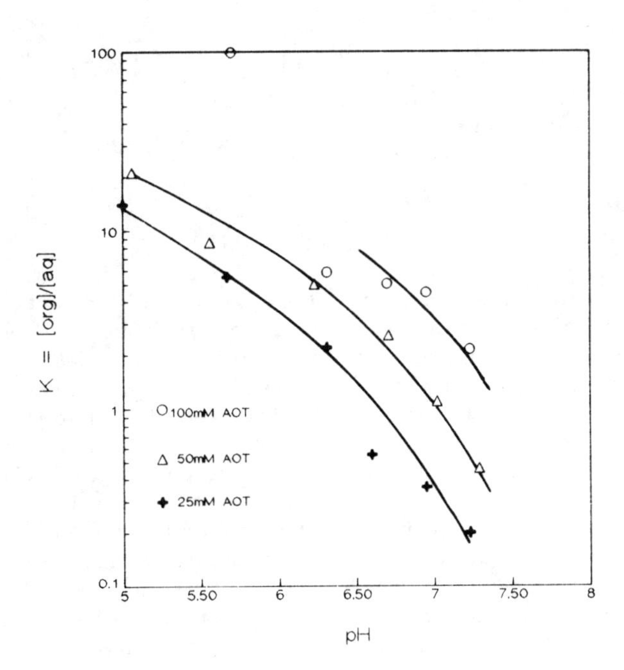

FIGURE 5. Effect of surfactant concentration on partition coefficients for ribonuclease-a.

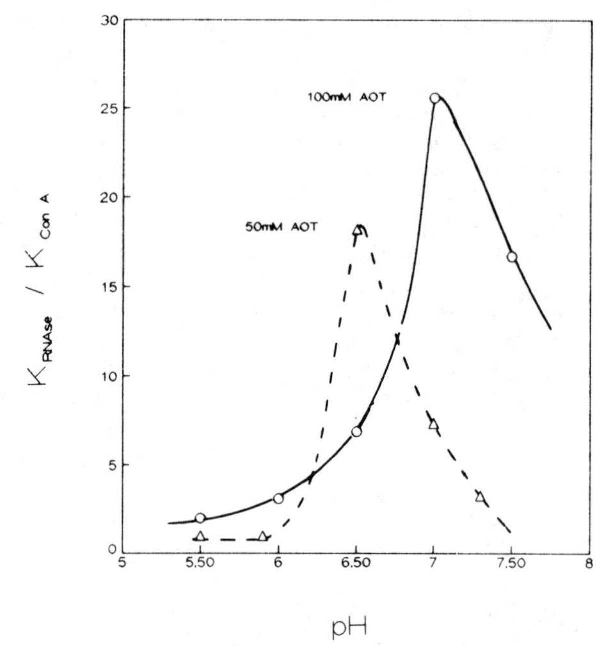

FIGURE 6. Extraction selectivity as a function of pH and AOT concentration.

AFFINITY PARTITIONING

Further enhancements in selectivity could be attained if it were possible to incorporate within the micellar extraction process a molecular recognition capability of the kind used in conventional affinity separations. For instance, if the affinity ligands used in chromatographic separations were to be grafted onto a long alkyl chain, the resulting amphipathic molecule would be positioned within the surfactant layer of the reversed micelle shell, with the hydrophilic ligand group protruding into the polar core. This ligand (frequently a substrate or inhibitor for the enzyme) would bind selectively to the target protein in the mixture, and pull the protein into the reversed micelle core. The long alkyl tail group would ensure that the affinity surfactant was anchored within the extractant phase and not stripped into the aqueous phase. The basic concepts are illustrated schematically in Figure 7, where it is clear that there is a need for only one affinity ligand surfactant per

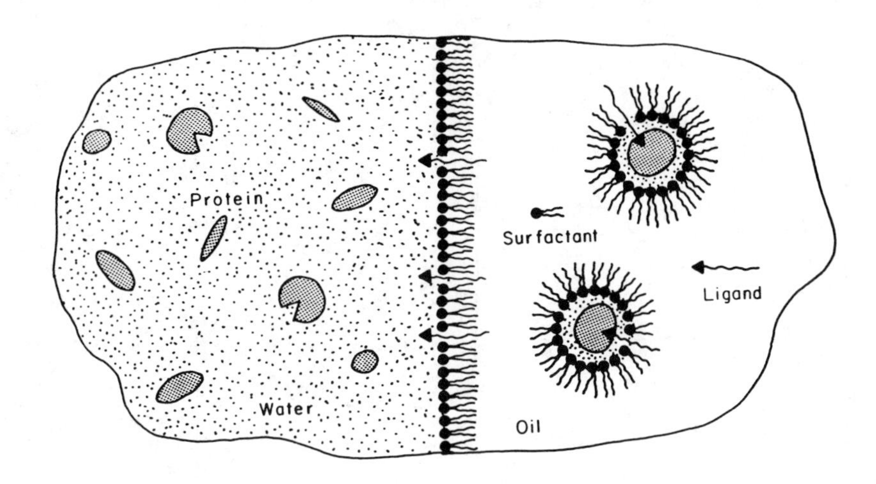

FIGURE 7. Principle of affinity partitioning in reversed micelles.

micelle, on average, since it is reasonable to assume single occupancy of these micelles.

Initial results have been obtained using the biological surfactant octyl-beta-D-glucopyranoside in AOT reversed micelles to enhance the extraction of concanavalin-A, a protein whose function is to bind carbohydrates. This surfactant is insoluble in isooctane, but low concentrations of the pyranoside are readily solubilised in normal 50 mM AOT solution, suggesting that it is behaving as a cosurfactant in the formation of mixed micelles. The aqueous solutions were bufferred, and low concentrations of calcium and magnesium salts were used, these heavy metal ions being required for concanavlin-a activity.

The affinity partitioning results are shown in Figure 8. The percent protein transferred as a function of pH is seen to depend strongly on the biosurfactant concentration, indicating that indeed the affinity surfactant does enhance the extraction of this protein. This was not an artifact resulting from the formation of mixed micelles having different solubilisation characteristics than the original micelles, since the presence of the biosurfactant did not affect the partitioning of ribonuclease-a, either in solution

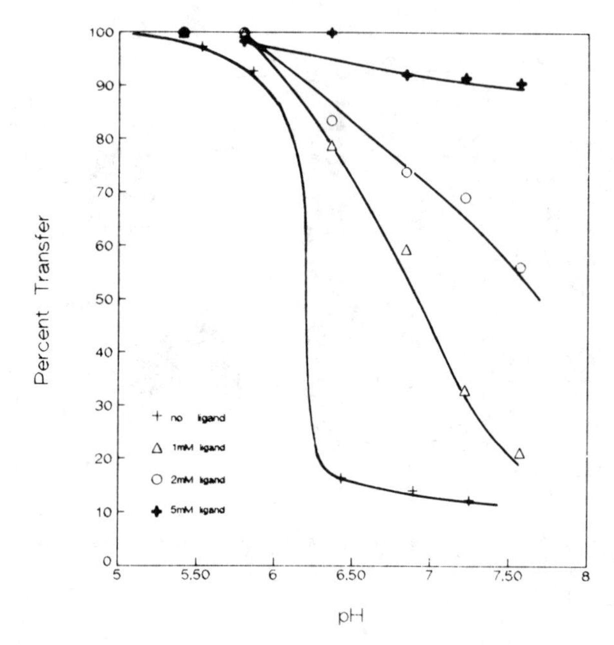

FIGURE 8. Effect of an affinity ligand on the partitioning of concanavalin-A to a reversed micelle phase.

by itself, or when in a mixture with the concanavalin-A. Thus it is apparent that added flexibility in the extraction of bioproducts from aqueous solutions can be provided by using affinity surfactants.

EXTRACTION OF AN EXTRACELLULAR ALKALINE PROTEASE FROM A FERMENTATION BROTH

All studies on protein recovery using reversed micelles reported to date have dealt with clean model solutions, and it has not been demonstrated that the solvents will be effective in the extraction of proteins from the complex solutions typical of real fermentation media. The separation of desired products from fermentation broths introduces additional considerations to those inherent in simple synthetic systems. In preliminary studies on the recovery of a detergent enzyme, an extracellular alkaline protease from Bacillus Sp. ATCC 21536, we have found that it is possible to extract and purify the desired protein using a multi-stage extraction procedure, although there is a tradeoff between the recovery of protein mass and the recovery of protein activity as the pH of the solution is varied, as shown in Figure 9. Lower pH's increase the driving force for protein transfer because of increasingly favourable electrostatic interactions between the protein and the surfactants, but below a pH of about 5.0 there is a dramatic falloff in protein activity. This is not unexpected, as the optimum pH for this protease is approximately 11.0.

The results shown in Figure 10 are indicative of the type of separation possible using staged extractions. The feed solution, adjusted to a pH of 5.4, was contacted twice with a fresh 200mM AOT-isooctane solution, and the resulting extractant phases were stripped of the extracted protein using a 0.1 M KCl, pH 11.1 solution. Approximately ten percent of the total protein feed mass was recovered in each extraction step, as determined by reversed phase HPLC, although the activity yield was greater, being approximately twenty percent of the initial feed activity. This increase in the specific activity of the recovered protein, by a factor of 2.2 for each stage, illustrates the ability of this system to selectively extract the desired protein from the

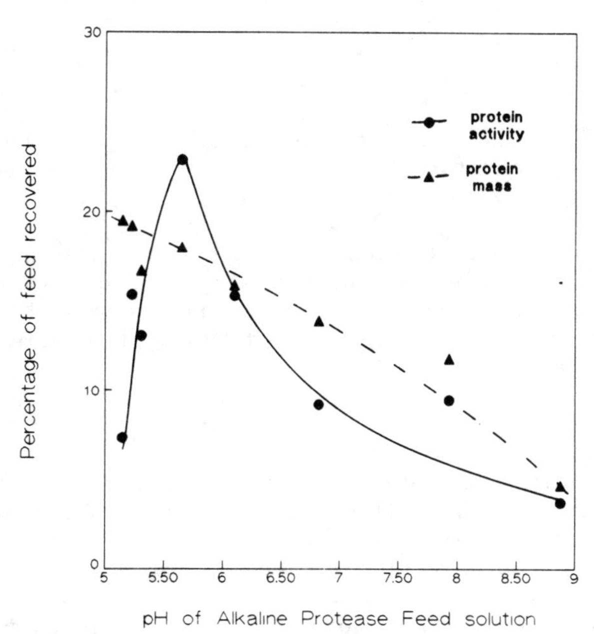

FIGURE 9. Effect of pH on recovery of total protein mass and alkaline protease activity from fermentation broth.

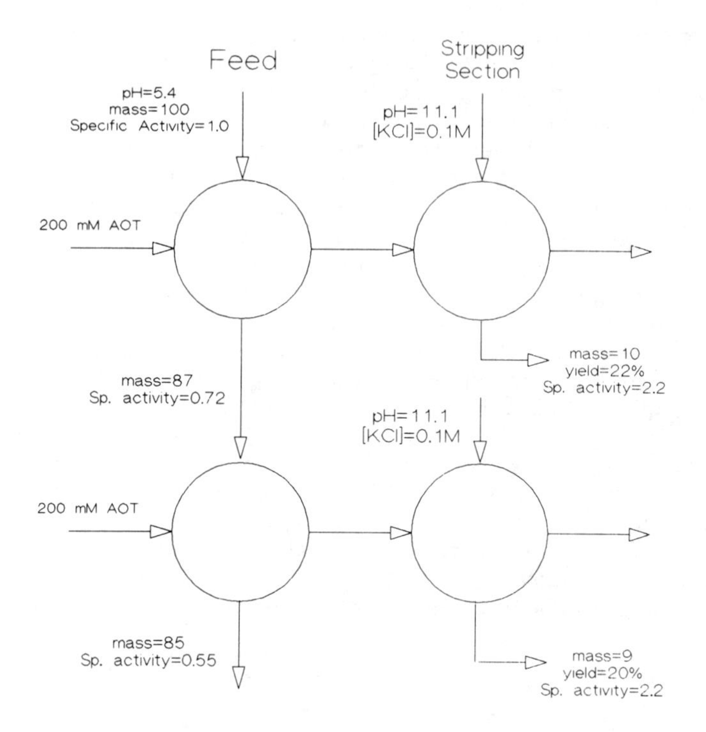

FIGURE 10. Staged extraction of an alkaline protease from a fermentation broth.

fermentation broth with no apparent loss of enzyme activity. With the addition of further stages, and the use of smaller stripping solution volumes, it will be possible to increase the product recovery, and to obtain a more concentrated product solution.

CONCLUSION

Reversed micelles show potential as effective extractants for the large-scale continuous recovery, concentration and purification of bioproducts from aqueous solutions and fermentation media. Selectivity of the process is influenced strongly by the electrostatic interactions between the charged protein surface residues and the ionic surfactant headgroups, and, as such, is very sensitive to variations in pH near the isoelectric point of the protein to be extracted. In addition to pH, the solubilisation of proteins can be affected by the concentration of surfactant used in the extractant solution, and this can be used to modify the selectivity for one protein relative to another in a protein mixture. Additional specificity can also be incorporated in protein extraction operations by using small concentrations of a second surfactant species having molecular recognition capabilities. In this manner, the advantages associated with other affinity separations can also be exploited in the reversed micellar extraction of proteins.

ACKNOWLEDGMENTS

We would like to thank Dr M. L. Yarmush of the Chemical Engineering Department at M.I.T. for his help in selecting an appropriate affinity ligand system for the study of affinity partitioning in reversed micelles, and Dr J. Cabral of the Instituto Superior Tecnico in Lisbon, Portugal, for his collaboration on the alkaline protease project, and for supplying the fermentation media used in this work.

REFERENCES

1. Luisi PL, Bonner FJ, Pellegrini A, Wiget P, Wolf R (1979). Micellar Solubilization of Proteins in Aprotic Solvents and Their Spectroscopic Characterization. Helv Chim Acta 62:740.
2. Meier P, Imre E, Fleschar M, Luisi PL (1984). Further investigations on the micellar solubilisation of bio-polymers in apolar solvents. In Mittal KL, Lindman B (eds): "Surfactants in Solution," New York: Plenum Press.
3. Goklen KE, Hatton TA (1985). Protein Extraction Using Reversed Micelles. Biotech Prog 1:1.
4. Goklen KE, Hatton TA (1986). Liquid Liquid Extraction of Proteins Using Reversed Micelles. Proc ISEC'86 Munich.
5. Goklen KE, Hatton TA (1987). Liquid-Liquid Extraction of Low Molecular Weight Proteins by Selective Solubilization in Reverse Micelles. Sep Sci Tech (in press).
6. Goklen KE, Hatton TA (1987). Separation of Proteins by Liquid Extraction: Selective Solubilization in Reverse Micelle Solutions. (in preparation).
7. Van't Riet K, Dekker M (1984). Preliminary Investigations on an Enzyme Liquid-Liquid Extraction Process. Proc 3rd Eur Con on Biotech Munich.
8. Dekker M, V'ant Riet K, Weijers SR (1986). Enzyme Recovery by Liquid-Liquid Extraction Using Reverse Micelles. Biochem Engr J (in press).

9. Leser ME, Wei G, Luisi PL, Maestro M (1986). Application of Reverse Micelles for the Extraction of Proteins. Biochem Biophys Res Comm 135:629.

Protein Purification: Micro to Macro, pages 131–148

THREE PHASE PARTITIONING (TPP) VIA t-BUTANOL: ENZYMES SEPARATION FROM CRUDES

Rex Lovrien, Craig Goldensoph, Paul C. Anderson and Bruce Odegaard

Biochemistry Department, Gortner Laboratory, University of Minnesota, St. Paul, Minnesota 55108

ABSTRACT. A three phase partitioning method using ammonium sulfate and t-butanol was developed for separating enzymes and proteins into a midlayer. Fourteen crude enzymes and proteins were investigated. Retrieval of most of the total activity and increases in specific activity was yielded for the majority of enzymes. In addition to partitioning proteins and saccharides from one another, TPP acts as an extraction method for routing apolar pigments into the t-butanol layer (cleaning). It works at room temperature and handles gram quantities of crude inputs in relatively small volumes.

INTRODUCTION

Three Phase Partitioning via t-butanol (TPP) was first used as an upstream method for isolating cellulases (1). Cellulases are multienzymes and include three classes. Retrieval of all three classes of enzymes in one TPP step, and earlier work in 1972 (2) showing how t-butanol enhances enzyme activities indicated that TPP might be expanded further. Two other laboratories have used TPP-t-butanol separation (3,4). Eighteen enzymes and proteins have been examined in their TPP behavior here. Occasionally enzymes are inactivated by TPP. However most of the enzymes we've investigated emerged with retention of their activity. Our work has dealt mainly with carbohydrase enzymes.

Figure 1 outlines TPP. Table 1 lists a typical protocol. Tertiary butanol is infinitely soluble in neat water. In ca. 30% saturation ammonium sulfate, t-butanol forms a second

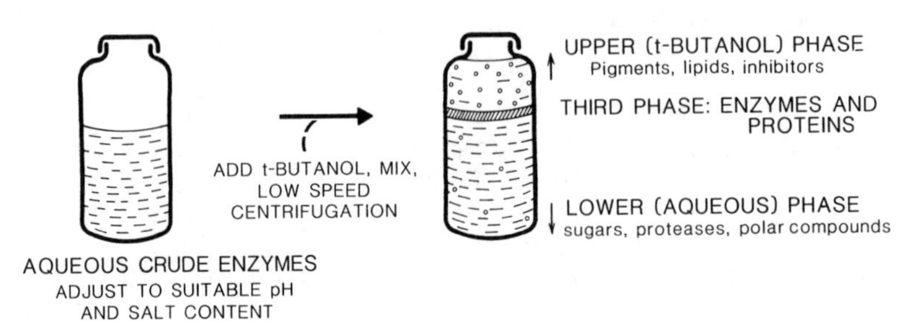

Figure 1. Means for carrying out TPP starting with crude enzymes in aqueous buffer.

Table 1
STEPS IN TPP t-BUTANOL, 20-30°C

Step	Time Needed
1. Adjust pH of crude aqueous sample.	1-5 min
2. Add ammonium sulfate to desired concentration.	1 min
3. Add t-butanol 1 ml./ml. aqueous phase, mix.	2 min
4. Centrifuge (200-2000 R.P.M.) or let settle.	2-5 min
5. Remove t-butanol layer.	2 min
6. Retrieve middle enzymes/proteins layer or penetrate it with a needle to pull out the lower aqueous layer and leave the enzymes/protein layer behind.	2 min

phase. If the aqueous phase contained enzymes to start with and was adjusted to a suitable pH and salt concentration, enzymes come out of the aqueous phase on addition of t-butanol. The phases usually develop cleanly on low speed centrifugation and sometimes by simple settling. The midphase often is a gel containing the desired enzymes plus t-butanol, water and salts. Such gels usually redissolve on adding aqueous buffer. For analysis and assay, redissolved enzymes were usually dialyzed to get rid of NH_4^+ (in determining microkjeldahl nitrogen), and remove reducing sugars and phenolics that interfere with protein analyses. Dialysis is not necessary for most applied uses.

Under some conditions certain enzymes do not come out in the midlayer in the first stage of TPP. Instead unwanted polymers and foreign molecules may be pushed out for discard, leaving the desired enzyme still in solution in the aqueous lower phase. On adding more salt or on shifting pH, the enzyme may then come out and form a midlayer for capture. Such a procedure is 'two stage TPP'. It is outlined using a crude *B. amyloliquifaciens* α-amylase as an example:

Crude α-Amylase pH 7.5, 25° → (First stage, 75% ammonium sulfate) → Midlayer ppt., includes all amylolytic and most proteolytic activity. Dialyze. → (Second stage, 30-40% ammonium sulfate, pH 7.5.) →
- Midlayer, 6X overall increase in S.A. α-Amylase
- Lower aqueous layer, 3X overall increase in S.A. protease

Nearly all our work with TPP dealt with very crude commercial preparations. Some crudes are simply lyophilized fermentation broths, or lysates of cells, or broken cells. Many crudes are very pigmented and laden with foreign polymers; starches, nucleic acids, etc. Two matters concerning crudes need realistic regard. (i) Some isolations steps ostensibly work, when dealing with simple mixtures of already-clean proteins. However the gross impurities in crude enzymes such as emulsifying agents quite derail some of these methods which have only been challenged with ideal, concocted test mixtures. (ii) Impurities such as phenolic glycosides, lipids, etc. grossly interfere with analyses and compromise the data from monitoring separations. A main question in development of any separations technique is not simply whether separation is possible for two or three common prepurified 'test proteins'. Rather, what do gross impurities (junk) in real crudes do to erstwhile separations procedures and erstwhile analyses?

TPP via t-butanol functions as an extraction method besides as a proteins partitioning method. Extraction is increasingly helpful as input crudes become increasingly dirty. TPP outlined in Figure 1 may appear as a hybridization of conventional salting out and ethanol/acetone precipitation. So in fact it is, but there are important differences in practice. Besides extraction, one can operate at room temperature using the t-butanol system. Conventional ethanol/acetone precipitation requires -5 to -20°C in most applications. Rather less ammonium sulfate is needed for TPP than in conventional salting out technique, depending on pH.

Partitioning via t-butanol also may appear as simply a variation of Albertsson's two phase water soluble polymers systems for proteins (5). However we start from a very different predicate than that of Albertsson systems concerning the putative need for all enzymes and proteins to be kept strictly aqueous. Most enzymes are not in strictly aqueous environments in vivo. There is no concrete reason for expecting that enzymes have to be separated or used in neat aqueous solvents in vitro (6,7,8). In both kinds of systems, t-butanol and aqueous polymers, the phase builders are mixed in the output, the desired enzyme. However, t-butanol is considerably easier to shed than aqueous polymers.

ANALYTIC METHODS

Enzyme total activities (T.A.) combined with total protein measurements (T.P.) give specific activity quotients (S.A.):

$$\text{Specific activity (S.A.)} = \frac{\text{Total activity (T.A.)}}{\text{Total protein (T.P.)}}$$

Carbohydrase activities are reported in (μmoles reducing sugar production)/(minute) and in (μmole substrate conversion)/(minute). Protease activities were obtained in units (mg. dry weight protein solubilized)/(minute) via conversion factors recently determined by us (9) for using spectrophotometric data from standard protein substrates such as azocasein. Relative activities before and after TPP separation of proteins like Bowman-Birk Inhibitor are expressed in an operational manner. The slope of BBI's inhibitory titration of trypsin gotten from trypsin's action on the synthetic substrate BAPNA measures BBI's activity (10). Peroxidase and lipase were monitored via dianisidine prochromogen and p-nitrophenyl palmitate ester respectively. Their activities are expressed in (μmoles substrate converted)/(minute) under pH and buffer conditions given in more detail below.

Impurities in crudes strongly interfere with total protein (TP) analyses. Accordingly four analytic methods were used: Spectrophotometric 'microkjeldahl' total nitrogen, biuret, Hartree-Lowry and bicinchoninic acid (BCA). Our slopes for calibrating T.P., slopes for reducing sugar and for neutral sugars analysis, and our calibrating materials and references are listed in Table 2.

We avoided Coomassie Blue (Bradford) analysis of T.P. because several papers render the Coomassie method dubious

TABLE 2[a]
ANALYSIS FOR CARBOHYDRATE AND TOTAL PROTEIN; SLOPES OF CALIBRATION PLOTS[a]

Method	Reference	Calbrating Compound	Measured Slope
Dinitrosalicylate (DNS)	(11)	Glucose	5.50 A_{575}//(cm)·(mg sugar)/(ml. f.a.v.)
Nelson-Somogyi reducing sugar	(12)	Glucose	$6.3x10^{-3}$ A_{520}//(cm)·(nanomole glucose)/(ml. f.a.v.)
Phenol-sulfuric acid neutral sugar	(13)	Mannose	$8.6x10^{-2}$ A_{485}//(cm)·(µgm sugar)/(ml. f.a.v.)
Biuret protein	(14)	Serum albumin	$2.3x10^{-4}$ A_{550}//(cm)·(µg protein)/(ml. f.a.v.)
Hartree-Lowry protein	(15)	Serum albumin	$1.7x10^{-2}$ A_{650}//(cm)·(µg protein)/(ml. f.a.v.)
Bicinchinonic acid protein	(16)	Serum albumin	$1.5x10^{-2}$ A_{562}//(cm)·(µg protein)/(ml. f.a.v.)
Colorimetric microkjeldahl nitrogen	(17)	Ammonium sulfate	1.3 A_{660}//(cm)·(µg nitrogen)/(ml. f.a.v.)

[a]f.a.v. = final assay volume.

(18,19). Foreign compounds in crudes cause T.P. analyses to vary by factors from X2 to X10 from one to another analysis (16). Since (S.A.) = (T.A.)/(T.P.) similar errors in S.A. accrue. Biuret analysis remains one of the more reliable methods albeit biuret is 50-100X less sensitive than Lowry and BCA analysis. Spectrophotometric 'microkjeldahl' total nitrogen (Koops _et al_. method (17)) is perhaps the best choice for T.P. if nucleic acid and NH_4^+ are not present. Streptomycin sulfate and dialysis, respectively were used to remove them. Weight absorption coefficients, $E^{1\%}_{280}$, were mainly based on dialyzed retentates of samples from TPP.

RESULTS

Table 3 lists crude enzyme and protein inputs for developing the TPP method, the aqueous phase best pH ranges, T.A. and S.A. values gotten. Many of the crudes were Sigma Company's crudest preparations. In parallel with a survey of eight commercial cellulases (9), crude carbohydrases sometimes are largely salt and other compounds that readily dialyze away. Some crudes are simply first cuts from ammonium sulfate salting out of fermentation broths. Dialyzable salts,

TABLE 3
ENZYMES INVESTIGATED IN TPP-t-BUTANOL PROCEDURES
(S.A. = SPECIFIC ACTIVITY, T.A. = TOTAL ACTIVITY).

Enzyme	Source	Conditions in aqueous layer, pH	Principal Results
Cellulase	*T. reesei*	5	95% recovery of all three activities.
Peroxidase	Horse radish	3.5-8	2X to 3X increase in S.A.
Protease	Pepsin	2-3	Retrieval of nearly all activity.
Protease	*B. subtilis*	8	5X inc. S.A., 3X inc. T.A.
α-Amylase	Barley malt	8-9	10X inc. S.A., retain T.A.
α-Amylase	*A. oryzae*	4-6	2X-3X inc. S.A.; retain T.A.
α-Amylase	*B. amyloliquifaciens*	7.5	6X-8X inc. S.A.; 2X-3X inc. T.A.
α-Amylase	*B. licheniformis*	4-6	4X inc. S.A.; 2X-3X inc. T.A.
Invertase	*S. cerevisiae*	4-6	75X inc. S.A.; 50X-100X inc. T.A.
β-Galactosidase	*A. oryzae*	4.5	8X-10X inc. S.A.; 6X inc. T.A.
β-Glucosidase	Almond	10.5	10X-20X inc. S.A. Retrieve 80% T.A.
Amyloglucosidase	*Rhizopus*	2.3	3X inc. S.A.; 2X-3X inc. T.A.
Lipase	*C. cylindracea*	>5	8X inc. S.A.; 9X inc. T.A.
Protease Inhibitor	Bowman-Birk, soybean	4	4X inc. S.A.; Retrieve T.A.

sugars, etc. commonly contribute 80-95% dry weight of the original sample. There is no malfeasance by the vendor, usually one receives crude enzymes having S.A. and T.A. within ±50% of vendor specifications.

Perceived S.A. values depend on how the T.P. denominator is measured and perceived. Dry weights of original crude and of dialyzed crude retentate, weight protein via BCA, microkjeldahl analysis were used to determine T.P. values and thence S.A.s. Table 3 is largely based on BCA, Hartree-Lowry and biuret protein. As before we recommend (9) determination and reporting on a basis of dry weight dialyzed retentate where possible for comparison to wet chemical analyses. Quantities entering $E^{1\%}_{280}$ values also are subject to large variations in how the denominators of $E^{1\%}_{280}$ quotients are determined. The absorbancies (numerators) of $E^{1\%}_{280}$ values vary because of pigments and nucleotide contaminants which strongly absorb light.

Table 3 lists several cases of increased T.A. values resulting from TPP. Increases in T.A. occur mainly through removal of inhibitors from enzymes, particularly carbohydrases.

Crudes often arrive with many reaction products from substrate reactions and often are product inhibited from the start, as in case of cellulases, cellobiose and glucose (20). Inhibitors can be beneficial under stresses imposed during separations however, for inhibitors often tighten and protect enzyme molecules (21). Thus amylases may be protected by starch and lower saccharides although inhibited by them. Requirements by enzymes such as α-amylase need for Ca^{++}, and time dependency in restoration of enzyme activity after stresses like TPP, often need attention. Since sulfate and sometimes phosphate is present in crudes and after TPP, and since Ca^{++} gets precipitated by these anions, assay of certain α-amylases requires replacement of sulfate-phosphate buffers with β-glycerophosphate or zwitterionic buffers such as HEPES or MES. Replacement was usually performed by dialysis against the appropriate buffer.

Usually there is a "window" or optimal range of ammonium sulfate in TPP aqueous phase below which the desired enzyme will not split out and above which too many foreign polymers also come out to form a midphase. In case of α-amylases such a window often is in the 40-50% salt saturation range. Table 3 lists pH ranges for optimal TPP operation but they are not necessarily optimal pHs for enzyme assay before, or after TPP. Assays were carried out at fixed pH by adding aliquots of samples to assay buffers having sufficient capacity to overwhelm the sample and send it to a standard pH. A few enzymes such as cellulases are sharply dependent on enzyme/substrate ratio because of limitations imposed by binding to solid surfaces (9), a matter needing attention in T.A. and S.A. measurement.

In assay and also for larger scale preparative TPP it is advantageous to mesh with other means for isolation. E.g. by using a heat shock step as Rudd et al. did with a lipase (3). Crudes from yeast contain much nucleic acid that clutters both TPP isolation, spectrophotometric and nitrogen analysis. We used streptomycin sulfate to help remove them beforehand (22) but it is likely Jendrisak's polyethyleneimine method (23) will replace it. Summarizing so far, results perceived in TPP or in any other isolations technique depends first on means for analysis. A very close second is dependence on how TPP distribution affects not only sought-for enzymes and proteins but on how intefering compounds get distributed. The need to consider two patterns of distribution - not simply one - becomes sharper as inputs become more crude.

α-Amylase, Barley Malt.

Input crude was Sigma Type VIII-A. It lost 90% dry weight on dialysis. Roughly 30-60% of the retentate is sugar (phenol-sulfuric acid), likely starch; the rest is protein (BCA analysis). Table 4 shows how protein and neutral sugar between midphase and lower aqueous phase on varying pH during TPP. The data may be added - horizontally - to compare sums of protein and carbohydrate partitioned into both phases, with total dry weight of dialyzed retentate before TPP. In the pH 4.0 case, 10 gm. original crude was started

TABLE 4
PARTITIONING BEHAVIOR AND DISTRIBUTION OF CRUDE, ONCE DIALYZED, BARLEY MALT α-AMYLASE; 1030 mg. DRY WEIGHT SAMPLE IN EACH TPP SEPARATION. (BEFORE TPP, S.A. = 85, T.A. = 148,000).

pH, Aqueous phase	Protein recovered		Neutral sugar recovered		S.A. µmole/min/mg Pr.		T.A. recovered µmole/min
	Mid	Lower	Mid	Lower	Mid	Lower	
3.0 (0.05 M glycine)	*	*	*	*	0	0	0
4.0 (.05 M acetate)	170	100	54	695	171	0	29,100
6.0 (0.05 M MES)	115	65	52	585	370	0	42,600
6.9 (0.05 M HEPES)	252	100	37	242	400	0	100,800
9.0 (0.05 M Tris)	193	92	101	250	850	0	164,000

*Under pH 3 conditions, all material denatured and did not redissolve.

with and dialysis left 1030 mg. dry weight retentate. The sum of neutral sugars in both phases was 749 mg. and of T.P.,290 mg. for an overall sum via wet analysis of 1019 mg., fortuitously close to the 1030 mg. input. At pH 6 the overall sum was 817 mg. output vs. 1030 mg. dry weight input. Table 4 shows how at all pH values TPP gave back an increase in S.A. although at pH 3-4 TPP gave poor return of T.A. Above pH 6 input T.A. is retrieved but not more. Figure 2 plots S.A. of both crude and TPP treated α-amylase exposed to varying pH. All activity measurements were carried out at pH 6 (with Ca^{++} added). The amylolytic substrate was Lintner potato starch developed by DNS colorimetry. Optimum separation in TPP occurred at pH 9 although that is not optimum for enzyme activity. A minor increase of T.A. at pH 9, 164,000 units output vs. 148,000 units input is not signi-

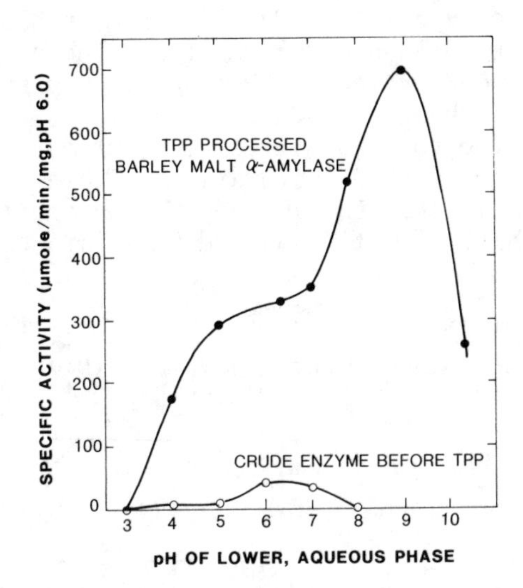

Figure 2. Dependency of specific activities of barley malt α-amylase on pH before and after TPP.

ficant but the ∿10X gain in S.A. probably is. All α-amylolytic activity entered the midlayer where it concentrated and was easily captured.

Pigments partitioned to the upper t-butanol phase although TPP is not outstandingly depigmenting for this enzyme. Weight absorption coefficients $E^{1\%}_{280}$ (gm/100 ml.)$^{-1}$ cm^{-1} reflect how u.v. absorbing materials, foreign proteins and pigments were shed. Original crude not dialyzed had an $E^{1\%}_{280}$ = 0.27, low because of much starch in it. After dialysis and TPP at pH 9, $E^{1\%}_{280}$ = 7.2 based on dry weight; corrected for polysaccharide still present $E^{1\%}_{280}$ = 14.4; based on BCA protein $E^{1\%}_{280}$ = 20.1. Pure proteins have $E^{1\%}_{280}$ values ∿5-25. Thus dialysis and TPP upgraded the crude to reasonable $E^{1\%}_{280}$ parameters consisent with Table 4's reflection of how foreign materials were removed. All dialyzing solvents included 1-10 mM $CaCl_2$ for this, and other α-amylases reported below. Assay temperatures were 50° for amylolysis, 37° for proteolysis.

α-Amylase, Aspergillus oryzae.

Sigma's no. A0273 Type X-A was the input crude. It lost 25% weight on dialysis and was densely pigmented. Dialyzed retentate was ca. 34% protein, 41% carbohydrate. The α-amylases from one cycle of TPP, also from Sephadex G100 chromatography were 5% neutral sugar and the enzymes are glycoproteins. One cycle of TPP moved much pigment into the t-butanol layer, changing $E_{280}^{1\%}$ from 126 to 15, i.e. from an unreasonably large to a normal value. Optimal salt concentration during TPP was 50% saturation. Table 5 lists results from varying pH in TPP, showing how the pH optimum is near 5 with a major shift of both T.A. and S.A. into the midlayer.

TABLE 5
ASPERGILLUS ORYZAE CRUDE α-AMYLASE DISTRIBUTION IN TPP/t-BUTANOL, 50% AMMONIUM SULFATE IN THE AQUEOUS PHASE; DEPENDENCE ON pH. ORIGINAL CRUDE INPUT: 85 mg TOTAL PROTEIN, 103 mg NEUTRAL CARBOHYDRATE (NOT DIALYZABLE), S.A. = 683 μmole/mg protein/minute; 58,300 TOTAL UNITS ACTIVITY

pH, layer		Protein output (mg)	Neutral sugar output (mg)	S.A., specific activity	Fold change of S.A., rel. to crude	Output total activity
4.0	Mid	40	5	1350	X2.0	54,000
	Lower	8	14	760		5,900
5.0	Mid	55	5	1756	X2.6	97,000
	Lower	14	13	197		2,800
6.0	Mid	27	4	2110	X3.1	57,000
	Lower	8	13	453		3,720
6.9	Mid	22	4	2330	X3.4	52,100
	Lower	10	11	360		3,410
9.0	Mid	25	5	1370	X2.0	33,900
	Lower	8	14	810		6,425
10.0	Mid	19	7	1190	X1.7	22,850
	Lower	8	17	1410		11,500

α-Amylase, Bacillus amyloliquifaciens.

Our input crude was Sigma's Type XI-A. The crude is 80-90% salt, roughly 10% protein and there is some insoluble starch. In 25 gms. crude there was 2300 mg. T.P., 1.8×10^6 units T.A., 2900 units of protease activity giving S.A. of α-amylase = 780 (μmoles reducing sugar)/(mg protein)·(min) and S.A. of protease = 1.3 (mg. azocasein solubilized)/(mg. protein)·(min) under the assay conditions used for the preceding α-amylases. Similar S.A. values were gotten via Hatree-Lowry and biuret T.P. assay. Two TPP programs were devised, one-stage and two-stage. One-stage TPP with 50% salt saturation has a broad optimum pH from 5 to 8 at 25° and increases amylolytic S.A. from ca. 800 to 3000 in the mid-layer but S.A. ≅ 1000 in the lower layer and protease is still largely in the midlayer. Two-stage TPP outlined above at pH

7-7.5 loses some α-amylase activity but sheds foreign protein such that most of the protease activity is shunted aside. Of 1.8×10^6 units T.A. crude, two stage TPP gave back 8.6×10^5 units T.A. in the midlayer and 1.1×10^6 T.A. in the lower layer. (Table 3 above refers to products from one-stage TPP). After two stage TPP, dialysis and freeze drying, 25 gms. input crude (nearly black) gave 2 gms. of white α-amylase powder, completely soluble with S.A. = 6900. Crude starting material had $E^{1\%}_{280}$ = 31 (above normal), $E^{1\%}_{300} \cong 12$ (far above normal) and A_{280}/A_{260} = 0.6 (pure proteins, 1.5-3).

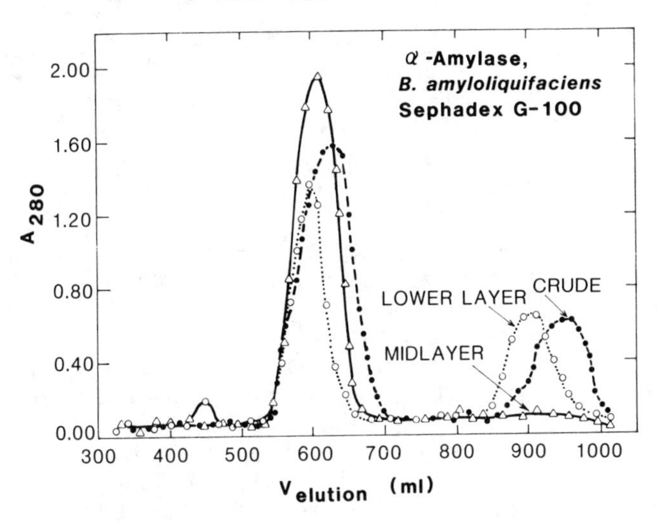

Figure 3. Chromatography from 2 stage TPP output.

After two stage TPP these values were 22, 0.3 and 2 respectively. Figure 3 illustrates how Sephadex G-100 chromatography elutes α-amylases in a first peak (M.W. ∿50,000) from crude and from two stage TPP. The protease second peak's M.W. ∿ 20-30,000 present in crude is largely transferred into the lower layer away from the midlayer in two stage TPP.

α-Amylase, Bacillus licheniformis

The crude, Type VII-A, Sigma no. A3403 is in solution, densely colored and had $E^{1\%}_{280} \cong 40-50$. TPP using 45-50% salt in the aqueous layer is mainly an extractive method, as we used up to six successive t-butanol renewals as outlined in Table 6. Fifty ml. of crude had 18 gm. total dry weight, 80-90% of which dialyzed away. It had 3.4 gm. protein

TABLE 6

SUCCESSIVE TPP EXTRACTION OF B. licheniformis α-AMYLASE[a]

Extraction number	Total protein, mg	Neutral sugar, mg	S.A., μmole maltose/(min)•(mg protein)	T.A., μmole maltose/(min)•(mg protein)
Zero (crude)	58	17	570	33000
1	11	2	3100	34000
2	7	2	2800	20200
3	8	2	3600	30300
4	4	2	4600	18900
5	4	1.5	4700	19600
6	5	1	4100	20600

[a]Based on 1 ml. crude liquid sample, pH 6.0, 50% ammonium sulfate in each extraction, Hartree-Lowry protein analysis.

(perhaps high because of pigment interference), 0.8 gm. polysaccharide, 3.3×10^4 units activity/ml. crude and S.A. ∿ 570 units/(mg. T.P.)·(min). If Ca^{++} is maintained the enzyme is tolerant to pH excursions in TPP from 4 to 9 but its activity and t-butanol extraction optimum both lie near pH 6. Several fold extracted material is "clean", not absorbing at 300 mμ. Its A_{280}/A_{250} ratio ≅ 2.5, $E^{1\%}_{280}$ = 26.4. Such a value seems high but is close to the average $E^{1\%}_{280}$ = 23 for four α-amylases listed by Stein et al. (24). Sephadex G-100 chromatographic peaks corresponding to α-amylase and protease in Figure 3 for B. amyloliquifaciens were obtained but the ∿25,000 M.W. peak for B. licheniformis was not active toward azocasein.

Invertase, Saccharomyces cerevisiae.

S. cerevisiae invertase increases its S.A.s by factors of from X10 to X100 on dilution from ca. 100 to 1 μgm/ml. This rather odd behavior needs be kept in account in regarding the results. Our crude was Sigma's grade V 'practical' which was leached 24 hrs. at 4°, pH 4.5-5 and may be largely yeast cells. Crude invertase before streptomycin sulfate had A_{280}/A_{260} ca. 0.4 to 0.5; after TPP a ratio ≅ 1.8. In TPP with ca. 50% salt, pH 4-6, the majority of enzyme remains in the lower aqueous layer and unwanted less active material entered the midlayer for discard. Crude enzyme had S.A. of 30-70 μmole sucrose/(min)·(mg T.P.) on dilution to 1-2 μgm protein/ml. (50° assay). TPP treated, dialyzed lower layer samples attained S.A. from 2000-6000, contain 25-30% carbohydrate and had $E^{1\%}_{280} = 20.6$ on dry weight protein basis. Such a coefficient aligns well with $E^{1\%}_{280} = 21.1$ calculated from literature giving 65 tyrosine and 33 tryptophan per 135,000 gm. protein (25).

β-Galactosidase, Aspergillus oryzae.

Grade XI, Sigma's no. G7138 was our input crude. S.A. and T.A. both increased 8X and 6X respectively in 40-70% ammonium sulfate in pH 3.5-8.5. Crude's S.A. = 2.2 μmole o-nitrophenylate/(min)·(mg. protein) at central assay pH = 4.5, 0.1 M citrate-P_i buffer, 5 mM o-NP-β-galactoside substrate. Increased T.A. presumably resulted from removal of inhibitors which however may occur also from dialysis in addition to TPP. Total protein recovered from mid- and lower phases after TPP summed to ±30% input crude protein. Modes of distribution of T.P. and of β-galactosidase activity switched on varying salt. Below 50-60% salt the enzyme tended toward the lower phase but above that the majority was captured in midphase. Both S.A. and T.A. so behave in pH 4.5-7.5 range. In 40% salt, $S.A._{mid}/S.A._{lower} = 0.3$; $T.A._{mid}/T.A._{lower} = 0.1$. Using 80% salt, the ratio of S.A.s was 2 and of T.A.s, ca. 20. Trypsin was deliberately added to β-galactosidase to test the ability of TPP to pull them apart under conditions leading to moderate autolysis and galactosidase destruction. In 50% salt pH 4.5-6, 80% galactosidase activity traveled to the lower layer and 90% of trypsin to midphase. Partial splitting is feasible and likely can be improved.

Amyloglucosidase, Three Sources of Crude.

Two crudes were Novo AMG 200 and Spirizyme 200L from *A. niger*; the third was Sigma's A7255 from *Rhizopus*. These enzymes behaved similarly in TPP giving 2 to 4X increase in S.A., 1 to 3X increase in T.A. in pH 2.5-4.5, one cycle of TPP, 60% salt. Starting crudes had S.A. of 100-400 μmole glucose (Lintner starch, 50°)/(min)·(mg. T.P. Hartree-Lowry, BCA). Crudes are only ca. 10% protein, strongly pigmented, $E^{1\%}_{280} \cong 40$, which TPP reduced to ca. 15 to 20. Proteases (azocasein) shift to the lower layer; redissolved midlayer, ultrafiltered, cycled again with TPP and freeze dried gave a near-white powder, S.A. = 1600-2000, readily soluble.

Bowman-Birk Inhibitor Protein (BBI) from Soybean for Trypsin.

Crude was donated by Dr. I.E. Liener. Ca. 70% ammonium sulfate, pH 5-7 was used in TPP giving an increase in S.A. of X4 measured by slopes of BBI-trypsin titration illustrated in Figure 4, also in BBI inhibition of α-chymotrypsin. Before TPP, BBI's $E^{1\%}_{280}$ = 13.4, A_{280}/A_{260} = 0.80. After TPP these were 12.7 and 1.22 respectively. Accounting for the M.W.s of BBI, 8000 and trypsin, 23000, Figure 4's abscissal intercept for TPP treated BBI gives an interaction stoichiometry to trypsin within ±30% of the theoretical 1:1 value.

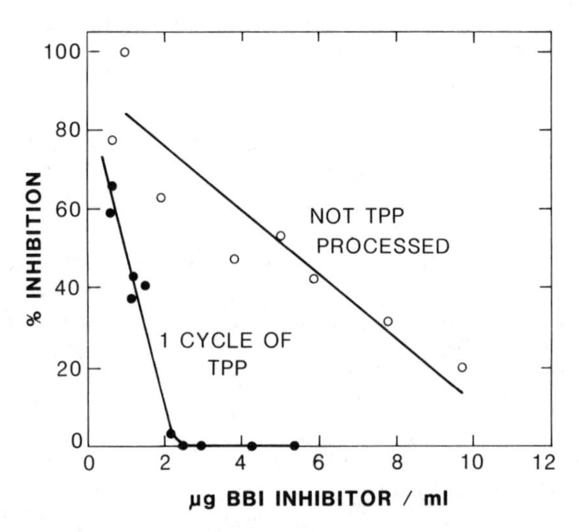

Figure 4. Inhibition titration of 10.7 μg trypsin/ml. with BBI inhibitor, pH 8.2 Tris buffer 0.05 M 0.001 M $CaCl_2$, BAPNA substrate for trypsin, 2 mg.; total assay volume 2.6 ml., 37°, 5 minute reaction, stopped with 30% acetic acid, read at 410 nm. Increasing slope measures increasing power of BBI to inhibit trypsin.

β-Glucosidase; Almond Meal; Almond Emulsin.

Almond meal is solvent extracted (defatted) ground almond and emulsin is a crude salting cut from an extract. Table 7 outlines our principal results worked out by Sigurd Assev (26). The feasible pH range for TPP is 4.5-10.5 but optimal pH for highest S.A. is pH ∿ 10. Maximal S.A.s were gotten via two stage TPP using 40% salt first, discarding the midphase, increasing to 60% salt and retaining the midphase. Overall increases in T.A. were not gotten. All assays were at pH 5.0, 0.05 M acetate, 37° using p-nitrophenyl-β-D glucoside. One cycle of TPP for emulsin in the alkaline

TABLE 7
TPP OF ALMOND MEAL AND ALMOND EMULSIN CRUDE
β-GLUCOSIDASE

Raw material	TPP
Defatted almond meal S.A. = 0.2 to 0.6	Yield (midphase): 60% S.A. = 8.1 ± 0.7
Almond emulsin S.A. = 8.4	Yield (midphase): 62% S.A. = 27 ± 4

region gave 3X increase in S.A. to reach 27 μmole PNP/ (min)·(mg protein), nearly half of maximal S.A.s ≅ 50-60 via chromatography (26).

Peroxidase, Horse Radish.

Sigma No. P8000 was the starting material, S.A. = 0.10 μmole chromogen/(mg. T.P.)·(min). Partitioning was developed around 50% salt, pH 5 to 7, but the enzyme mainly remains in the lower layer. Some foreign material enters the midlayer for discard. Lower layer peroxidase gained 2X to 4X increase in S.A. over crude.

Lipase, Candida cylindracea.

Crude Sigma Type VII lipase, S.A. ≅ 2 x 10^{-8} moles p-nitrophenylate release from the palmitate/(min)·(mg. protein), in TPP using 35-50% salt at pH 5.5 or greater gave readily captured midlayer with S.A. ∿ 15 to 30 x 10^{-8} units at 25°. Total activities increased 1.5-2X. The principal gain occurs from shunting foreign protein to the aqueous layer while precipitating lipase to the middle.

Cellulase: Trichoderma reesei; Aspergillus niger.

Cellulases are multienzymes, the first enzymes isolated via TPP-t-butanol (27). Increases in S.A. and T.A. were not gotten. However TPP concentrates and decolorizes crude cellulases. TPP has been 'tuned' in the 30-40% salt ranges to attempt splitting β-glucosidase from the endo- and exo-enzymes. About 50% of β-glucosidase can be so separated at pH 5, 25°.

Alkaline Protease, Bacillus subtilis.

The source was Gist-Brocades, U.S.A. Assay via azocasein was carried out at pH 8. In first stage TPP, 43 gm. $(NH_4)_2SO_4$ was added per 100 ml. liquid crude plus 1 volume t-butanol. From 11000 mg. total crude protein having S.A. = 5.5 azocasein units (9) and T.A. = 63000 units, a midlayer was captured having S.A. = 29, T.A. ∿ 210,000 units. Upon a second TPP cycle, S.A. ≅ 43. Hence TPP retains and usually enhances S.A. and T.A.

DISCUSSION

Macromolecular charge, Z, the relationship between isoelectric points of proteins and the pH of optimal TPP probably control partitioning to some extent. Why proteins float in TPP (midlayer) and sink in conventional salting out is unknown. Likely there is connection to t-butanol binding, solvent exclusion and salt anion binding to protein molecules with attendant effects on molecular buoyancies. However these matters need careful thermodynamic analysis. Meanwhile, the empirical useful features offered by TPP-t-butanol are: (i) Rapid, (ii) Easy to experiment with, (iii) Can be scaled up - and down, (iv) Concentrative, (v) Depigmenting, (vi) Useable upstream, (vi) Flexible in that pH, temperature kind of salt and of alcohol, reducing agents, cofactors can be controlled, (vii) Works at room temperature, (viii) Does not use polymers needing removal later, (ix) Not expensive, (x) t-Butanol is high boiling, much less flammable than C_1 and C_2 cosolvents.

ACKNOWLEDGEMENT

This work was supported by the Biological Process Technology Institute of the University of Minnesota, and General Mills Company.

REFERENCES

(1) Anderson PC (1979) Enzymatic preparation of protoplasts from corn coleoptile tissue. PhD Thesis, University of Minnesota, St. Paul, p. 69

(2) Tan K, Lovrien R (1972) Enzymology in aqueous-organic cosolvent binary mixtures. J Biol Chem 247:3278

(3) Rudd EA, Mizuno N, Brockman HL (1987) Isolation of two forms of carboxylester lipase from porcine pancreas. Biochim Biophys Ac, in press

(4) Niehaus WG, Dilts RP (1982) Purification and characterization of mannitol dehydrogenase from A. parasiticus. J Bact 151:243

(5) Albertsson P-E (1986) Partition of Cell Particles and Macromolecules, New York: Wiley-Interscience p. 1

(6) Butler L (1979) Enzymes in nonaqueous solvents. Enzyme Microb Tech 1:253

(7) Kazandjian R, Dordick J, Klibanov A (1986) Enzymatic analyses in organic solvents. Biotech Bioeng 28:417

(8) Clegg J (1978) Hydration dependent metabolic transitions and the state of water in Artemia, in Crowe J, Clegg J, "Dry Biological Systems" New York: Academic Press p 117

(9) Lovrien RE, Gusek T, Hart B (1985) Cellulase and protease specific activities of commercially available cellulase preparations. J App Biochem 7:258

(10) Turner R, Liener IE, Lovrien R (1975) Equilibria of Bowman-Birk Inhibitor Association with Trypsin and α-chymotrypsin. Biochemistry 14:275

(11) Miller GL (1959) Dinitrosalicylic acid reagent for determination of reducing sugar. Analyt Chem 31:426

(12) Spiro RG (1966) Analysis of sugars in glycoproteins. Meth Enzymol 8:7

(13) Dubois M, Gilles KA, Hamilton JK, Rebers P, Smith F (1956) Colorimetric method for determination of sugars and related substances. Analyt Chem 28:350

(14) Watters C (1978) One-step biuret assay for protein in presence of detergent. Analyt Biochem 88:695

(15) Hartree F (1972) Determination of protein; modification of the Lowry method. Analyt Biochem 48:422

(16) Smith P, Krohn I, Hermanson G, Mallia A, Gartner F, Provenzano M, Fujimoto E, Goeke N, Olson B, Klenk D (1985) Measurement of protein using biscinchinonic acid. Analyt Biochem 150:76

(17) Koops J, Klomp H, Elgersma RH (1975) Rapid determination of nitrogen by colorimetric estimation of ammonia following accelerated digestion. Neth Milk Dairy J 29:169

(18) Pierce J, Suelter CH (1977) An evaluation of the Coomassie Brilliant Blue G250 dye-binding method for quantitative protein determination. Analyt Biochem 81:478

(19) Jernejc K, Cimerman A, Perdih A (1986) Comparison of different methods for protein determination in _A. niger_ mycelia. App Microbiol Biotech 23:477

(20) Mandels M (1982) Cellulases. Ann Rep Ferment Proc 5:35

(21) Citri N (1973) Conformational adaptability in enzymes. Adv Enz 37:488

(22) Suelter CH (1985) A Practical Guide to Enzymology, New York: Wiley Interscience p 66

(23) Jendrisak J (1987) Use of polyethyleneimine in protein purification. J Cell Biochem, Supp 11C abstracts:166

(24) Stein EA, Hsiu J, Fisher EH (1964) Alpha-amylases as calcium metalloenzymes. I. Preparation of calcium free apoamylases. Biochemistry 3:56

(25) Neumann NP, Lampen JO (1967) Purification and properties of yeast invertase. Biochemistry 6:468

(26) Assev SM (1986) Isolation and characterization of β-glucosidase in almond meal. Thesis, Univ. of Minnesota p 14

(27) Odegaard BH, Anderson PC, Lovrien R (1984) Resolution of the multienzyme cellulase complex of _T. reesei_ QM9414. J App Biochem 6:156

Protein Purification: Micro to Macro, pages 149–162

IMMOBILIZED METAL ION AFFINITY CHROMATOGRAPHY OF PROTEINS

Eugene Sulkowski

Department of Molecular and Cellular Biology,
Roswell Park Memorial Institute, Buffalo, New York 14263

ABSTRACT Some new observations and new insights into IMAC of proteins are reported: (a) the perturbation of the microenvironment of the active site histidines of α-chymotrypsin and pancreatic ribonuclease can be probed by IDA-Cu^{2+} chromatography; (b) the overall electric charge of the IDA-Me^{2+} chelate is negative, a conclusion derived from the analysis of the chromatographic behavior of cytochrome c and calmodulin; (c) the pH-dependence of the binding of bovine serum albumin on IDA-Ni^{2+} is consistent with the participation of histidine side chains in the coordination to immobilized metal; (d) the elution of bovine pancreatic ribonuclease from IDA-Cu^{2+} with imidazole concentration gradient as a function of pH from 6 to 8 has been studied; and (e) human tumor necrosis factor (TNF) is retained on IDA-Cu^{2+}, whereas human lymphotoxin (LT) is retained also on IDA-Co^{2+}, IDA-Ni^{2+} and IDA-Zn^{2+}. This chromatographic behavior of human TNF and human LT can be rationalized in terms of the peculiarities of histidine distribution (histidine clusters).

INTRODUCTION

Since its introduction by Porath et al. (1), immobilized metal ion affinity chromatography (IMAC) has gained wide acceptance, both on micro and macro scales, as an useful and on occasion of choice, technique for the isolation of proteins (Table 1).

Moreover, due to its *modus operandi*, IMAC may serve as a facile probe of the surface topography of potential

TABLE 1
IMAC OF PROTEINS[a]

PROTEIN	Me++	REFERENCE
1. Human serum proteins	Zn, Cu	J. Porath et al., Nature 258, 598 (1975)
2. Lactoferrin	Cu	B. Lonnerdal et al., FEBS Lett. 75, 89 (1977)
3. α_2-SH glycoprotein	Zn	J.P. Lebreton, FEBS Lett. 80, 351 (1977)
4. Human fibroblast interferon	Zn	V.G. Edy et al., J. Biol. Chem. 252, 5934 (1977)
5. α_2-Macroglobulin	Zn	T. Kurecki et al., Anal. Biochem. 99, 415 (1979)
6. Plasminogen activator	Zn	D.C. Rijken et al., Biochim. Biophys. Acta 580, 140 (1979)
7. Lysozyme	Cu	A.R. Torres et al., Biochim. Biophys. Acta 576, 385 (1979)
8. Nucleoside diphosphatase	Zn, Cu	I. Ohkubo et al., Biochim. Biophys. Acta 616, 89 (1980)
9. Dolichos b. lectin	Ca	C.A.K. Borrebaeck et al., FEBS Lett. 130, 194 (1981)
10. Non-histone proteins	Cu	H. Kikuchi et al., Anal. Biochem. 115, 109 (1981)
11. Human serum albumin	Cu	H. Hanson et al., J. Chromat. 215, 333 (1981)
12. Human fibrinogen	Zn	M.F. Scully et al., Biochim. Biophys. Acta 700, 130 (1982)
13. Phosphotyrosyl-protein phosphatase	Zn	D. Horlein et al., Biochemistry 21, 5577 (1982)
14. Superoxide dismutase	Cu	B. Lonnerdal et al., J. Appl. Biochem. 4, 203 (1982)
15. Human serum proteins	Ni	J. Porath et al., Biochemistry 22, 1621 (1983)
16. Nitrate reductase	Zn	M.G. Redinbaugh et al., Plant Physiol. 71, 205 (1983)
17. EGF, shrew	Ni	T.T. Yip et al., FEBS Lett., 187, 345 (1985)
18. rMuIFN-β	Cu	S. Matsuda et al., J. Interferon Res. 6, 519 (1986)
19. Carboxypeptidases (Aspergillus niger)	Cu	S. Krishnan et al., J. Chromat. 370, 315 (1986)
20. Superoxide dismutase, Catalase	Cu	M. Miyata-Asano et al., J. Chromat. 370, 501 (1986)

[a]Selected examples.

electron donor residues ("Porath triad"): cysteine, histidine, tryptophan. This has been explored, in particular, for histidine residue (2).

Full realization of the heuristic value of Porath's seminal contribution (1) is only now beginning to be appreciated, mostly as a result of the studies with model proteins. Acquisition of new observations in this area is still very much in demand.

Even though many of the experimental conditions for IMAC have been established (1,3), there is a need for their further refinement.

This communication reports some new experimental data pertaining to both aforementioned areas of IMAC studies.

METHODS

Chelating Sepharose 6B was purchased from Pharmacia. All experiments were performed on 0.9x8 cm columns (bed volume, ca 5 mL). Proteins, 5 to 10 mg samples, in 5 mL of an equilibration solvent, were applied and the columns were developed by means of a peristaltic pump at a flow rate of 1 mL per 5 min, at room temperature. Some other details are given in figure legends.

RESULTS

Microenvironment of a Ligand (Histidine).

It is intuitively plausible that the microenvironment of an electron donor grouping will have an impact on its coordination to immobilized metal ion. This notion has been experimentally tested for an imidazoyl (histidine) donor grouping on the surface of α-chymotrypsin and ribonuclease.

<u>α-Chymotrypsin</u>. Conversion (zymogen activation) of chymotrypsinogen (Figure 1) to α-chymotrypsin results from

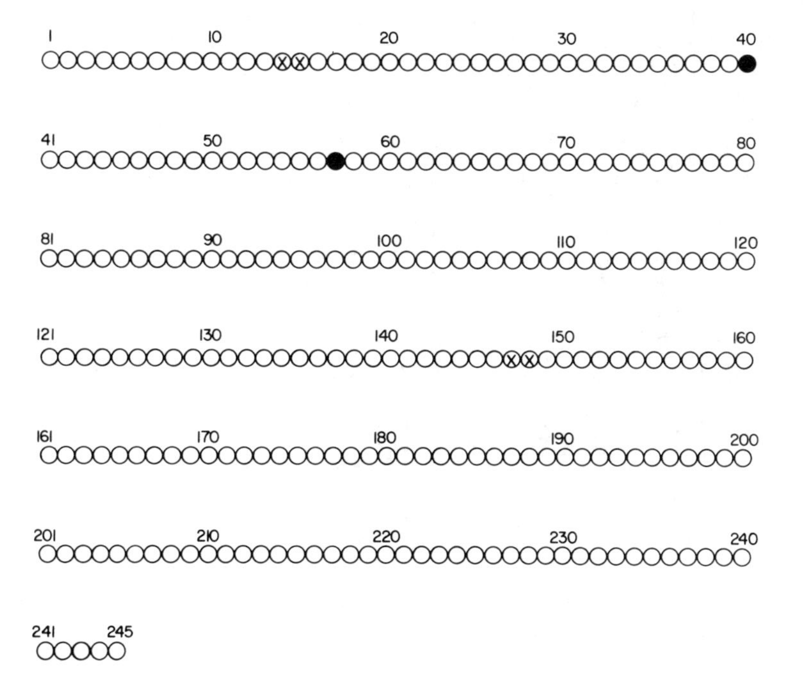

FIGURE 1. Distribution of histidine in α-chymotrypsinogen A and α-chymotrypsin. ○ , an amino acid residue; ⊗, amino acid residues removed from chymotrypsinogen during its activation; ● , histidine residue.

the proteolytic cleavage and ensuing removal of two dipeptides. The conformational changes occurring during this conversion are localized around the catalytic site and thus perturb the microenvironment of histidine 57 (4). An additional perturbation can be accomplished by the modification of serine 195 (active site serine) with diisopropylphosphofluoridate (DIPF).

Figure 2 illustrates the chromatography of α-chymotrypsinogen A, α-chymotrypsin and α-chymotrypsin (DIPF treated). Clearly, they display different affinity for IDA-Cu^{2+}. On a practical plane, IDA-Cu^{2+} chromatography can be quite useful in the additional purification (mutual contamination) of these commercially available "pure" proteins.

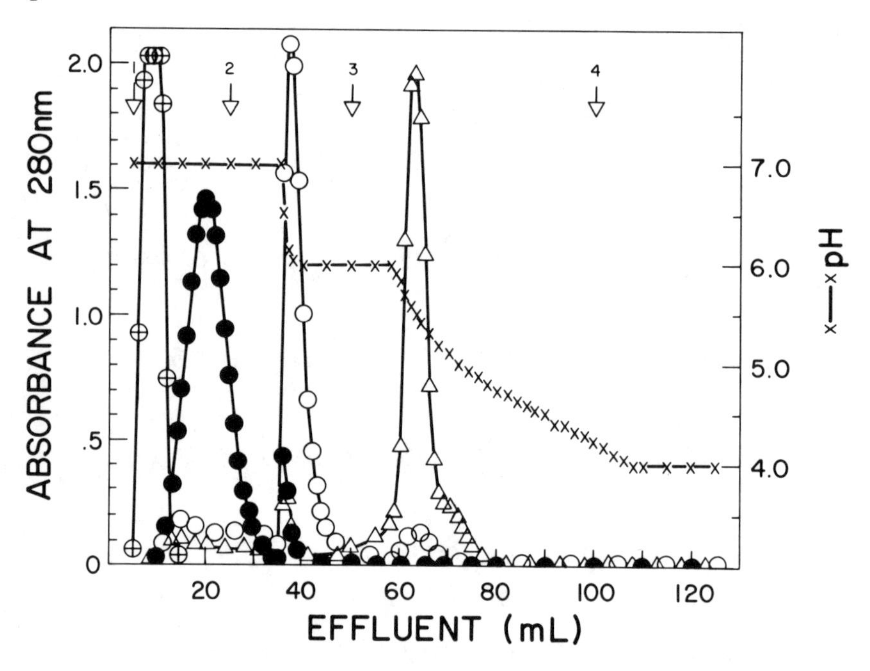

FIGURE 2. Chromatography of α-chymotrypsinogen A and α-chymotrypsin. ⊗ , chromatography of α-chymotrypsin on a column uncharged with copper; ●, IDA-Cu^{2+} chromatography of α-chymotrypsinogen A; ○, IDA-Cu^{2+} chromatography of α-chymotrypsin; △, IDA-Cu^{2+} chromatography of α-chymotrypsin treated with DIPF. All buffers, 0.02M sodium phosphate, pH 7.0 and 0.1M acetate, pH 6 and pH 4, contain 1M sodium chloride.

Ribonuclease. Bovine pancreatic ribonuclease may be non-glycosylated (RNase A) or glycosylated (RNase B, C and D). RNase B carries an oligosaccharide moiety ($GlcNAc_2Man_{4-5}$) on asparagine 34 (5). The enzymatic activity (relative specific activity) of RNase B (96%) toward small molecular size substrate (cytidine 2':3' phosphate) is almost the same as that of RNase A (100%). However, RNase B displays only 84% of RNase A activity toward macromolecular RNA substrate. This difference is apparently caused by the oligosaccharide moiety hindering the access of macromolecular substrate to the active site of the enzyme (5). Could this saccharide moiety hinder the access of IDA-Cu^{2+} (molecular leash) to the histidines 12 and 119 of the enzyme's active site? Figure 3 illustrates the weaker retention of RNase B vis a vis RNase A on IDA-Cu^{2+} sorbent. This would be the anticipated result consistent with the steric hindrance hypothesis (5).

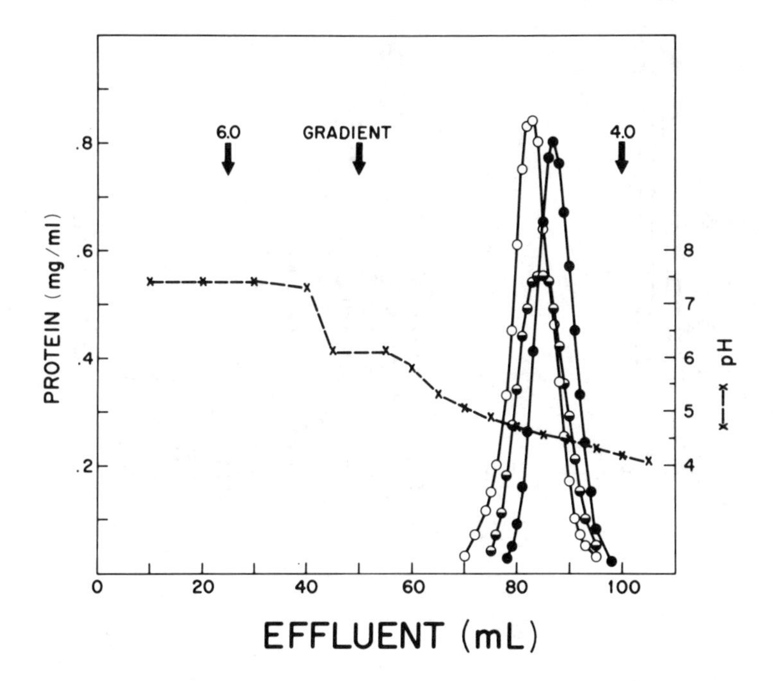

FIGURE 3. IMAC of bovine pancreatic ribonuclease on IDA-CU^{2+}. ○ , RNase B; ◒ , mixture of RNase B and RNase A (1:1); ● , RNase A. Constant background of 1M sodium chloride.

Adsorption of a Protein on IDA-Me^{2+}: Salt Effect.

Chromatography of proteins on IDA-Me^{2+} (Co, Ni, Cu, Zn) is usually performed in the presence of an electrolyte in order to quench electrostatic interactions (1-3). In order to ascertain the effects of salt (NaCl) on the IDA-Cu^{2+} chromatography two proteins: horse heart cytochrome c (pI 10.6) and calmodulin (pI 4.1) were selected. Both proteins display single histidine residues available for coordination (6,7).

Figure 4 illustrates the chromatography of cytochrome c. This protein is not retained on a column, uncharged with Cu^{2+}, when applied in 1.0M sodium chloride or 0.1M

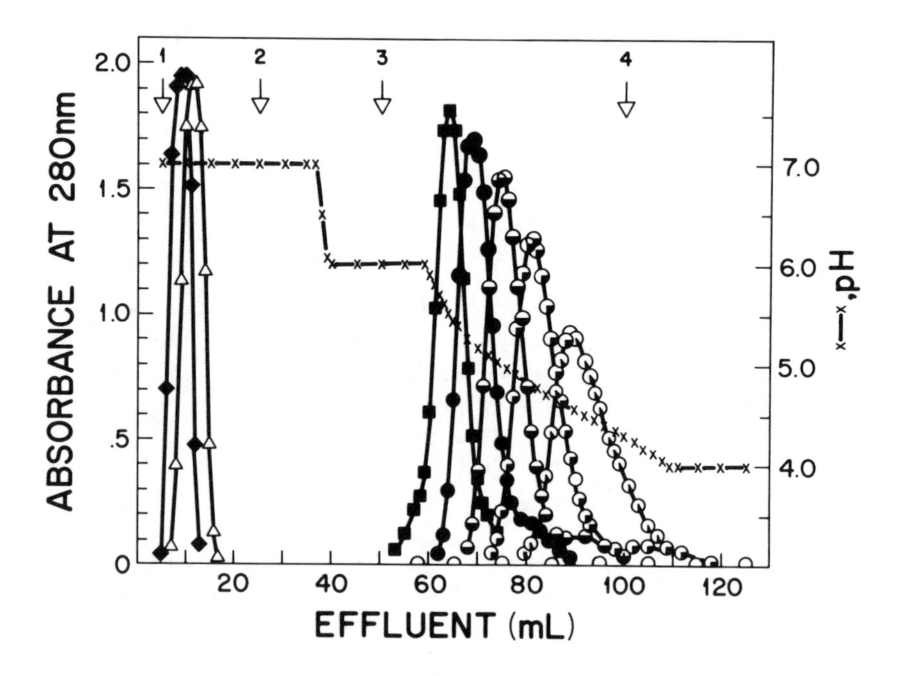

FIGURE 4. Chromatography of horse heart cytochrome c. ♦, cyt c on a column uncharged with copper in 20 mM phosphate (1.0M NaCl); Δ, cyt c on a column uncharged with copper in 20 mM phosphate (0.1M NaCl); ○, IDA-Cu^{2+} chromatography of cyt c in 20 mM phosphate and 0.1M acetate buffers (no sodium chloride); IDA-Cu^{2+} chromatography in the presence of sodium chloride: ■, 2M; ●, 1M; ◒, 0.50M; ◔, 0.25M; and ◔, 0.10M.

sodium chloride in 20 mM sodium phosphate, pH 7.0. Cytochrome c is retained on an IDA-Cu^{2+} column, when applied in 20 mM phosphate buffer, pH 7.0, and could not be recovered with a falling pH gradient developed with 0.1M sodium acetate. However, inclusion of 0.1M sodium chloride in the acetate buffers resulted in the elution of cytochrome c. Moreover, an increase in salt concentration (0.25M-0.50M-1.0M-2.0M) facilitated the recovery of cytochrome c by shifting upwards the pH of elution by about full unit, elution being effected at pH 5.5 (2M NaCl) rather than at pH 4.5 (0.1M NaCl).

Chromatography of calmodulin is illustrated in Figure 5. This protein is only weakly retained on IDA-Cu^{2+} when applied in 20 mM sodium phosphate, pH 7.0. However, inclusion of 0.1M sodium chloride in both phosphate and acetate buffers, results in good retention and subsequent elution when pH is lowered to ca 5.2. Additional increase in salt concentration (0.5M-1.0M) results in still stronger

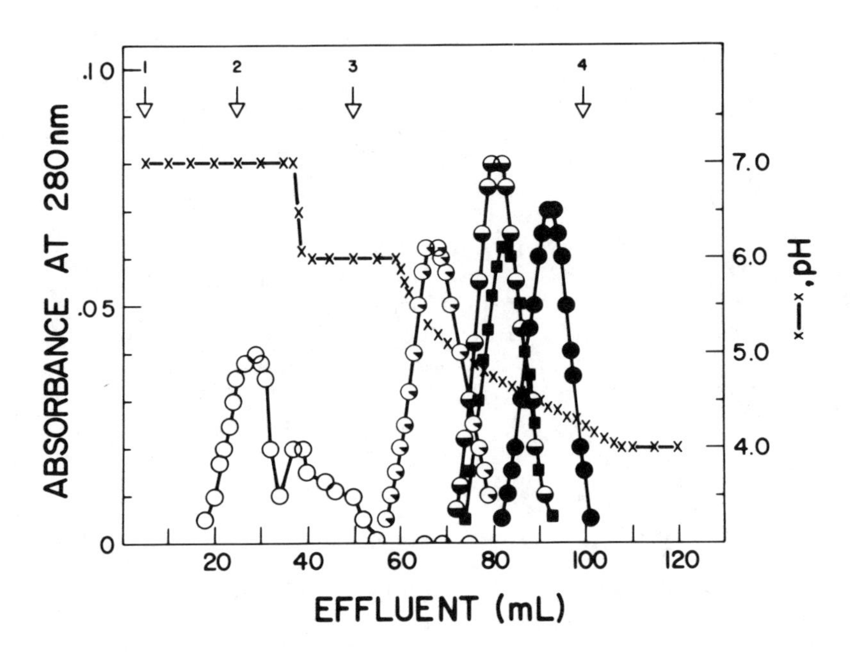

FIGURE 5. IMAC of calmodulin on IDA-Cu^{2+}. ○ , no sodium chloride; ◔ , 0.1M sodium chloride; ◒ , 0.5M sodium chloride; ● , 1.0M sodium chloride; and ■ , 2.0M sodium chloride.

retention: at 1.0M sodium chloride the elution occurs at pH 4.4 (the reversal of this trend at 2.0M sodium chloride may be due to dependence of calmodulin conformation on salt concentration.

Clearly, the effect of salt on the chromatography of cytochrome c and calmodulin on IDA-Cu^{2+} is opposite: salt facilitates the elution of cytochrome c and enhances the adsorption of calmodulin. As sodium chloride is expected to quench an electrostatic interaction one must conclude that this interaction is of *attractive* nature in the case of cytochrome c and of *repulsive* nature in the case of calmodulin. In view of the pI values of these proteins, it seems reasonable to account for their disparate behavior by postulating that the overall charge of IDA-Cu^{2+} is a negative one.

The modulation of the primary type of interaction, i.e., coordination, by an electrolyte can be employed as an experimental parameter, especially when protein stability depends substantially on the pH value of the solvent.

Adsorption of a Protein on IDA-Me^{2+}: pH Effect.

Bovine serum albumin was chromatographed on IDA-Ni^{2+} equilibrated with 20 mM sodium phosphate (1M sodium chloride) at varying pH (Figure 6). The retention of albumin was quite strong at pH 7.0, resulting in some leakage of the protein only after ca 50 mL of the solvent have passed (ten total bed volumes of the column). The retention of albumin at pH 6.0 is, by contrast, barely discernible. The progressive change of pH between 6 and 7 influences the interaction of albumin with IDA-Ni^{2+} in a progressive fashion, with the abrupt transition being at ca pH 6.5. This chromatographic behavior of albumin on IDA-Ni^{2+} indicates the involvement of an amino acid residue with pK_a 6.5. It is reasonable to assume that this residue is a histidine (1).

Affinity Elution with Imidazole.

The chromatography of proteins on IDA-Me^{2+} (Co, Ni, Cu, Zn) whose retention may be due to histidine residues (imidazoyl side chains) can be readily done by inclusion of imidazole in the eluants. Whether or not this can be also accomplished when the electron donor grouping is tryptophan or cysteine has not been reported as yet.

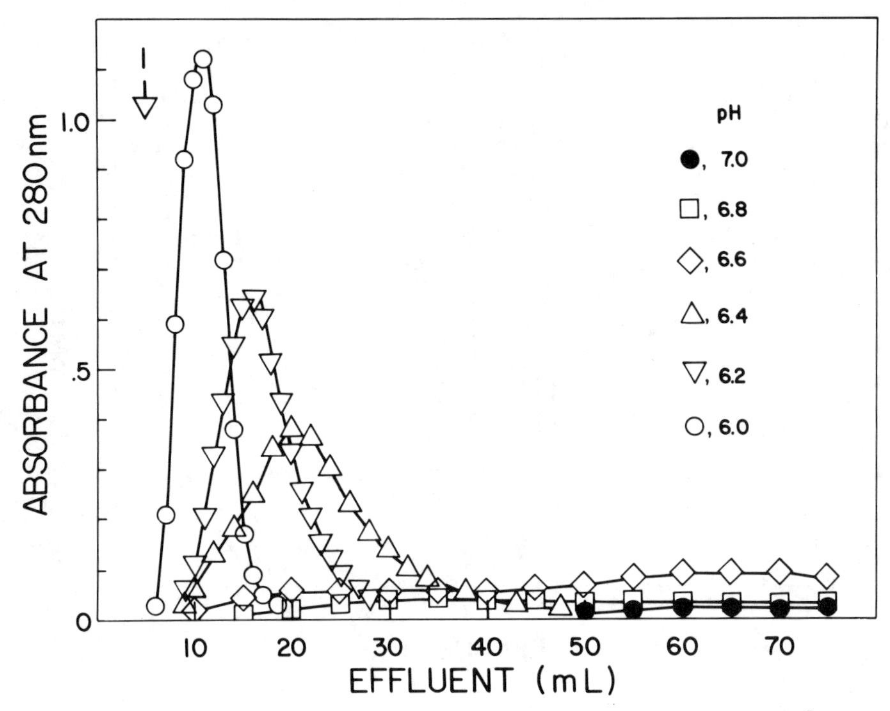

FIGURE 6. IMAC of bovine serum albumin on IDA-Ni^{2+}. Isocratic elution with 20 mM sodium phosphate (1M NaCl), at indicated pH.

Figure 7 illustrates the chromatography of bovine pancreatic ribonuclease on IDA-Cu^{2+} with an imidazole concentration gradient in 20 mM sodium phosphate (1M NaCl) at varying pH, within pH 6 to pH 8 range. Both the adsorption of ribonuclease and its elution from IDA-Cu^{2+} depends on the state of protonation of histidine residues ($pK_a \sim 6.5$) and imidazole (pK_a 6.95). One can, therefore, anticipate that such a "calibration" experiment may differ, so far as the concentration range of imidazole is concerned, from one protein to another depending on the pK_a values of their histidine residues, which can vary considerably. The experiment with ribonuclease indicates, however, that the elution of a protein from an IDA-Me^{2+} sorbent can be accomplished within a relatively narrow range of imidazole concentration provided that the experiment is performed between pH 6 and pH 8.

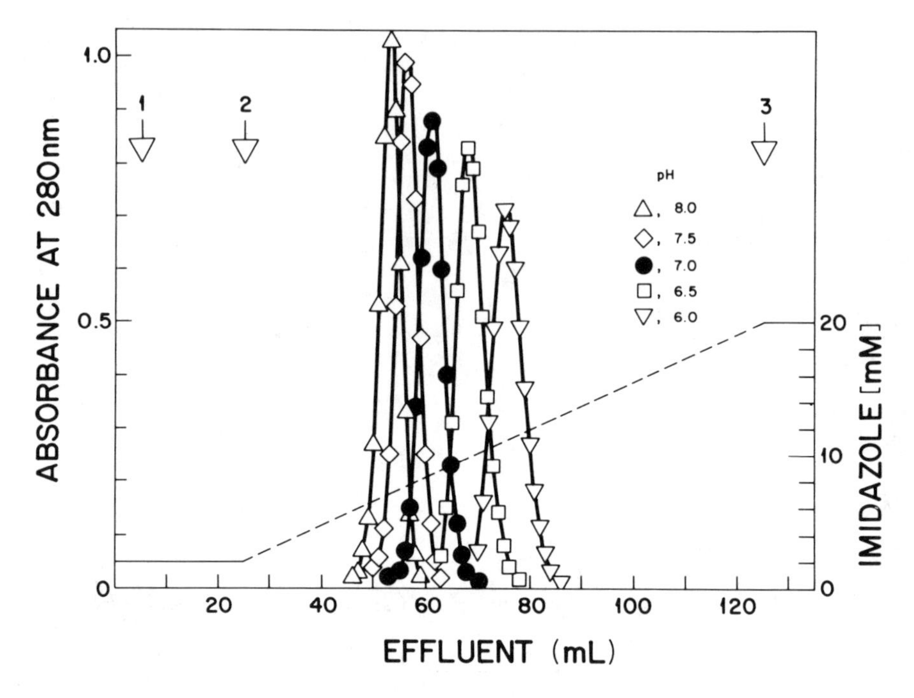

FIGURE 7. Elution of bovine pancreatic ribonuclease from IDA-Cu^{2+} with imidazole gradient at indicated pH. Constant background of 20 mM phosphate buffer and 1M sodium chloride.

Selectivity of Me^{2+} Recognition.

A coordination bond, established between single histidine residue and IDA-Cu^{2+}, is sufficiently strong to retain protein molecule on a column. The binding of a protein to an IDA-Ni^{2+} column seems to require more than one histidine residue. The requirements for protein retention on IDA-Co^{2+} and IDA-Zn^{2+} are even more stringent. The presently available data indicate that the presence of two closely spaced histidines is necessary for the retention (2). This is the case when two histidines, in an α-helical segment of a polypeptide, are separated by two or three other amino acid residues, as for example in sperm whale myoglobin and human fibroblast interferon, respectively (2). The close orientation of two histidines, originating from distant parts of a polypeptide, as in the case of human serum transferrin (8), is also sufficient to result in retention of this protein on IDA-Co^{2+} and IDA-Zn^{2+} sorbents. This is not, however, the case with pancreatic ribonuclease (9), even though histidines 12 and 119 are closely spaced (10).

Our understanding of the selectivity of protein retention on IDA-Me^{2+}, even though limited and for the most part phenomenological at present, has allowed a prediction of the chromatographic behavior of human tumor necrosis factor (TNF) and human lymphotoxin (LT).

The histidine content of TNF (11) and LT (12) is quite different (Figure 8). Conspicuously, LT displays two clusters of histine residues which may have α-helical configuration (Chou Fasman analysis). Therefore, it was reasonable to anticipate that LT should be retained on IDA-Co^{2+} and IDA-Zn^{2+} whereas TNF should not. This, in fact, was confirmed experimentally (W. v. Muenchhausen and E. Sulkowski, unpublished observations), see Table 2.

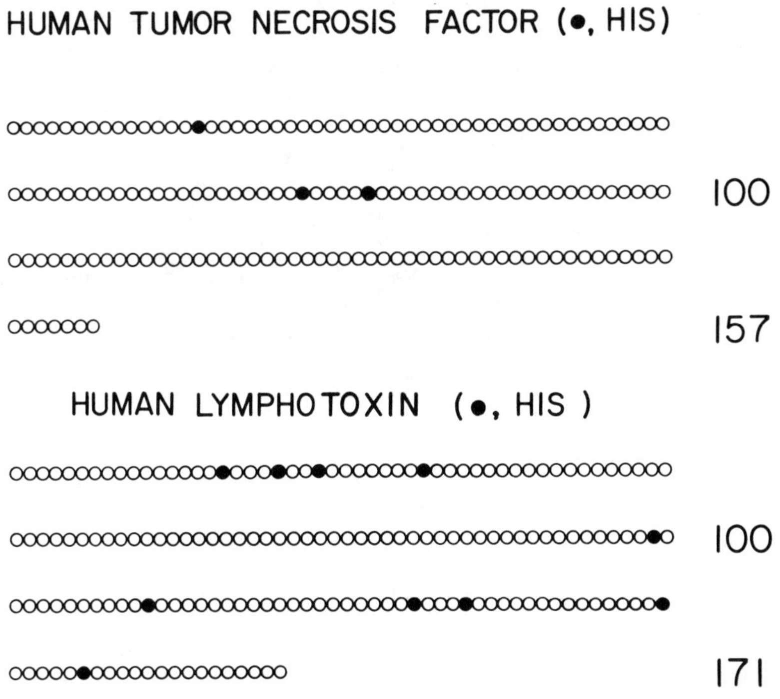

HUMAN TUMOR NECROSIS FACTOR (●, HIS)

100

157

HUMAN LYMPHOTOXIN (●, HIS)

100

171

FIGURE 8. Distribution of histidine in human tumor necrosis factor and human lymphotoxin.

TABLE 2
AFFINITY OF TNF AND LT FOR IDA-Me^{2+}[a]

P / Me^{2+}	Co	Ni	Cu	Zn
TNF	–	–	+	–
LT	+	+	+	+

[a]Binding in 20 mM sodium phosphate (1M NaCl), pH 7.0.

DISCUSSION

IMAC may serve, in addition to other well established physical methods, as a facile probe of the topography of histidine residues of a protein molecule. In particular, the following facets of histidine topography can be explored by IMAC:

1. Localization: surface or interior.
2. Multiplicity.
3. Microenvironment.
4. Clustering.

A brief account of published and unpublished observations is in order at this junction.

1. Proteins which lack histidine in their composition are not retained on IDA-Cu^{2+}: pancreatic trypsin inhibitor (9) and Peking duck lysozyme. Proteins, such as tuna cytochrome c, which have histidines but not accessible to the solvent (His 18 and His 26), are not retained on IDA-Cu^{2+}. By contrast, horse heart cytochrome c, which in addition to His 18 and His 26 has His 33, accessible to the solvent, is retained on IDA-Cu^{2+} (2).

2. Proteins displaying more than one histidine on their molecular surfaces are retained stronger on IDA-Cu^{2+}: Candida k. cytochrome c, Ring-necked pheasant and California quail lysozymes.

3. Single histidine lysozymes (hen, Bobwhite quail) display sufficiently different affinity for IDA-Cu^{2+} which indicates an influence of amino acid residues other than those vicinal to His 15 in the polypeptide, i.e., residues brought about to the vicinity of His 15 by protein folding. Glycosylated and non-glycosylated species of mouse interferons can be separated on IDA-Cu^{2+} (9).

4. Proteins with clusters of histidines: human fibroblast interferon (His 93 and His 97, α-helical assignment), sperm whale myoglobin (His 113 and His 116, α-helix), human serum transferrin (two histidines involved in the coordination to Fe^{3+}), are all retained on IDA-Co^{2+} and IDA-Zn^{2+}.

Further refinement of IMAC as a surface topography probe of histidine residues may be advanced by studies on protein engineering, co-crystallization of proteins with IDA-Me^{2+} and x-ray analysis, computer modelling, etc.

IMAC was suggested previously (2,9) as a tool to ascertain the native format of some recombinant DNA protein products. Perhaps this will occur in the near future.

ACKNOWLEDGEMENT

I am grateful to Marcia Held for her excellent secretarial assistance in preparing this manuscript.

REFERENCES

1. Porath J, Carlsson J, Olsson I, Belfrage G (1975). Metal chelate affinity chromatography, a new approach to protein fractionation. Nature 258:598.
2. Sulkowski E (1985). Purification of proteins by IMAC. Trends Biotechnol 3:1.
3. Porath J, Olin B (1983). Immobilized metal ion affinity adsorption and immobilized metal ion affinity chromatography of biomaterials. Serum protein affinities for gel-immobilized iron and nickel ions. Biochemistry 22:1621.
4. Stryer L (1975) Biochemistry, Second Edition, Chapter 8, W.H. Freeman and Company, p 157.
5. Baynes JW (1976). Effect of glycosylation on the in vivo circulating half-life of ribonuclease. J Biol Chem 251:6016.
6. Dickerson RE, Takano T, Eisenberg D, Kallai OB, Samson L, Cooper A, Margoliash E (1971). Ferricytochrome c. I. General features of the horse and bonito proteins at 2.8A resolution. J Biol Chem 246:1511.
7. Klee CB, Crouch TH, Richman PG (1980). Calmodulin. Annu Rev Biochem 49:489.
8. Chasteen ND (1983). The identification of the probable

locus of iron and anion binding in the transferrins. Trends Biochem Sci 8:272.

9. Sulkowski E, Vastola K, Oleszek D, von Muenchhausen W (1982). Surface topography of interferons: a probe by metal chelate chromatography. In Gribnau TCJ, Visser J, Nivard RJF (eds): "Affinity Chromatography and Related Techniques", Amsterdam: Elsevier, p 313.
10. Crestfield AM, Stein WH, Moore S (1963). Properties and conformation of the histidine residues at the active site of ribonuclease. J Biol Chem 238:2421.
11. Pennica D, Nedwin GE, Hayflick JS, Seeburg PH, Derynck R, Palladino MA, Kohr WJ, Aggarwal BB, Goeddel DV (1984). Human tumour necrosis factor: precursor structure, expression and homology to lymphotoxin. Nature 312:724.
12. Aggarwal BB, Henzel WJ, Moffat B, Kohr WJ, Harkins RN (1985). Primary structure of human lymphotoxin derived from 1788 lymphoblastoid cell line. J Biol Chem 260:2334.

Protein Purification: Micro to Macro, pages 163–175

FROM AFFINITY CHROMATOGRAPHY TO HPAC

Karin Ernst-Cabrera and Meir Wilchek

Department of Biophysics, The Weizmann Institute of Science
Rehovot, 76100 Israel

ABSTRACT Affinity Chromatography has become the method of choice for the purification of biologically active macromolecules. Its expansion, which has continued especially for industrial approaches, has mainly been due to developments in the field of biotechnology. The factors which are required to adapt this technique for industrial usage are discussed in this paper. Among these factors are the choice of carriers, procedures for activation and coupling of ligands, and the choice of ligands as well as conditions for elution.

INTRODUCTION

Since its introduction in 1968 (1), affinity chromatography has become a fundamental tool for the purification of biologically active materials. Indeed, many different proteins, including antibodies, antigens, enzymes, receptors, lymphokinases etc., have been purified as indicated in several reviews (2-6). The success of this method is essentially due to its high efficiency, which is based on the inherent biospecificity of a ligand covalently attached to a chromatographic support which can then interact with its complementary partner.

Despite its wide use as a purification method in research laboratories, affinity chromatography has only recently been adapted for large-scale purification. The need for efficient large-scale purification methods is obvious if one follows the development in the field of biotechnology. Production of therapeutic substances (such as vaccines, human insulin, interferon, drugs etc., obtained by recombinant DNA-technology and by monoclonal antibodies), requires industrial scale methods to isolate the desired product at a high degree of purity. A very impressive example of this kind is the purification of interferon in large quantities on immobilized antibody columns, which has led to a very large increase in the specific activity of the interferon (up to 5000 fold) (7).

Nevertheless, the use of affinity chromatography on the industrial scale is still in the beginning stages of development, since several factors so far have limited its use. Among the factors which have to be considered in adopting this method in industry are the following:

1) The availability of a suitable carrier material which should have (besides certain other physical properties) the ability to bind high amounts of ligands.
2) Methods for the immobilization of ligands are required which lead to a stable linkage of ligand to the carrier thus avoiding leakage of the ligand and therefore contamination of the desired product.
3) The ligand to be chosen has to be sufficiently specific and available in large quantities at resonable prices.
4) Efficient procedures are needed in order to elute the adsorbed compound without loss of biological activity and without affecting the affinity matrix in order to provide a chromatographic system which can be used many times.

A relatively new development which offers high flow rates and more rigid affinity matrices required in large scale affinity columns is high performance affinity chromatography (HPAC). This technique combines the biospecificity of affinity chromatography with the high speed and resolution obtained in high performance liquid chromatography (HPLC) (8-10). The versatility of this method has already been demonstrated with the purification of enzymes, antigens, antibodies, receptors, glycoproteins etc., within very short time periods. However, while this technique seems to be promising, particularly for industrial application, the lack of suitable affinity supports of high capacity has thus far prevented the possibility of a major breakthrough in this area.

CARRIERS

An ideal hypothetical carrier for affinity chromatography and HPAC would posses the following characteristics:

a) The matrix should have a hydrophilic surface to prevent nonspecific adsorption of contaminants.
b) Mechanical strength is required providing a chromatographic process of high flow rate.
c) High porosity of the particles should provide a large surface area for ligand substitution and offer good flow properties.
d) The matrix should be stable to a wide range of solvents which may be used during the chromatographic process.
e) A reasonable amount of functional groups on the surface should be available for immobilization of the ligands.
f) Stable chemical bonding of the ligand is necessary to avoid leakage and contamination of the purified product.

Many different matrices have already been used for the purpose of immobilizing proteins (11). Table 1 gives a summary of the most important carriers. These can be classified into 3 main groups: 1) polysaccharides, 2) synthetic polymers and 3) inorganic materials.

TABLE 1
COMMONLY USED CARRIERS FOR IMMOBILIZATION OF LIGANDS

Polysaccharides	Agarose (Sepharose) Dextran (Sephadex) Cellulose Alginate Carrageenan
Synthetic Polymers	Acrylic copolymers Polystyrene Silicone Nylon Vinyl polymers Polyamino acids
Inorganic Materials	Silica Alumina Magnetic Particles Zirconium oxide

Agarose, a polysaccharide based on galactose (commercially available as Sepharose), is still the most commonly used carrier which fulfills many of the above-listed requirements. However, its relatively low mechanical stability and high price oftentimes prevents its consideration for large-scale purification. Therefore, other supports have been and are being developed.

Silica gel, an inorganic material with high mechanical stability, has recently experienced increased usage as a carrier for HPAC (13). The time factor (i.e. the use of high flow rates for the purification in short time periods), one of the main advantages of HPAC (13,14) over conventional chromatography, makes silica to a promising matrix for industrial use. The introduction of affinity supports based on silica for HPAC was accomplished by Ohlson et al (15), who immobilized anti human serum albumin and N^6-(6-aminohexyl)-AMP to a silica derivative containing aldehyde groups. Other silica derivatives (epoxide-, diol silica) have been used to couple ligands (16-19), but all show relatively low capacities. A comparison of different silica derivatives with respect to coupling yields obtained by two different ligands, trypsin and bovine serum albumin, are given in Table 2.

TABLE 2
COUPLING OF BOVINE SERUM ALBUMIN AND TRYPSIN TO DIFFERENT SILICA DERIVATIVES

Silica derivative	Albumin (mg/g)	Trypsin (mg/g)
Epoxide silica	1.5	4.1
Diol Silica	3.4	39.8
Aldehyde silica	15.0	49.2
Primary Hydroxyl Silica		
p-Nitrophenyl carbonate	47.0	69.7
N-Succinimide carbonate	8.0	96.9
Imidazoyl carbonate	30.8	50.9
Tosylate	-	14.6

Primary hydroxyl silica, developed in our group (20,21) provides a carrier of high capacity following activation of the resin and coupling of the ligands (see Tab 2). Primary hydroxyl silica can be synthesized according to the three pathways outlined in Fig 1. It can be activated with common reagents in high yields, as it will be discussed below. Primary hydroxyl silica has been successfully applied for the purification of antibodies, enzymes and an enzyme inhibitor using HPAC (20,21).

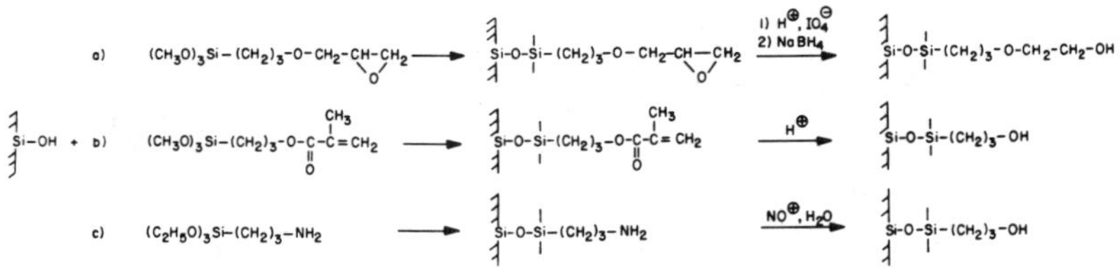

$$\text{a)}\quad (CH_3O)_3Si-(CH_2)_3-O-CH_2-\underset{\diagdown O \diagup}{CH-CH_2} \longrightarrow Si-O-Si-(CH_2)_3-O-CH_2-\underset{\diagdown O \diagup}{CH-CH_2} \xrightarrow[2)\ NaBH_4]{1)\ H^{\oplus},\ IO_4^{\ominus}} Si-O-Si-(CH_2)_3-O-CH_2-CH_2-OH$$

$$Si-OH\ +\quad \text{b)}\quad (CH_3O)_3Si-(CH_2)_3-O-\underset{O}{\underset{\|}{C}}-\overset{CH_3}{\overset{|}{C}}=CH_2 \longrightarrow Si-O-Si-(CH_2)_3-O-\underset{O}{\underset{\|}{C}}-\overset{CH_3}{\overset{|}{C}}=CH_2 \xrightarrow{H^{\oplus}} Si-O-Si-(CH_2)_3-OH$$

$$\text{c)}\quad (C_2H_5O)_3Si-(CH_2)_3-NH_2 \longrightarrow Si-O-Si-(CH_2)_3-NH_2 \xrightarrow{NO^{\oplus},\ H_2O} Si-O-Si-(CH_2)_3-OH$$

FIGURE 1. Schematic pathways for the synthesis of primary hydroxyl silica.

ACTIVATION PROCEDURES AND COUPLING

The preparation of an affinity support is a two-step process if silica-based matrices are chosen which have hydroxyl groups for anchoring the ligand: Activation of the hydroxyl groups in the first step is followed by the coupling of the desired ligand. In the case of matrices containing epoxide or aldehyde functions direct coupling is possible, leading to an amine derivative and a Schiff base.

Despite several disadvantages, the CNBr method is still the most commonly used to activate hydroxyl groups (22,23). Studies on the mechanism of activation with CNBr (23) revealed that the active species is mainly a cyanate ester which links proteins to the carrier via their free amino groups (Fig 2,I). The resulting isourea derivatives are known to be unstable and labile to nucleophilic attacks. Therefore leakage of the ligand often contaminates the purified product. In addition, isourea bonds serve as centers for nonspecific binding due to their ion-exchange properties.

In order to overcome these problems, new activation methods have been developed, which, after coupling of the ligand, lead to stable matrices without introduction of charge. Among the many potential activating reagents, we have chosen chloroformates (Fig 2,II) (24), carbonyldiimidazole (Fig 2,III) (25) and tosylchloride (Fig 2,IV) (26), which result in stable carbamate- and amine derivatives (see Fig 2).

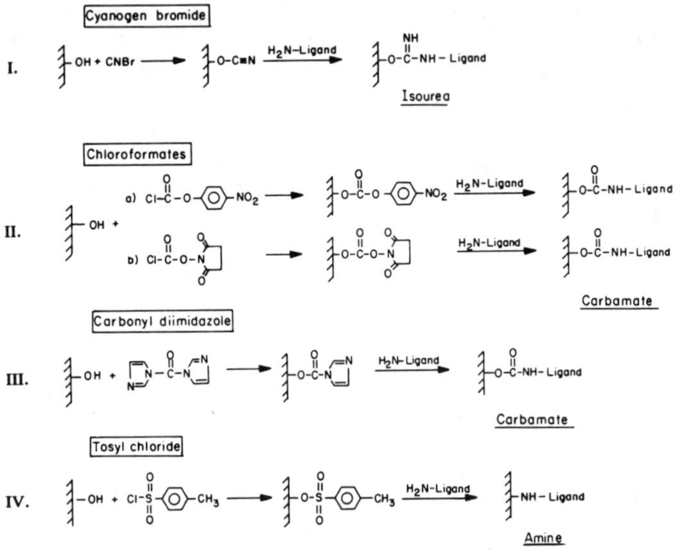

Cyanogen bromide

I. —OH + CNBr ⟶ —O-C≡N —H_2N-Ligand⟶ —O-C(=NH)-NH-Ligand

Isourea

Chloroformates

II. —OH + a) Cl-C(=O)-O-C6H4-NO_2 ⟶ —O-C(=O)-O-C6H4-NO_2 —H_2N-Ligand⟶ —O-C(=O)-NH-Ligand

b) Cl-C(=O)-O-N(succinimide) ⟶ —O-C(=O)-O-N(succinimide) —H_2N-Ligand⟶ —O-C(=O)-NH-Ligand

Carbamate

Carbonyl diimidazole

III. —OH + Im-C(=O)-Im ⟶ —O-C(=O)-Im —H_2N-Ligand⟶ —O-C(=O)-NH-Ligand

Carbamate

Tosyl chloride

IV. —OH + Cl-S(=O)(=O)-C6H4-CH_3 ⟶ —O-S(=O)(=O)-C6H4-CH_3 —H_2N-Ligand⟶ —NH-Ligand

Amine

FIGURE 2. Different methods of activation.

Using chloroformates, we have obtained very high yields of coupled ligands (see Table 2). For our studies we have used two chloroformates of different reactivity, p-nitrophenyl chloroformate and N-hydroxysuccinimide chloroformate leading respectively to active p-nitrophenyl carbonate- and N-succinimide carbonate-containing silica derivatives (Fig 2, IIa and b). A systematic study of the optimal conditions for attaching the ligand to the carrier, indicated that a pH between 6 and 7 results in the best yields (21). Coupling at pH 8 reduced the coupling capacity. This can be explained by considering the effect of two competitive reactions which occur during the coupling of ligands in aqueous solution: a) The aminolysis, due to the reaction of the ligand with the active resin and b) hydrolysis of the activated carbonate resin. At lower pH values the hydrolysis reaction is suppressed thereby leaving enough active carbonate groups available for the aminolysis (27). The importance of the two processes during coupling was demonstrated by performing the reaction with two differently activated resins at a constant pH = 6: With the N-succinimide carbonate resin (which are much more sensitive to hydrolysis due to the good leaving character of the N-succinimide group) the coupling process is already finished after 5 hours. With the p-nitrophenyl carbonate resin on the other hand, the coupling continues even after 36 hours in agreement with the fact that this resin is less reactive, and therefore, the occurrence of hydrolysis is much less (see Fig 3).

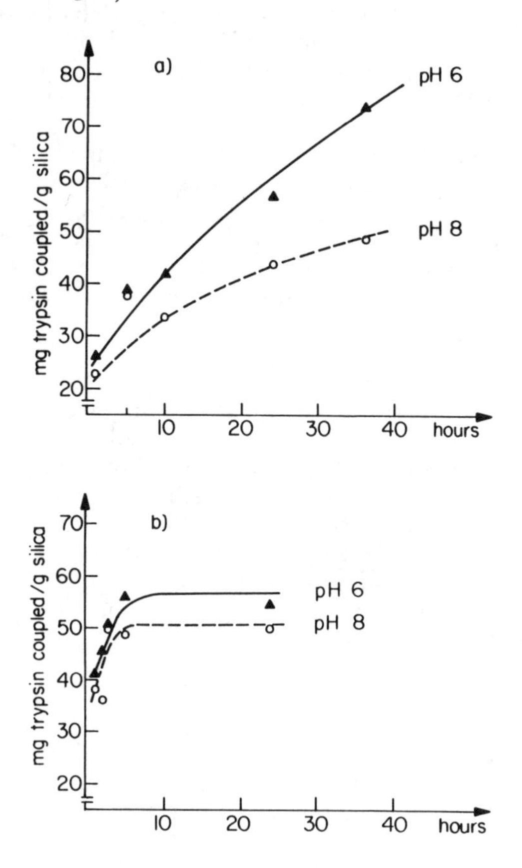

FIGURE 3. Coupling of trypsin to a) p-nitrophenyl carbonate silica and b) N-succinimide carbonate silica as a function of time and pH of the coupling buffer.

LIGANDS

Numerous ligands have been coupled to different matrices for the purification of a large variety of compounds using affinity chromatography. A representative list is given in reference 2.

Affinity ligands are divided into two groups: General and specific ligands. General ligands show a specific affinity towards a group of compounds. Immobilized adenosine monophosphate (AMP) (28,29), for example has been used for the purification of a variety of related dehydrogenases. This ligand is a part of the cofactor nicotinamide adenine dinucleotide (NAD) which can be recognized by a large number of different enzymes.

Lectins comprise another group of general ligands which have a high specificity towards particular sugar residues. This class of compounds has found widespread applicability for the purification of glycoproteins and glycolipids from membrane extracts (30). The best known lectin for this purpose is Concanavalin A which specifically recognizes D-mannose and D-glucose residues on macromolecules.

Triazine dyes (31) have gained importance especially in large-scale purification procedures. This group of ligands is easy to synthesize in large quantities and therefore quite low in cost. The additional property of high chemical stability makes these ligands almost ideal for industrial usage. Triazine dyes show a broad affinity for numerous proteins. They bind at the nuleotide binding sites of many enzymes as well as hydrophobically to many other proteins. Cibacron Blue for example is now routinely used for the one-step large-scale purification of serum albumin (32). Many different forms of immobilized triazine dyes are now available commercially.

In contrast to the general ligands are ligands which bind mainly to one particular solute. Among the specific ligands are enzymes, hormones, receptors, antigens and many others. Monoclonal antibodies are ligands with exquisite specificity, which are gaining more and more importance since they are available in large quantities through hybridoma technology (33). Therefore, monoclonal antibodies offer the possibility of large-scale purifications, and very likely, affinity chromatography will, in the future, gradually be overshadowed by immunoaffinity chromatography.

Immunoaffinity chromatography

The antibody-antigen interaction is of relatively high affinity with a constant of 10^6 to 10^{11} M^{-1}. The specificity is based on the fact that the antibody only recognizes one epitope on the antigen. As a result, many antibodies of different specificity are obtained against one antigen. Polyclonal antibodies, used for affinity-based purifications, generally show such tight binding to the adsorbed product that elution can usually take place only under harsh conditions. Such conditions (i.e. very high or very low pH, as well as chaotropic agents such as KSCN or guanidine HCl), used for the dissociation of the formed affinity complex, oftentimes denature the protein (34). The use of monoclonal antibodies which exhibit lower dissociation constants for the desired antigen thereby allowing milder conditions for elution, may eventually solve these problems.

Another problem which often occurs using harsh desorbing methods is the contamination of the purified antigen with the ligand (antibody). The major cause of this difficulty is the low stability of the antibody linkage to the support. Most of the immunosorbents in use today are still prepared by the CNBr-activation method for antibody immobilization. The resulting isourea linkages between ligand and support are very labile towards the above-described elution conditions thus contaminating the product with the antibody. Thus, it would be advisable to employ other activation methods which would result in more stable attachment of the antibody to the support. Alternatives are described in the section on activation procedures and coupling.

Monoclonal antibodies are beeing rapidly adapted for purification purposes in all fields of biology and biotechnology. Large-scale purifications are beeing reported more and more in the literature, interferon (7) and urokinase (35) are only two impressive examples were precious material was obtained in large quantities using monoclonal antibody-based immunoaffinity chromatography.

ELUTION

As already indicated above (see section on immunoaffinity chromatography) the dissociation of a immobilized affinity complex is often a serious problem. Although it can be the most important step in the purification of a compound using affinity chromatography, relatively little has been devoted to improve the elution conditions.

General strategies (for example, alterations in solvent, buffer, pH or ionic strength) have been used for the desorption of solutes having only low affinity towards the ligand. Biospecific elution is also possible when low-molecular-weight inhibitors are available which have a similar affinity towards the desired ligand. In this case, the inhibitor and the adsorbed molecule compete for binding sites on the ligand thus displacing the desired molecule upon saturation of all possible sites on the carrier.

Harsher conditions are required when the immobilized affinity complex is characterized by an especially high binding constant as is many times the case with antigen-antibody interactions. Chaotropic reagents like KSCN, urea or guanidine are then necessary to dissociate the affinity complex. This is often accompanied by the denaturation of the purified molecule and concomitant destruction of the affinity matrix. A possible way of solving these problems is the systematic construction of ligands which are on the one hand specific enough but on the other hand of such low binding strength that mild conditions are sufficient for elution of the desired molecule.

CONCLUDING REMARKS

Until very recently, researchers who have applied affinity chromatography have been satisfied with Sepharose as the carrier and CNBr as the reaction for coupling of ligands. From time to time new carriers and new coupling methods have been introduced, and, in most cases, were not extensively adopted by the scientific community.

The advent of the new "biotechnology era" has underscored the fact that the purification of genetically engineered proteins, for example, is the current bottleneck in most preparations for reasons discussed above. Therefore, the field of downstream processing has recently gained high priority, and it is our opinion that we will be witnessing in the not-too-distant future new purification methods and a wide range of new carriers (both synthetic and natural). In conjunction with this new revolution, affinity chromatography and high performance affinity chromatography will also flourish upon the careful design and development of new activation methods and new ligands. These new ligands will hopefully be derived from a better understanding of the forces involved in the formation of binding or recognition sites. Until then, the best approach currently available is immunoaffinity chromatography or some combination between affinity and nonspecific interactions such as a mixture of ion exchange and hydrophobic interactions.

REFERENCES

1. Cuatrecasas P, Wilchek M, Anfinsen C (1968). Selective enzyme purification by affinity chromatography. Proc Natl Acad Sci USA 61:636.
2. Wilchek M, Miron T, Kohn J (1984). Affinity chromatography. In Jacoby WB (ed): "Methods in Enzymology 104: Enzyme Purification and Related Techniques," Orlando: Academic Press, Inc, p 3.
3. Parikh I, Cuatrecasas P (1985). Affinity chromatography. Chem & Eng News 26:17.
4. Walters RR (1985). Affinity chromatography. Anal Chem 57:1099A
5. Birch IR, Hill CR, Kenney AC (1985). Affinity chromatography: Its Role in Industry. In Cheremisinoss PN, Ovellette RP (eds): "Biotechnology Applications and Research," Lancaster: Techcromic Publishing Company, Inc, p 594.
6. Wilchek M (1984). Affinity chromatography. Angew Makrom Chem 123/124:175.
7. Secher DS, Burke DC (1980). A monoclonal antibody for largescale purification of human leukocyte interferon. Nature 285:446.
8. Larsson PO, Glad M, Hansson L, Mansson MO, Ohlson S, Mosbach K (1983). High-performance liquid affinity chromatography. In Giddings JC, Grushka E, Cazes J, Brown PR (eds): "Advances in Chromatography 21," New York: Marcel Dekker, p 41.
9. Dean PDC, Walker JNB (1986). Recent advances in high-performance liquid affinity chromatography columns.

Biochem Soc Trans 13:1984.
10. Lowe CR (1984). New developments in downstream processing. J Biotech 1:3.
11. Taylor RF (1985). A comparison of various commercially available liquid chromatographic supports for immobilisation of enzymes and immunoglobulins. Anal Chim Ac 172:241.
12. Weetall HH (1985). Enzymes immobilized on inorganic supports. Trends Biotech 3:276.
13. Southern D, Hollis D (1986). Fast affinity chromatography. Biotech 4:17.
14. Parker M (1986). Fast affinity chromatography. Internat Labmate:17.
15. Ohlson S, Hansson L, Larsson P-O, Mosbach K (1978). High-performance liquid affinity chromatography (HPLAC) and its application to the separation of enzymes and antigens. FEBS Let 93:5.
16. Lowe CR, Glad M, Larsson P-O, Ohlson S, Small DAP, Atkinson T, Mosbach K (1981). High-performance liquid affinity chromatography of proteins on Cibacron Blue F3G-A bonded silica. J Chromatogr 215:303.
17. Borchert A, Larsson P-O, Mosbach K (1982). High-performance liquid affinity chromatography on silica bound concanavalin A. J Chromatogr 244: 49.
18. Walter RR (1983). Minicolumns for affinity chromatography. Anal Chem 55:1395.
19. Crowley SC, Walters RR (1983). Determination of immunoglobulins in blood serum by high-performance affinity chromatography. J Chromatogr 266:157.
20. Ernst-Cabrera K, Wilchek M (1986). Silica containing primary hydroxyl groups for high-performance affinity chromatography. Anal Biochem 159:267.
21. Ernst-Cabrera K, Wilchek M (1987). Coupling of ligands to primary hydroxyl-containing silica for HPAC: Optimization of conditions. J Chromatogr in press.
22. Porath J, Axen R, Ernback S (1967). Chemical coupling of proteins to agarose. Nature 215:1491.
23. Kohn J, Wilchek M (1982). Mechanism of activation of sepharose and sephadex by cyanogen bromide. Enzyme Microb Technol 4:161.
24. Wilchek M, Miron T (1982). Immobilization of enzymes and affinity ligands onto agarose via stable and uncharged carbamate linkages. Biochem Intern 4:629.
25. Bethell GS, Ayers JS, Hearn MTW, Hancock WS (1981). Investigation of the activation of various insoluble polysaccharides with 1,1'-carbonyldiimidazole and of the

properties of the activated matrices. J Chromatogr 219:361.
26. Nilsson K, Mosbach K (1984). Immobilization of ligands with organic sulfonylchlorides. In Jacoby WB (ed): "Methods of Enzymology 104: Enzyme Purification and Related Techniques," Orlando: Academic Press,Inc, p 56.
27. Ernst-Cabrera K, Wilchek M. Kinetic studies on the coupling of ligands to chloroformate activated carriers. In preparation.
28. Clonis YD, Lowe CR (1980). Affinity chromatography on immobilized nucleotides. Eur J Biochem 110:279.
29. Brodelius P, Larsson P-O, Mosbach K (1974). The synthesis of three AMP-analogues: N^6-(6-aminohexyl)-adenosine 5'-monophosphate, N^6-(6-aminohexyl)-adenosine 2',5'-biphosphate, and N^6-(6-aminohexyl)-adenosine 3',5'-biphosphate and their application as general ligands in biospecific affinity chromatography. Eur J Biochem 47:81.
30. Lotan R, Nicholson GL (1979). Purification of cell membrane glycoproteins by lectin affinity chromatography. Biochim Biophys Acta 559:329.
31. Dean PDG, Quadri F (1983). Affinity chromatography on immobilized dyes. In Scouten WH (ed): "Solid Phase Biochemistry," New York: John Wiley & Sons, Inc, p 79.
32. Harvey MJ, Brown RA, Rott J, Lloyd D, Lane RS (1983). In Curling JM (ed): "Separation of Plasma Proteins," Uppsala: Pharmacia Fine Chemicals AB, p 79.
33. Koehler G, Milstein C (1975). Continuos cultures of fused cells secreting antibody of predefined specificity. Nature 256:495.
34. Murphy RF, Iman A, Hughes AE, McGucken J, Buchanan KD, Conlon JM, Elmore DT (1976). Avoidance of strongly chaotropic eluents for immunoaffinity chromatography by chemical modification of immobilized Ligand. Biochim Biophys Acta 420:87.
35. Herion P, Bollen A (1983). Purification of urokinase by monoclonal antibody chromatography. Biosci Report 3:373.

Protein Purification: Micro to Macro, pages 177–195

CONTROLLED PORE GLASS CHROMATOGRAPHY OF PROTEINS

Eugene Sulkowski

Department of Molecular and Cellular Biology
Roswell Park Memorial Institute, Buffalo, New York 14263

ABSTRACT An understanding of the interaction between a protein and glass surface (controlled pore glass, silicic acid) is of significant theoretical and practical interest. In order to gain an insight into this interaction, chicken egg white lysozyme was chromatographed on controlled pore glass using several classes of potential eluants. Lysozyme was displaced from controlled pore glass with: a) an electrolyte and an organic solute (ethylene glycol, dimethyl sulfoxide, acetone, acetonitrile) used as co-solvents; b) various alkylamines; c) natural and alkylated amino acids, and d) alkylureas. As a result, it is proposed that tetramethylammonium chloride be used as a standard solvent in the chromatography of proteins on unmodified glass (controlled pore glass, silicic acid). The present study provides an experimental evidence that the simultaneous suppression of both electrostatic and hydrophobic interactions, is necessary and sufficient, to prevent or abrogate an interaction between a protein and glass surface.

INTRODUCTION

Controlled pore glass (CPG) and silicic acid (SA) have become sorbents of choice in the initial isolation of many proteins important in human therapy: interferons (1-3), tumor necrosis factor (4), lymphotoxin (5), interleukin 2 (6) and erythropoietin (7).

Physico-chemistry of CPG is discussed by its inventor, W. Haller, in a comprehensive review (8), while the use of CPG for the purification of proteins is the subject of a recent review by T. Mizutani, pioneer and major contributor to the chromatography of proteins on glass (9). The reader is directed to both articles for the authoritative presentations of basic observations and principles. The remainder of the literature, some exceptions notwithstanding (3,7,10, 11,12), is mostly of the phenomenological character.

Controlled pore glass was introduced as a medium for permeation chromatography (13). However, its present use is mainly for the adsorption chromatography (8,9). Bock et al. (14) chromatographed a variety of proteins on CPG using a plethora of eluants and, in particular, chaotropic salts. Braude et al. (10) studied the mechanism of binding to CPG and concluded that it involved electrostatic and hydrophobic interactions.

METHODS

Controlled pore glass, CPG00350, mesh 120/200, Lot 05B08, was purchased from Electro-Nucleonics, Fairfield, NJ. It had a mean pore diameter, 381A and surface area, 64.7 M^2 per gram. All experiments were performed on a column with packed bed, 0.6x15 cm, i.e., ca 4 mL. Flow rate was 1 mL per 3.6 min (ca 60 $mL/cm^2/hr$). Protein solution, 10 mL, containing 1 mg/mL, in 50 mM sodium phosphate (PB), pH 7.4, was applied and the column was rinsed with 20 mL of 50 mM PB, pH 7.4. All eluants contained 50 mM PB, pH 7.4. Experiments were performed at $4^{o}C$. Solvents were applied on the columns by means of a satellite pump, mp13 GJ-4, Gilford.

RESULTS

Elution of Lysozyme from CPG with NaCl or Ethylene Glycol.

Lysozyme binds to CPG when applied in 50 mM sodium phosphate buffer (PB), pH 7.4. It remains bound on the column (Figure 1) upon a prolonged wash with PB (25 total column volumes, 100 mL). However, lysozyme can be readily displaced from the column with an electrolyte (sodium chloride). Apparently, the quenching of electrostatic attractive forces alone is sufficient to reverse the binding of this protein to glass surface. However, ethylene glycol (EG) can displace lysozyme from CPG column, as well (Figure 1). Apparently, the enhancement of electrostatic interactions by ethylene glycol (low dielectric constant) is more than compensated for by its function as a solute decreasing

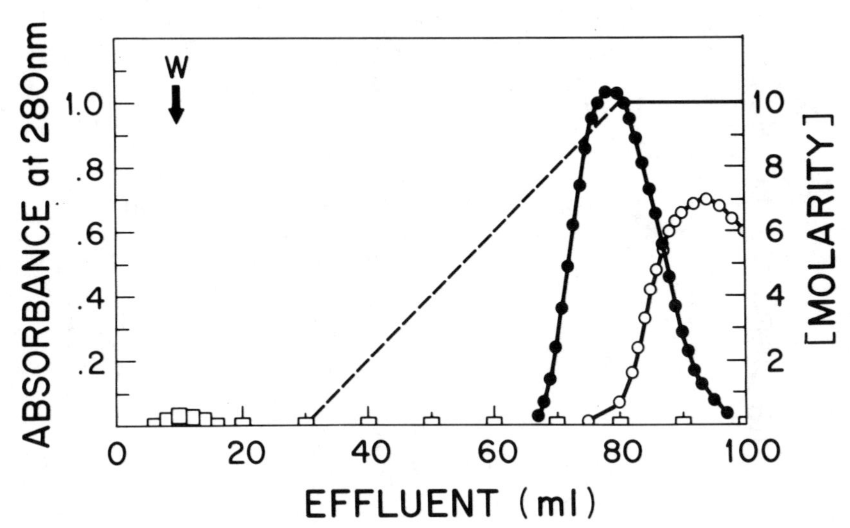

FIGURE 1. Chromatography of lysozyme on CPG. □ , 50 mM sodium phosphate, pH 7.4. ● , sodium chloride gradient, 0 to 1M, in 50 mM sodium phosphate, pH 7.4. o , ethylene glycol gradient, 0 to 10M, in 50 mM sodium phosphate, pH 7.4.

hydrophobic interactions. Thus, the interaction of lysozyme with CPG may involve both, coulombic and hydrophobic forces. If so, then the simultaneous use of an electrolyte and an organic solvent (a polarity reducing agent) should be the most efficient to displace lysozyme from CPG.

Elution of Lysozyme from CPG with Sodium Chloride and Ethylene Glycol (co-solvents).

The elution of lysozyme from CPG by linear concentration gradients of sodium chloride and ethylene glycol, run concurrently, is illustrated in Figure 2. Lysozyme was recovered with 0.5M NaCl and 5M EG (30%); 0.6M NaCl and 3M EG (18%); 0.7M NaCl and 1.7M EG (10%), respectively. Clearly, NaCl and EG, when used as co-solvents, are much more efficient as eluants than when used separately (Figure 1). Moreover, the concentration of EG necessary for the recovery of lysozyme from CPG can be significantly lowered by just a moderate increase in NaCl concentration.

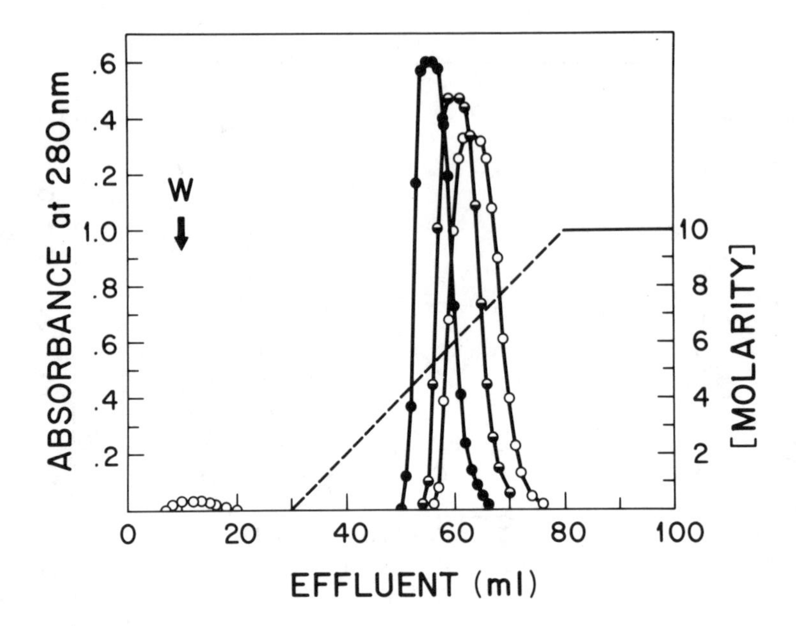

FIGURE 2. Chromatography of lysozyme on CPG. ● , sodium chloride gradient, 0 to 1M; ethylene glycol gradient, 0 to 10M. ◒ , sodium chloride gradient, 0 to 1M; ethylene glycol gradient 0 to 5M; o , sodium chloride gradient, 0 to 1M; ethylene glycol gradient, 0 to 2.5M. All solvents contained 50mM PB, pH 7.4.

Elution of Lysozyme from CPG with Sodium Chloride and Organic Solvents (co-solvents).

Figure 3 illustrates a series of experiments in which a linear concentration gradient (0 to 0.5M) of sodium chloride was developed alone or concurrently with an identical gradient (0 to 0.5M) of an organic solvent: dimethyl sulfoxide, acetone and acetonitrile. Elution of lysozyme was followed by absorption at 280 nm or by colorimetry (acetone). All organic solvents failed to displace lysozyme when used alone (Figure 3). Lysozyme was recovered with sodium chloride alone. However, the elution of lysozyme with an electrolyte and an organic solvent was more efficient than with an electrolyte alone.

In order to probe further the participation of electrostatic and hydrophobic interactions, the efficiency of various alkylamine chlorides as eluants was studied.

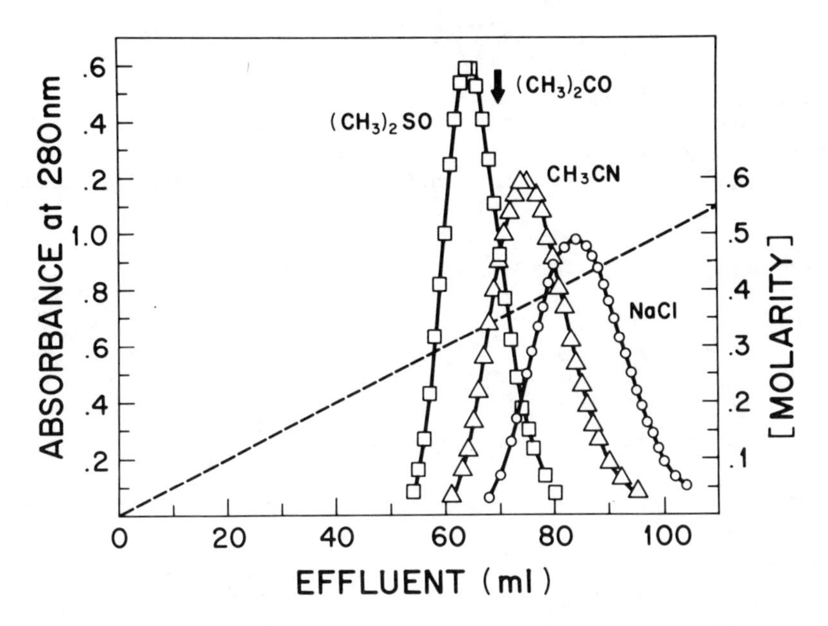

FIGURE 3. Chromatography of lysozyme on CPG. □, dimethyl sulfoxide $(CH_3)_2SO$. ⬇, Acetone, $(CH_3)_2CO$. Δ, Acetonitrile, CH_3CN. o, sodium chloride, NB: only elution phase of the experiments is illustrated; the adsorption and wash phases of the experiment are identical to those shown earlier (Figures 1 and 2).

Elution of Lysozyme from CPG with $(CH_3)_n$ NCl's.

In an orienting experiment, lysozyme was eluted from CPG with: sodium chloride, ammonium chloride and tetramethylammonium chloride (TMAC). Since TMAC was found to be the most efficient, a relationship between the extent of alkylation of ammonium ion and its eluting power was further studied.

The elution of lysozyme by ammonium chloride and alkylated ammonium chlorides from CPG is shown in Figure 4. The elution of lysozyme by NH_4Cl (Figure 4) is more efficient than that by NaCl (Figure 3), reflecting a higher affinity of $NH_4^+ > Na^+$ for glass surface. The increasing alkylation of ammonium ion, in turn, resulted in an enhanced efficiency of elution of lysozyme (Figure 4); tetramethylammonium chloride is more efficient as an eluant (it displaces protein at a lower concentration) than ammonium chloride; this correlation is graphically presented in Figure 5.

Elution of Lysozyme from CPG with $[CH_3(CH_2)_n]_4$ NCl's.

Tetramethylammonium chloride is the lowest homologue in the family of so-called "hydrophobic electrolytes". It was thus of interest whether or not the higher homologues: tetraethyl-, tetrapropyl-, and tetrabutyl-ammonium chlorides would function as still more efficient eluants. To this end, lysozyme was eluted with appropriate salts. The following values (molarity of salt) were obtained:

Salt	Mol
$N^+M_4Cl^-$	0.16
$N^+E_4Cl^-$	0.16
$N^+P_4Cl^-$	0.22
$N^+B_4Cl^-$	0.27

Clearly, tetramethylammonium chloride ($N^+M_4Cl^-$) was as efficient as tetraethylammonium chloride ($N^+E_4Cl^-$); however, $N^+P_4Cl^-$ was less efficient and $N^+B_4Cl^-$ was the least efficient.

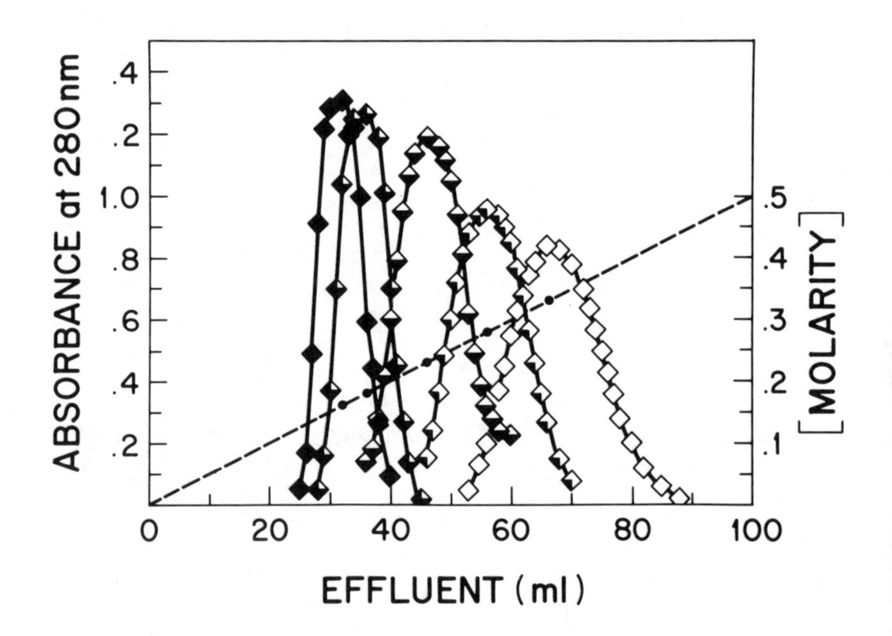

FIGURE 4. Elution of lysozyme from CPG by alkylamines. ◇, ammonium chloride. ◇, methylammonium chloride. ◇, dimethylammonium chloride. ◇, trimethylammonium chloride. ◆, tetramethylammonium chloride.

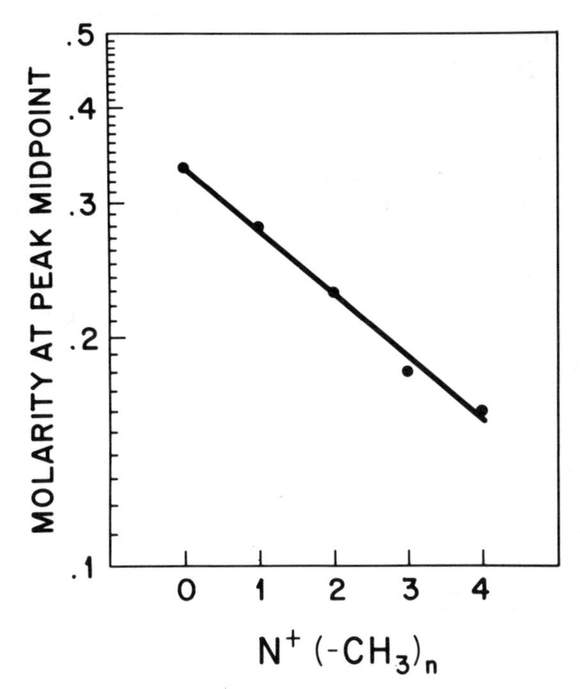

FIGURE 5. Positive correlation between the extent of alkylation (methylation) of ammonium ion and its efficiency (molar concentration) as eluant.

Figure 6 illustrates the results of a series of experiments on the elution of lysozyme from CPG by methylamine, ethylamine, propylamine and butylamine, all neutralized with hydrochloric acid to near neutral pH and adjusted to pH 7.4 in the presence of 50 mM phosphate pH 7.4. As expected (see Figures 4 and 5), methylammonium chloride was found to be more efficient as an eluant than ammonium chloride. However, the eluting powers of ethylammonium-, propylammonium- and butylammonium-chlorides were found to be identical. Therefore, it may be concluded that an increased length of a single alkyl substituent, from methyl through butyl, does not impart greater eluting efficiency on the alkylammonium cation.

This is in contrast to the data presented in Figure 5. Apparently, the tetrahedral disposition of alkyl substituents around a central nitrogen ion (N^+) meets better the requirements for generating an affinity for the glass surface.

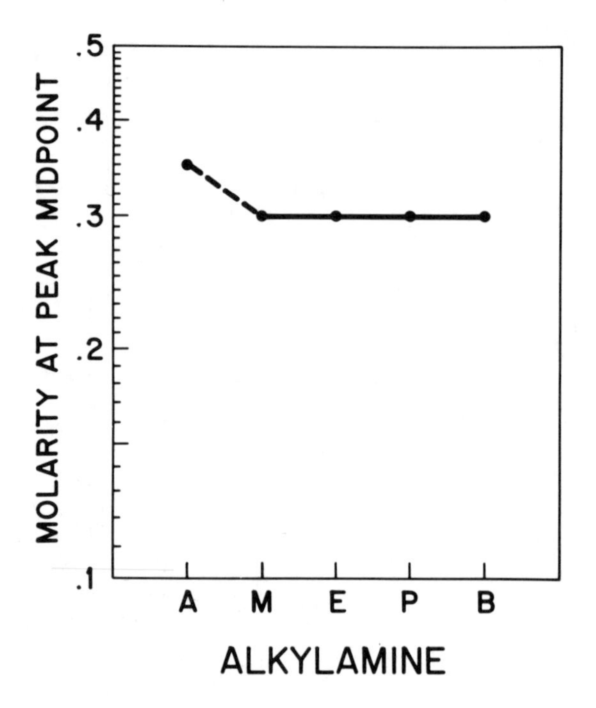

FIGURE 6. The lack of correlation between the length of alkyl substituent of ammonium ion and its eluting power. A, ammonium chloride. M, methylammonium chloride. E, ethylammonium chloride. P, propylammonium chloride. B, butylammonium chloride.

Elution of Lysozyme from CPG with Amino Acids.

All solutions of amino acids were supplemented with equimolar concentration of sodium chloride. The selection of amino acids was determined, in part, by their solubility. Otherwise, it was of interest to assess the eluting power of acidic, basic and neutral amino acids.

Figure 7 records (arrows) the peak tube positions of the elution profiles of lysozyme. L-glutamic acid was found to stabilize the binding of lysozyme on CPG, by comparison to glycine; L-lysine and L-arginine facilitated the displacement of lysozyme. Preferential displacement of lysozyme from CPG by lysine (arginine) corroborates well the earlier findings on the role of lysine in the binding of a protein to glass (10) as well as studies on the binding of amino acids to controlled pore glass (9).

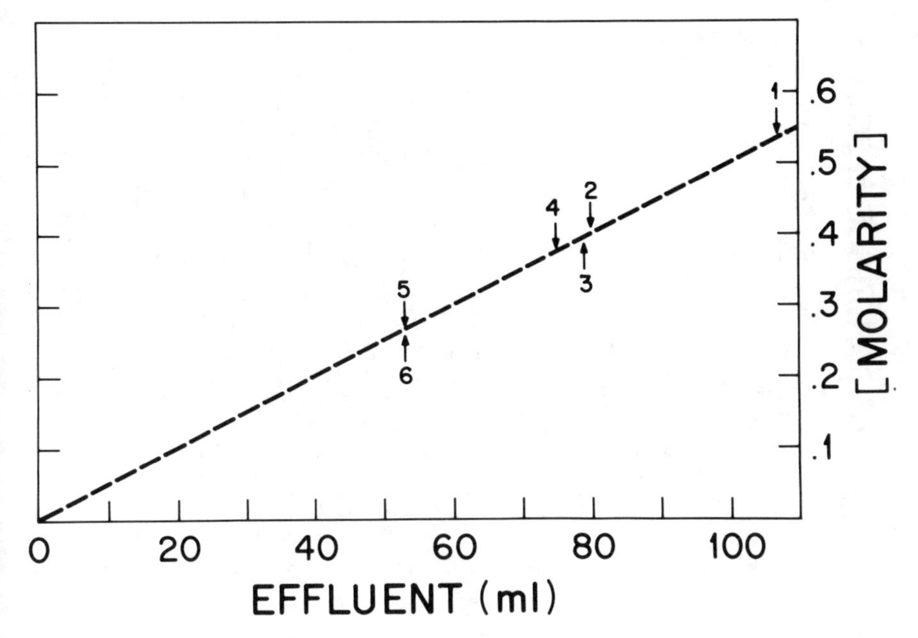

FIGURE 7. Elution of lysozyme from CPG with amino acids. 1, L-glutamic acid. 2, glycine. 3, L-alanine. 4, L-valine. 5, L-arginine. 6, L-lysine.

Elution of Lysozyme from CPG by Glycine and N-Alkylated Glycines.

In view of the data on the alkylamines (Figures 4 and 5), it was of considerable interest to explore the efficiency of N-alkylated derivatives of glycine as eluants. It could be anticipated that progressive methylation of α-amino group should produce a series of solutes with an increasing elution power toward lysozyme bound on CPG. This anticipation is justified by the results illustrated in Figure 8. N-methylglycine (sarcosine) was found to be better eluant than glycine; N,N-dimethylglycine was, in turn, better than sarcosine; fully methylated derivative, betaine, was proven to be the most efficient eluant of this series. Again, as in the case of progressive methylation of ammonium ion (Figure 5), there was a linear relationship between the extent of alkylation (methylation) of glycine and its efficiency as an eluant (Figure 9).

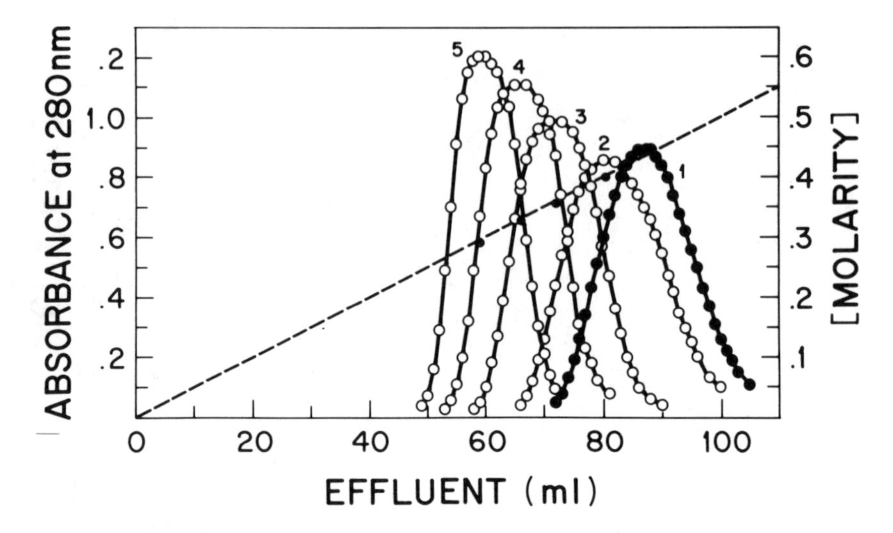

FIGURE 8. Elution of lysozyme from CPG by glycine and its N-methylated derivatives. 1, sodium chloride. 2, glycine. 3, sarcosine. 4, N,N-dimethylglycine. 5, betaine.

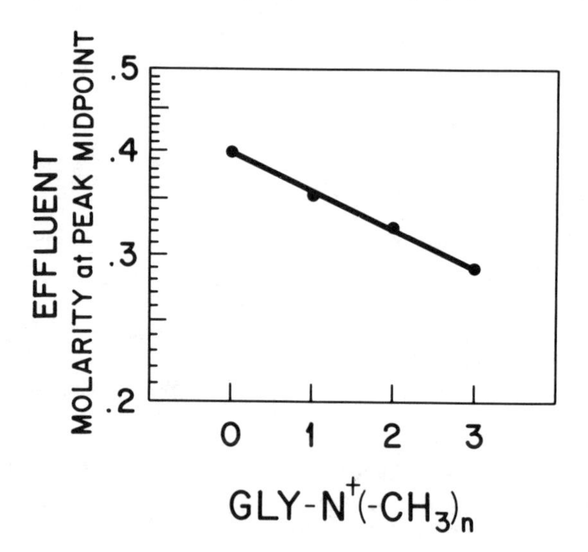

FIGURE 9. Positive correlation between the extent of alkylation of glycine and its efficiency as an eluant. 0, glycine; 1, N-methylglycine; 2, N,N-dimethylglycine; 3, betaine.

Elution of Lysozyme from CPG by Alkyl Chain Amino Acids.

In order to assess the influence of the overall hydrophobicity of an α-amino acid, a series of amino acids with an increasing alkyl side chain was studied:

Glycine	H -C α
L-Alanine	H_3C -C α
L-α-amino-n-butyric acid	H_3C-CH_2 -C α
L-Norvaline	$H_3C-CH_2-CH_2$ -C α

The elution power of L-norvaline was found to be the same as that of L-valine. The eluting power of L-α-amino-n-butyric acid was found to be intermediate between L-alanine and L-norvaline (not illustrated). Thus, the contribution of the alkyl chain of an α-amino acid towards its eluting power was found to be experimentally discernible but quantitatively small. This is in contrast to the findings illustrated in Figure 9, i.e., the effect of an increase in the alkylation of amino group of an α-amino acid.

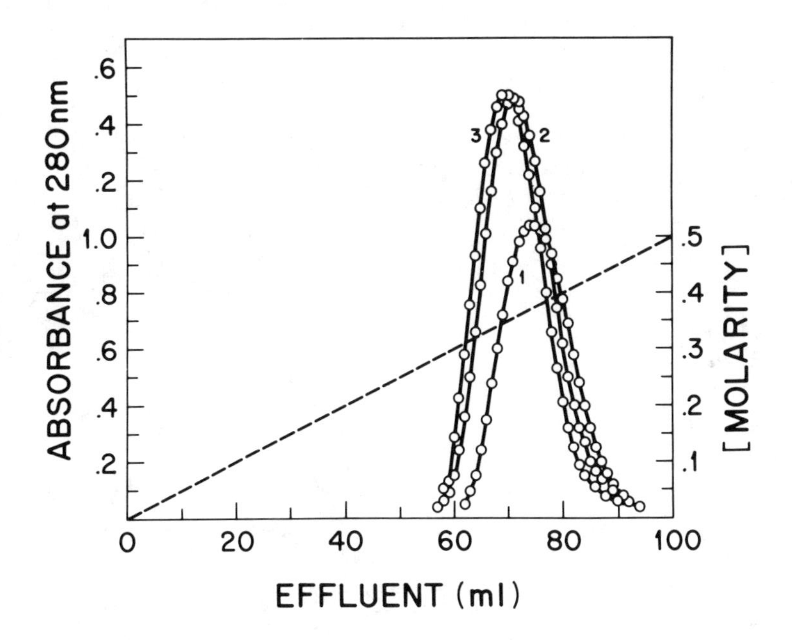

FIGURE 10. Elution of lysozyme from CPG with α-, β-, γ-amino acids. 1, α-aminobutyric acid. 2, β-aminobutyric acid and 3, γ-aminobutyric acid. All solutions of amino acids contained equimolar sodium chloride.

Elution of Lysozyme from CPG with α-, β-, and γ-Amino Acids.

In order to assess the influence of the separation of electric charges in an amino acid (zwitterion) on its elution power toward proteins bound to CPG, two short series of experiments were performed. Firstly, lysozyme was displaced from CPG with linear concentration gradients of L-alanine (2-amino-propionic acid) and β-alanine (3-amino-propionic acid). It was found that β-alanine was a better eluant than L-alanine (not illustrated). Secondly, lysozyme was eluted from CPG with α-, β-, and γ-butyric acids (Figure 10). Although it was found again that the disposition of an amino group in β position resulted in an improvement of the elution power of an aminobutyric acid, the additional separation of charges by placing amino group in γ-position had but a negligible effect. Perhaps, a still further separation of charges would remain inconsequential (not tested).

Elution of Lysozyme from CPG with Trimethyl-glycine (betaine) and Trimethyl-β-alanine.

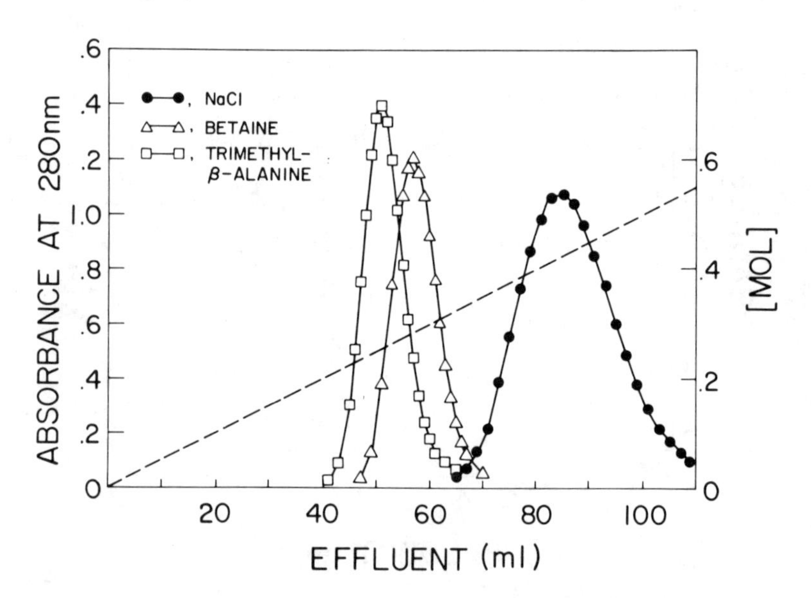

FIGURE 11. Elution of lysozyme from CPG with trimethylglycine and trimethyl-β-alanine. ● , sodium chloride. Δ , trimethylglycine. □ , trimethyl-β-alanine. Solutions of amino acids contained equimolar sodium chloride.

A $CH_3-\overset{CH_3}{\underset{CH_3}{N^{(+)}}}-CH_2-COO^{(-)}$ **B** $CH_3-\overset{CH_3}{\underset{CH_3}{N^{(+)}}}-CH_2-CH_2-COO^{(-)}$

Clearly, betaine (A) is less efficient than trimethyl-β-alanine (B).

Elution of Lysozyme from CPG with Alkyl Ureas.

The role of the hydrophobicity of an eluant molecule was studied by using alkylureas as eluants for lysozyme. The results are illustrated in Figure 12. In a preliminary experiment, it was found that urea itself, at 5 Molar concentration, was not able to displace lysozyme from CPG. However, when urea was supplemented with sodium chloride, the desorption of lysozyme was facilitated (Figure 12). The effect of an increasing alkyl arm was also observed.

sym-Dimethylurea and sym-diethylurea were found marginally more efficient than urea. Tetramethylurea was the most efficient, even more so than butylurea (not illustrated).

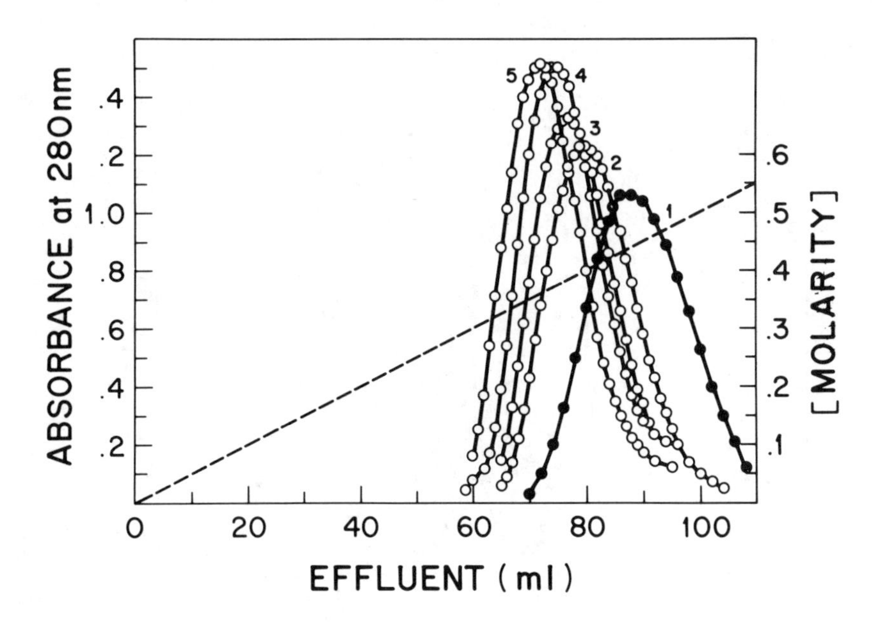

FIGURE 12. Elution of lysozyme from CPG with alkylureas. 1, sodium chloride. 2, urea. 3, methylurea and ethylurea. 4, propylurea. 5, butyl urea. All urea solutions contained equimolar sodium chloride.

Elution of Lysozyme from CPG with Selected Solutes.

In order to arrive at good comparison of the eluting power of some solutes employed in this study a series of experiments was performed on a single chromatographic column. The experimental series started with elution of lysozyme by sodium chloride, proceeded through various eluants as indicated by the sequential numbering and was finished by elution of lysozyme with sodium chloride. The elution profiles of lysozyme with sodium chloride, 1 and 6, indicate that the column did not suffer any drift in its chromatographic properties (affinity for lysozyme, capacity, mass transfer).

Clearly, tetramethylammonium chloride emerged as the most efficient eluant of lysozyme from CPG.

The selection of lysozyme as a model protein of this study was dictated by its high pI (11.2) and moderate surface hydrophobicity. The chromatography of proteins on CPG cannot, however, be limited to proteins sharing these properties with lysozyme. It was thus imperative to select some other proteins and evaluate their binding to CPG as well as to assess the eluting power of tetramethylammonium chloride, our eluant of choice.

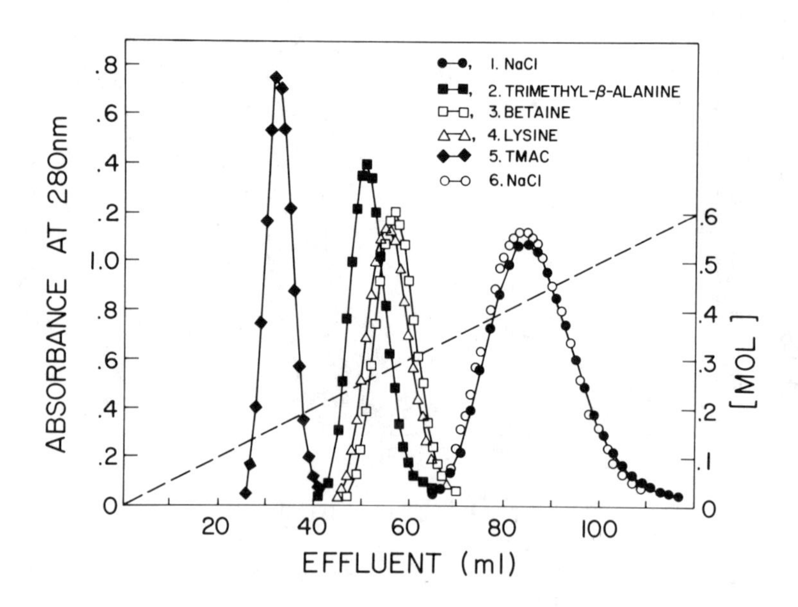

FIGURE 13. Chromatography of lysozyme on CPG. The sequence of experiments is indicated by numbers.

Chromatography of Model Proteins on CPG: Elution with TMAC.

Selected proteins were chromatographed on CPG under standard solvent conditions: adsorption in 50 mM sodium phosphate, pH 7.4; elution with linear concentration gradient of tetramethylammonium chloride, 0 to 0.5M, in 50 mM PB, pH 7.4. Proteins, monomeric and oligomeric, varied in size, shape, isoelectric point and surface hydrophobicity. A perusal of Figure 14 indicates that there is no simple correlation between the strength of binding and any single physical parameter. A combination of two parameters is more predictive, however.

Clearly, a notion that only basic proteins bind to glass surface (negative charge) is contradicted by the data. The binding of a protein seems rather to be critically dependent on its surface hydrophobicity. Conspicuously, the majority of studied proteins was recovered at relatively low concentration, less than ca 0.25M, of TMAC.

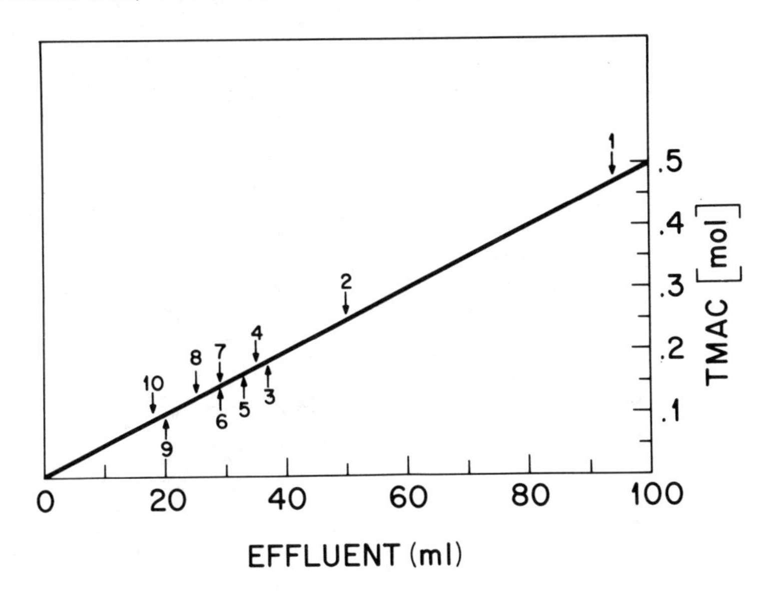

FIGURE 14. Chromatography of proteins on CPG. Peak tubes of the elution profiles are indicated by arrows. 1, Fibrinogen (pI 5.5). 2, Avidin (pI 10.1). 3, Hemoglobin (pI 6.8). 4, Cytochrome c (pI 10.6). 5, Lysozyme (pI 11.2). 6, Pancreatic trypsin inhibitor (pI 10.5). 7, Wax bean lectin (pI 5.5). 8, Myoglobin (pI 6.9). 9, β-lactoglobulin (pI 5.0). 10, Pancreatic ribonuclease A (pI 7.8).

DISCUSSION

The putative involvement of electrostatic and hydrophobic forces (10,11) in the binding of a protein to unmodified glass surface (CPG, SA) may be probed, as described in Figures 1, 2 and 3, by the modulation of the macroscopic properties of the mobile phase. To this end, the mobile phase may be rendered high in an electrolyte, low in polarity or both. The efficient elution of a protein from CPG requires the simultaneous presence of salt (NaCl, NH_4Cl) and a polarity reducing agent, such as ethylene glycol (Figure 2). The use of sodium chloride and ethylene glycol as co-eluants is now quite common in the purification of many proteins, although rarely an effort is made to use them in experimentally justifiable proportions (Figure 2). Attention should be paid, in particular, to the use of ethylene glycol at minimal necessary concentration: ethylene glycol is itself a moderate protein denaturant.

Mobile phase can also be modified, aside from admixing desirable solvents, by introdution of a solute of hybrid characteristics, carrying both electrostatic and hydrophobic properties. Ammonium ion can be rendered hydrophobic, by its alkylation, according to the scheme:

$$H-\overset{H}{\underset{H}{\overset{|+}{\underset{|}{N}}}}-CH_3 \qquad\qquad H-\overset{H}{\underset{H}{\overset{|+}{\underset{|}{N}}}}-CH_3$$

$$CH_3-\overset{H}{\underset{H}{\overset{|+}{\underset{|}{N}}}}-CH_3 \qquad\qquad H-\overset{H}{\underset{H}{\overset{|+}{\underset{|}{N}}}}-CH_2-CH_3$$

$$H-\overset{H}{\underset{H}{\overset{|+}{\underset{|}{N}}}}-H$$

$$CH_3-\overset{H}{\underset{CH_3}{\overset{|+}{\underset{|}{N}}}}-CH_3 \qquad\qquad H-\overset{H}{\underset{H}{\overset{|+}{\underset{|}{N}}}}-CH_2-CH_2-CH_3$$

$$CH_3-\overset{CH_3}{\underset{CH_3}{\overset{|+}{\underset{|}{N}}}}-CH_3 \qquad\qquad H-\overset{H}{\underset{H}{\overset{|+}{\underset{|}{N}}}}-CH_2-CH_2-CH_2-CH_3$$

The results illustrated in Figures 4 and 5 indicate clearly that only the construction of a hydrophobic "shell" leads to an increased efficiency in the elution of lysozyme from CPG. The construction of a hydrophobic "tail" fails to enhance the eluting power of the eluant beyond the effect of the first methyl unit (Figure 6).

Apparently, an increase in the overall hydrophobicity of a potential eluant molecule has its stereochemical caveat: tetrahedral disposition of the alkyl substituents around a central positive charge is obligatory. Hydrophobic shell is more effective than hydrophobic tail in dislodging protein molecule from the glass surface.

The decreasing efficiency in the elution of lysozyme from CPG in the series: tetraethyl-, tetrapropyl-, and tetrabutylammonium chlorides may be understood in terms of the coulombic effect of the central positive ion (N^+) which is progressively removed from the idealized glass surface due to an increase in the thickness of a hydrophobic shell.

The sterochemical requirement for the tetrahedral disposition of the alkyl substituents was found also in force in the alkylated glycine series (Figures 8 and 9). Again, the efficiency of elution of lysozyme from CPG correlated with the extent of alkylation of the α-amino group of glycine.

The fully alkylated glycine, i.e., betaine, can be formally considered as carboxylated tetramethylammonium ion. The introduction of one negative charge into tetramethylammonium ion results in a decrease of its eluting efficiency from 0.16M (TMAC) to 0.29M (betaine).

The separation of charges is better in trimethyl-β-alanine than in betaine. This may result in a higher affinity of trimethyl-β-alanine vis a vis betaine for glass surface (lesser electrostatic repulsion with silanols) and ensuing better eluting efficiency for lysozyme (Figure 11).

The displacement (elution) of proteins from glass surface ought to be, and on occasion must be, accomplished without an irreversible perturbation of their conformations. TMAC meets this requirement: it is known to stabilize the tertiary structure of proteins (15).

The literature search revealed that TMAC was used earlier to prevent adsorption to unmodified silica of some non-ionic and cationic polymers (16). The use of TMAC in recovery of human IFN-α (3,11,12) was recently extended to recovery of IFN-γ from CPG (17).

It remains to be seen whether the NaCl/EG co-solvents and TMAC may be combined to result in the recovery of a protein from glass when each eluting modality on its own is not acceptable.

In a closing comment, one can wonder whether the use of TMAC could not restore CPG to its original intended use (13). Hopefully, biotechnology field will make good use of unmodified glass as a chromatographic medium.

The main virtue of the naked glass (CPG, SA) is its price (cheap). This virtue is not shared by the chemically derivatized glass (covered surface), its Victorian cousin.

ACKNOWLEDGEMENT

I am grateful to Marcia Held for her excellent secretarial assistance in preparing this manuscript.

REFERENCES

1. Edy VG, Braude IA, De Clercq E, Billiau A, De Somer P (1976). Purification of interferon by adsorption chromatography on controlled pore glass. J Gen Virol 33:517.
2. Georgiades JA, Langford ML, Stanton GJ, Johnson HM (1979). Purification and potentiation of human immune interferon activity. IRCS Med Sci 7:559.
3. Chadha KC, Sulkowski E (1981). Chromatography of human leukocyte interferon on controlled-pore glass. Prep Biochem 11:467.
4. Aggarwal BB, Kohr WJ, Hass PE, Moffat B, Spencer SA, Henzel WJ, Bringman TS, Nedwin GE, Goeddel DV, Harkins RN (1985). Human tumor necrosis factor. Production, purification, and characterization. J Biol Chem 260: 2345.
5. Aggarwal BB, Henzel WJ, Moffat B, Kohr WJ, Harkins RN (1985). Primary structure of human lymphotoxin derived from 1788 lymphoblastoid cell line. J Biol Chem 260: 2334.
6. Pauly JL, Ovak GM, Russell CW (1984). Isolation of interleukin 2 (IL-2) from human and mouse lymphocyte culture supernatants by batch adsorption onto silicic acid. J Immunol Methods 75:73.
7. Zucali JR, Sulkowski E (1985). Purification of human urinary erythropoietin on controlled-pore glass and silicic acid. Exp Hematol 13:833.
8. Haller W (1983). Application of controlled pore glass

in solid biochemistry. In Scouten WH (ed): "Solid phase biochemistry: analytical and synthetic aspects," John Wiley and Sons, Inc., p 535.

9. Mizutani T (1985) Adsorption chromatography of biopolymers on porous glass. J Liquid Chromat 8:925.
10. Braude IA, Edy VG, De Clercq E (1979). Mechanism of binding of mouse interferon to controlled pore glass. Biochim. Biophys Acta 580:15.
11. Sulkowski E (1982). Purification and characterization of interferons. Tex Rep Biol Med 41:234.
12. Chadha KC, Sulkowski E (1982) Adsorption of human alpha (leukocyte) interferon on glass: contributions of electrostatic and hydrophobic forces. J Interferon Res 2:229.
13. Haller W (1965) Chromatography on glass of controlled pore size. Nature 206:693.
14. Bock HG, Skene P, Fleischer S (1976). Protein purification: adsorption chromatography on controlled pore glass with the use of chaotropic buffers. Science 191:380.
15. von Hippel PH, Wong K-Y (1964) Neutral salts: the generality of their effect on the stability of macromolecular conformations. Science 145:577.
16. Buytenhuys A, Van Der Maeden FPB (1978). Gel permeation chromatography on unmodified silica using aqueous solvents. J Chromat 149:489.
17. Friedlander J, Fischer DG, Rubinstein M (1984). Isolation of two discrete human interferon-γ (immune) subtypes by high-performance liquid chromatography. Anal Biochem 137:115.

Protein Purification: Micro to Macro, pages 197–206

PURIFICATION OF FLP RECOMBINASE USING SEQUENCE-SPECIFIC DNA AFFINITY CHROMATOGRAPHY

Cynthia A. Gates, Leslie Meyer-Leon, Janet M. Attwood, Elizabeth A. Wood, and Michael M. Cox

Department of Biochemistry, University of Wisconsin-Madison, Madison, Wisconsin 53706

ABSTRACT FLP recombinase mediates site-specific recombination at two sites within the yeast 2 micron plasmid. The gene encoding FLP recombinase has been cloned and expressed in *E. coli*. The recombination site is well defined and consists of three 13 bp repeats with the first and second separated by an 8 bp spacer. The FLP recombinase cleaves at the boundaries of the spacer. The protein binds to all three repeats. Using this information, a sequence-specific DNA Sepharose resin was synthesized for affinity chromatography of the recombinase. The immobilized ligand consists of a duplex DNA polymer containing multiple 13 bp repeats ligated in a head-to-tail orientation. After ammonium sulfate fractionation, cation exhange chromatography, and non-specific DNA Sepharose chromatography, FLP recombinase was purified to 95% purity using the sequence-specific DNA Sepharose resin.

INTRODUCTION

The 2 micron plasmid is an autonomously replicating, 6318 base pair (bp) circular DNA molecule present in many strains of yeast. The plasmid encodes a site-specific recombination system that inverts unique sequences separating two 599 bp inverted repeats. The gene encoding the protein that mediates this recombination is designated FLP and is located on the 2 micron plasmid (3). This recombination event results in amplification of the plasmid copy number (1,2). The FLP system is relatively simple and provides an accessible model for studies of eukaryotic site-specific recombination.

The gene encoding FLP recombinase has been cloned and expressed in E. coli and is the only yeast protein required for recombination. Using partially purified protein, an in vitro assay was developed and used to demonstrate that the reaction promoted by FLP recombinase has very simple requirements (4,5). In vivo studies revealed that only 65 bp within the 599 bp repeat are required for recombination (6).

The prominent features of the FLP recombination site are three 13 bp repeats as illustrated in Figure 1. Two of the

Figure 1: FLP recombination site

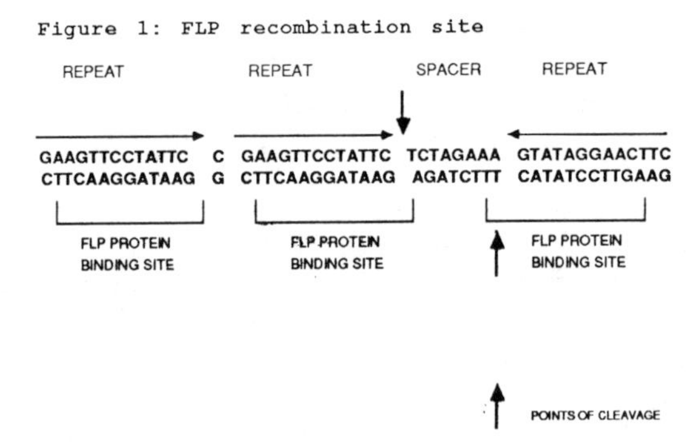

repeats are inverted relative to one another and are separated by an 8 bp repeat containing an XbaI restriction site. A third repeat flanks the second repeat, separated from its neighbor by one base pair, and is in the same orientation as the neighboring sequence (6). FLP recombinase binds to all three repeats and cleaves at the boundaries of the spacer. However, the third repeat is not required for cleavage or recombination (7,8).

The recombination site used by FLP recombinase is well defined. However, little progress has been made in kinetic and physical analysis of the protein and its interaction with its substrate. This has been hampered, in part, by the lack of purified FLP recombinase. The most successful purification scheme yielded protein at only 50% purity (9) although this has since been revised to 5% purity (10). Since FLP protein is known to bind the third repeat, but leave the DNA uncleaved, this information was used to design an affinity resin in which the immobilized ligand consists of repeating units of the 13 bp repeat ligated in head-to-tail orientation. We now report purification of FLP recombinase to 95% purity using this sequence-specific DNA affinity resin.

RESULTS

Design and Synthesis of Specific and Non-Specific DNA Sepharose Resins

An affinity ligand was constructed in which the 13 bp repeat was linked to a neighboring repeat and separated by a single base pair; this head-to-tail orientation is similar to that of the second and third repeats in the wild type recombination site. The construction of this ligand is illustrated in Figure 2, and details of the kinasing,

Figure 2: Design and construction of DNA affinity ligand

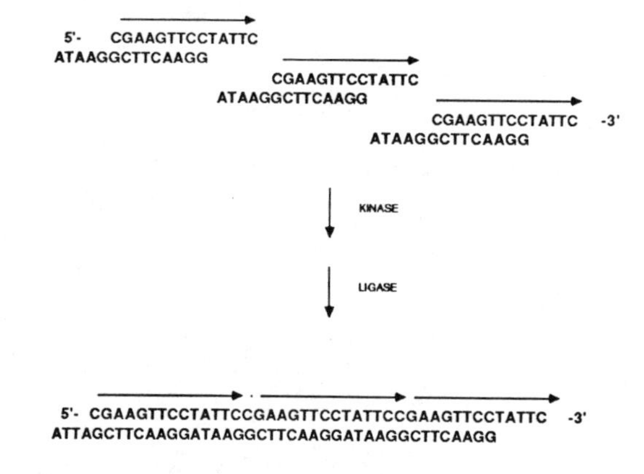

annealing, and ligation procedure are described elsewhere (11). The resulting polymer consisted of 5 to 20 repeating units of the 13 bp sequence. The duplex DNA ligand (5.7 mg) was coupled to cyanogen bromide (CNBr) activated Sepharose 4B (4.5 g dry weight, washed with 4 L 1 mM HCl and 1 L 10 mM sodium phosphate, pH 8.0) purchased from Pharmacia according to the procedure of Arndt-Jovin et al. (12). Approximately 2.3 mg of the DNA ligand was coupled to the resin. This was estimated by measuring the absorbance (at 260 nm (A_{260})) of the coupling solution and subsequent washes of the resin after 18 hours of reaction time at 25°C. This represents a 40% efficiency of coupling, with a concentration of 59 nmole of 13 bp site bound per gram of Sepharose. However, the actual amount of site accessible to the recombinase is uncertain since duplex DNA binds to Sepharose with multi-point attachments (12). Thus, some of the ligand is unavailable to the protein for binding.

A non-specific DNA Sepharose resin was synthesized by coupling sonicated calf-thymus DNA to CNBr activated Sepharose 4B. The Sepharose (70 g, suction dried) was activated using the procedure of Kohn and Wilchek (13), and the DNA was immobilized on the resin using the procedure of Arndt-Jovin et al. (12). The amount of DNA immobilized on the resin was estimated by measuring the A_{260} of the coupling solution and subsequent washes. Approximately 0.45 mg of DNA was bound per gram of Sepharose, representing a 30% binding efficiency.

Purification of FLP Recombinase

The source of FLP recombinase was *E. coli* C600K$^-$ containing the plasmid pMMC20; the construction of this plasmid and growth of the bacteria are described in detail elsewhere (11). The early stages of purification were as follows: (1) lysis of the cells by sonication and subsequent high speed (150,000 x G) centrifugation; (2) ammonium sulfate fractionation; (3) removal of the ammonium sulfate from the sample using a desalting resin; and (4) separation of FLP recombinse activity from proteins and nucleic acids using a sodium chloride gradient on a cation exchange resin (BioRex 70 from BioRad). The details of these steps and the assay conditions of FLP recombinase activity are described elsewhere (11). Protein concentrations were determined using the method of Warburg and Christian (14).

All the following steps were carried out at 0 to 5°C. Protein in fractions containing FLP recombinase activity eluted from the cation exchange resin was precipitated by addition of ammonium sulfate to 80% saturation at 5°C. Ammonium sulfate was removed on a BioGel P6-DG column (100 ml bed volume) equilibrated in 5 column volumes of 25 mM N-2-hydroxyethylpiperazine-N'-2 ethane sulfonic acid (HEPES) buffer, pH 7.0 at 5°C (14% anion), containing 1 mM EDTA and 0.1 M NaCl. This buffer henceforth will be referred to as Buffer H(0.1); the value of the number in parentheses will be varied according to the molar concentration of NaCl used at any given step. Fractions containing protein, as determined using a UV monitor with a set wavelength of 280 nm, were pooled and loaded onto the non-specific DNA

Sepharose column (70 ml bed volume) equilibrated in approximately 7 column volumes of Buffer H(0.1). The column was run using a peristaltic pump set to give a flow rate of 1 ml/min. After all of the sample was loaded the column was washed with 1 column volume of Buffer H(0.1). FLP recombinase activity was eluted with a step wash of Buffer H(0.35).

The protein sample containing activity eluted from the non-specific DNA Sepharose column was diluted with Buffer H(0) to give a sodium chloride concentration of 0.25 M as determined by conductivity measurements. This sample was then loaded onto the specific DNA Sepharose column (15 ml bed volume; equilibrated with 10 column volumes of Buffer H(0.25)). A peristaltic pump was used and set to give a flow rate of 1 ml/min. The protein sample was recycled over the resin at 1 ml/min for approximately 40 hours. Given that the volume of the sample load was 40 ml, the sample passed through the column at least 60 times during the 40 hour recycling. Next, the column was washed with 4 column volumes of Buffer H(0.4) to remove contaminating protein. FLP recombinsae activity ws then eluted with a step wash of Buffer H(1.0).

The protein concentration, units of activity, yield and purification factors for the above procedures are presented in TABLE 1. The protein components of the lysate, cation exchange chromatography pool, the non-specific DNA Sepharose 0.35 M NaCl step wash, and the specific DNA Sepharose 1.0 M NaCl step wash were separated by SDS-polyacrylamide gel electrophoresis (SDS-PAGE); the results are shown in Figure 3, and details are given in the figure legend.

DISCUSSION

The procedure described above yields FLP recombinase at approximately 95% purity as judged by a densitometric scan of the SDS-PAGE gel. Recycling the protein sample over the specific DNA Sepharose column increased both purity and yield, possibly because the prolonged exposure of FLP recombinase to the resin allows the recombinase to displace non-specific DNA binding proteins from the sequence-specific DNA ligand. The entire procedure requires four to five days to complete. However, the recombinase activity appears to be stable after elution from the non-specific DNA Sepharose column. The recombinase eluted from the specific DNA

Sepharose column is stable with no loss of activity observed for 5 days at 5°C or several months at -70°C. FLP recombinase is reported here to be purified 127-fold. Similar purifications have resulted in 600 to 1000-fold purification, depending on the efficiency of cell lysis and the initial amount of ativity.

Major losses of activity are observed at two stages of the purification. Removal of ammonium sulfate from the

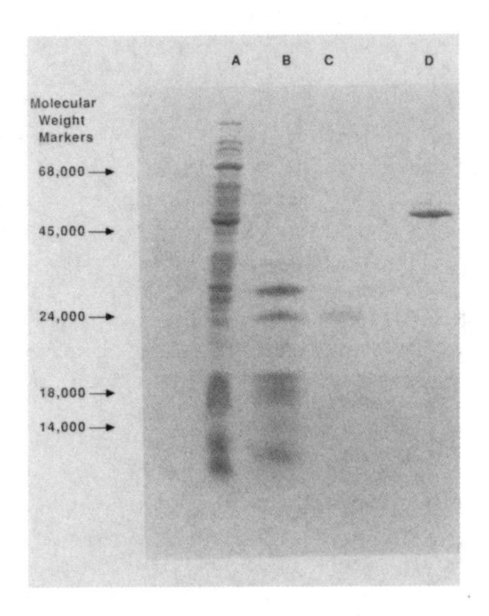

Figure 3. SDS polyacrylamide gel electrophoresis of FLP recombinase fractions

Protein in lanes A-D was denatured by heating to 100°C for 15 minutes in the presence of an SDS/reducing buffer before loading onto an SDS-polyacrylamide gel containing 11% crosslinking components in the separating gel and 6% in the stacking gel (21). The gel was stained, fixed, and destained according to published procedures (22). The molecular weight marker proteins (not shown) were bovine serum albumin (66,000), ovalbumin (45,000), trypsinogen (24,000), beta-lactoglobulin (18,000), and lysozyme (14,000). Lane A: 50 micrograms of the crude lysate. Lane B: 30 micrograms of the protein eluted from the BioRex-70 resin. Lane C: 15 micrograms of protein from the Buffer H(0.35) wash of the non-specific DNA Sepharose column. Lane D: 21 micrograms of protein eluted from the sequence-specific DNA Sepharose column with the Buffer H(1.0) wash; the major band with a MW of approximately 45,000 is FLP recombinase.

resuspended protein after the first ammonium sulfate fractionation (11) results in a large loss of activity. Similarly, activity is lost in the desalting step prior to the non-specific DNA Sepharose column. Use of a heparin-agarose column instead of the non-specific DNA Sepharose

Table I

Fraction	Total Protein (mg)	A(280)/A(260)	Total Units* (x1000)	Specific Activity (Units/mg x 1000)	Purification (x-fold)	Percent Recovery
Crude Extract	1670	0.5	8350	5	1	100
90% Ammonium Sulfate	960	0.5	4800	5	1	57
Biorex-70 Pool	73	1.3	1400	19	4	17
Non-Specific DNA-Sepharose Pool	19	1.3	400	21	4	5
Specific DNA-Sepharose Pool	0.63	1.3	400	63	127	5

*A unit of FLP recombiase activity is defined as the minimum amount of protein required to produce products in an agarose gel after a 60 minute reaction (see ref. 11)

resin results in improved yield of recombinase activity, but lower (85%) purity (11). Loading the protein sample eluted from the BioRex column directly onto the specific DNA Sepharose resin equilibrated in Buffer H(0.35) also results in a higher yield in activity, but with only 75 to 80% purity (15). Apparently, a step using non-specific DNA as a ligand prior to the specific DNA Sepharose column is a requirement for higher purity. Efforts are underway to improve the overall yield of pure recombinase using this information (15).

DNA binding proteins have been purified using two different strategies. The first was a procedure suggested by Alberts (17) in which the protein of interest is eluted from a non-specific cation exchange resin or a non-specific DNA affinity resin with short DNA fragments containing the binding site of the protein. This approach was used in an attempt to purify FLP recombinase. Although FLP recombinase was purified to 99% homogeneity by elution from heparin-agarose with a DNA fragment containing half the spacer and one adjacent repeat, half of the protein was covalently bound to the eluting DNA fragment and thus inactive (16). This, coupled with the expense of the procedure, led to the search for an alternative approach.

The second strategy, affinity chromatography in which the ligand that specifically binds to an enzyme is immobilized to a support of some type, has long been used by protein chemists. Purification of some DNA binding proteins has been accomplished by using resins of non-specific DNA linked to cellulose matrices. However, the procedures used to immobilize the DNA to the cellulose matrix do not result in stable, covalent bonds (12). Furthermore, the resins do not specifically separate all DNA binding proteins. A number of DNA binding proteins, FLP recombinase among them, recognize and bind to specific sequences of DNA, much like a dehydrogenase binds to a pyridine nucleotide substrate, whether free in solution or immobilized on a resin. By covalently linking the DNA binding site to a Sepharose resin, an effective affinity resin may be synthesized which separates the protein of interest from other DNA binding proteins. This approach is cost effective since the affinity resin can be used for extended periods of time with little or no leakage of the ligand from the resin. The affinity resin designed and synthesized for the purification of FLP recombinase was successful in producing protein of high purity. While the work described herein was in progess, similar procedures using sequence-specific DNA affinity chromatography were published for the purifications of Nuclear Factor I (18) and for transcription factor SpI (19,20). The use of affinity resin with covalently linked DNA containing specific recognition sequences has the potential for purification of a variety of DNA binding proteins.

ACKNOWLEDGEMENTS

This work was supported by grant #GM37835 from the National Institutes of Health and a March of Dimes Basil O'Connor Starter Research Grant 5-451. M.M.C. is supported by NIH Research Career Development Award AI00599. L.M-L. was supported by NIH Predoctoral Training Grant GM07215.

REFERENCES

1. Futcher AB (1986). Copy number amplification of the 2μm cirle plasmid of Saccharomyces cerevesiae. J Theor Biol 119:197.
2. Volkert FC, Broach JR (1986). Site-specific recombination promotes plasmid amplification in yeast. Cell 46:541.
3. Broach JR (1982). In Strathern JN, Jones EN, Broach JR (eds): "The Molecular Biology of the Yeast Saccharomyces I. Life Cycle and Inheritance," Cold Spring Harbor, NY, Cold Spring Harbor Laboratory, p. 445.
4. Meyer-Leon L, Senecoff JF, Bruckner RC, Cox MM (1984). Site-specific genetic recombination promoted by the FLP protein of the yeast 2-micron plasmid in vitro. Cold Spring Harbor Symp Quant Biol 49:797.
5. Sadowski PD, Lee DD, Andrews BJ, Babineau D, Beatty L, Morse MJ, Proteau G, Vetter D (1984). In vitro systems for genetic recombination of the DNAs of bacteriophage T7 and yeast 2-micron circle. Cold Spring Harbor Symp Quant Biol 49:789.
6. Broach JR, Guarascio VR, Jayaram M (1982). Recombination within the yeast plasmid 2μ circle is site-specific. Cell 29:227.
7. Andrews BJ, Proteau GA, Beatty LG, Sadowski P (1985). The FLP recombinase of the 2μm circle DNA of yeast: interaction with its target sequences. Cell 40:795.
8. Senecoff JF, Bruckner RC, Cox MM (1985). The FLP recombinase of the yeast 2-μm plasmid: characterization of its recombination site. Proc Natl Acad Sci USA 82:7270.
9. Babineau D, Vetter D, Andrews BJ, Gronostajski RM, Proteau GA, Beatty LG, Sadowski PD (1985). The FLP protein of the 2-micron plasmid of yeast: purification of the protein from Escherichia coli cells expressing the cloned FLP gene. J Biol Chem 260:12313.
10. Andrews BJ, Beatty LG, Sadowski PD (1987). Isolation of intermediates in the binding of the yeast plasmid 2-micron circle to its target sequence. J Mol Biol 193:345.
11. Meyer-Leon L, Gates CA, Attwood JM, Wood EA, Cox MM (1987). Purification of the FLP site-specific recombinase by affinity chromatography and re-examination of the basic properties of the system. submitted.

12. Arndt-Jovin D, Jovin T, Bahr W, Frischauf A-M, Marquadt M (1975). Covalent attachment of DNA to agarose. Eur J Biochem 54:411.
13. Kohn J, Wilchek M (1982). A new approach (cyano-transfer for cyanogen bromide activation of sepharose at neutral pH which yields activated resins free of interfering nitrogen derivatives. Biochem Biophys Res Comm 107:878.
14. Warburg O, Christian W (1942). Isolierung und kristallisation des garungsferments enolase. Biochem Z 310:384.
15. Gates CA, Cox MM. unpublished results.
16. Meyer-Leon L, Attwood JM, Cox MM. unpublished results.
17. Alberts BM (1984). The DNA enzymology of protein machines. Cold Spring Harbor Symp Quant Biol 49:1.
18. Rosenfeld PJ, Kelly TJ (1986). Purification of nuclear factor I by DNA recognition site affinity chromatography. J Biol Chem 261:1398.
19. Kadonaga JT, Tjian R (1986). Affinity purification of sequence-specific DNA binding proteins. Proc Natl Acad Sci USA 83:5889.
20. Briggs MR, Kadonaga JT, Bell SP, Tjian R (1986). Purification and biochemical characterization of the promoter-specific transcription factor, SpI. Science 234:47.
21. Laemmli UK (1970) Nature 227:680.
22. Weber K, Osbern M (1986) J Biol Chem 244:4406.

Protein Purification: Micro to Macro, pages 207–215

RAPID ISOLATION OF ACETYLCHOLINESTERASE FROM SNAKE VENOM

Jeffrey R. Deschamps
Naval Research Laboratory, Laboratory for the Structure of Matter, Code 6030, Washington, D.C., 20375-5000

Sam Morris
Beckman Instruments, Route 108, Columbia, MD.

ABSTRACT Acetylcholinesterase (EC 3.1.1.7) was isolated by batch affinity chromatography from the venom of *Naja naja kaouthia* and *Naja naja atra* using procainamide-HCl linked to a high-performance affinity matrix. The binding capacity of the derivatized affinity support was 1,000 units cholinesterase/ gram support. Affinity purified acetylcholinesterase (AChE) was analyzed for purity and specific activity.

A single step affinity purification of cobra venom AChE yields about a 340 fold increase in purity. The purified enzyme has a specific activity of over 2600 U/mg, and about 50% of the activity applied to the column is recovered in the product. Size-exclusion HPLC reveals a nearly homogeneous product with a molecular weight of 67 kDa and higher molecular weight contaminants.

INTRODUCTION

Acetylcholinesterase is primarily associated with neuronal cells, but has been isolated from a number of sources including neuronal tissue, erythrocytes, and various organs (1,2). AChE hydrolyzes the esterlinkage in acetylcholine releasing acetate and choline at the postsynaptic membrane thus restoring the polarized state, and excitability, of the postsynaptic membrane (3). The proper function of AChE is necessary for normal neuronal transmission.

On the basis of enzymatic activity, substrate specificity, and the effect of various inhibitors, the enzymes isolated from various sources have been classified either as 'true cholinesterases', or 'pseudocholinesterases'.

'True cholinesterases' hydrolyze acetylcholine more rapidly than other choline derivatives. 'Pseudocholinesterases', sometimes called plasma cholinesterase or butrylcholinesterase, are less specific and hydrolyze succinylcholine more rapidly than the true cholinesterases do. One property common to all of the true cholinesterases is substrate inhibition.

Soluble forms of AChE have been found in several elapid venoms and their specificity studied (4-9). A comparison of eel electroplax AChE, a membrane-bound 'true cholinesterase' with the soluble cholinesterase found in elapid venom revealed many similariteis, including substrate inhibition, between the enzymes (4). The soluble cholinesterase found in elapid venom is therefore a convenient model of the membrane bound cholinesterases.

Acetylcholinesterase is present in concentrations as high as 1.7 mg/g dry venom (*Naja naja oxiana*). Many isoenzymes of AChE have been observed in AChE isolated from elapid venom, as many as 15 different isoenzymes of AChE have been reported in one venom (8). The large number of isoenzynmes may be the result of proteolytic activity as proteases are also in the venom; it is therefore essential that the purification be carried out quickly so as to minimize contact with the proteases.

Previously AChE has been isolated from cobra venom by a combination of chromatographic techniques. The final purification step is generally affinity chromatography. Recently, a high performance affinity procedure for the isolation of acetylcholinesterase has been described (B.P. Doctor, unpublished; Sam Morris, personal communication). This new affinity procedure was modified for batch chromatography and tested for use in isolating elapid venom AChE.

EXPERIMENTAL METHODS

Preparation of affinity support

A Beckman Ultraffinity-EP column capacity kit was activated with Procainamide-HCl, U.S.P. grade (Unit Dose Laboratories, Inc.) as follows: 2 ml of a solution containing 5 mg/ml procainamide-HCl in 1 M potassium phosphate, pH 6.8 was injected into the vial containing the affinty support (0.1g). The vial was agitated until all of the affinity support was suspended. The suspension was set aside at room temperature for 24 hours. The derivatized affinity support was washed 3 times with 1 ml distilled water followed by 1 ml 95% ethanol. The derivatized affinity support is stored in 95 % ethanol.

Binding tests

Initial tests of binding were performed by pipeting a known quantity of AChE activity, from *Naja naja kaouthia* venom fraction, into a tube containing the affinity support, suspending the affinity support in the AChE containing solution, centrifuging, and measuring the amount of activity left in solution. The support was prepared for the next experiment by washing with 1 ml of 100 mM decamethonium bromide (Aldrich Chemical Co.), to remove bound enzyme, followed by washing with 3 1 ml aliquots of 0.1 M phosphate buffer (pH 7.0) to re-equilibrate the affinity support.

Enzyme Purification

Naja naja kaouthia AChE was purified by placing 0.5 ml of cobra venom fraction into the tube containing the affinity support and suspending the affinty support. This mixture was allowed to sit for one minute, centifuged, and the affinty support washed with 1 ml of 0.1 M phosphate buffer (pH 7.0). The product was then eluted from the support with 1 ml of 100 mM decamethonium bromide, and after centrifugation the affinity purified AChE decanted. This procedure was repeated five times and the affinity-purified AChE pooled. The purified enzyme was then concentrated in a centricon (Amicon Inc.), and washed with 10 ml of 50 mM phosphate buffer (pH 7.0), in 2 ml aliquots, to remove the decamethonium bromide. *Naja naja atra* AChe was purified from 1 ml of a solution containing 3.1 mg dry venom (Sigma Chemical) as described above. The washing of non-binding or weakly associated protein was extended to four 1 ml aliquots of phosphate buffer.

Characterization of Affinty Purified AChE

Cholinesterase activity was determined using a colorimetric method (10) with a Shimadzu UV-260 spectrophotometer. The purified AChE was treated with 25 and 250 micromolar decamethonium bromide and the inhibition of enzymatic activity determined. Total protein was determined by the Bradford dye binding assay (11), and specific activity calculated from these data.

The affinity purified AChE was further charaterized by high-performance size-exclusion chromatography on a TSK

2000-SW column (Toya Soda Inc.) and on a Superose-12 column (Pharmacia Inc.), and by reversed-phase HPLC on an Ultrashere (Beckman Inst. Inc.). The HPLC system consisted of a Beckman model 336 binary gradient HPLC equiped with a Kratos Spectraflow-783 variable wavelength detector.

RESULTS AND DISCUSSION

Binding Study

From the binding assays the capacity of the affinity support in the test kit was estimated to be about 1000 units. Thus about 10,000 units could be bound on a small Ultraffinity-EP column. As the amount of enzyme applied to the affinity support approaches the maximum binding capacity, the amount of enzyme retained approaches 50 to 60 per cent of the activity applied.

TABLE 1.

Summarized results on the binding of cobra venom acetylcholinesterase to the procainamide derivatized 'Ultra-affinity' support.

UNITS Applied	UNITS Bound	per cent Bound
1.77	1.71	96
17.7	15.9	90
17.7	16.2	92
17.7	16.0	90
24.4	14.3	59
186	99.0	53
373	101	27

Affinity Chromatography

The batch purification of 3 ml of a cobra venom fraction, containing about 1110 units AChE, resulted in a 53 per cent yield of affinity purfied AChE (see Table 2). The product had a specific activity of 193 units/mg which represents a 12 fold purification.

TABLE 2.
BATCH AFFINTY PURIFICATION OF ACETYLCHOLINESTERASE FROM *Naja naja kaouthia* VENOM FRACTION

SAMPLE	FRACTION VOLUME(ul)	ACTIVITY U/ul(a)	TOTAL Units	PROTEIN (ug/ul)	SPECIFIC ACTIVITY
Crude AChEase	3000(b)	0.37	1110(c)	23.9	15.4
affinity AChEase	1000	0.60	600	3.1	193

(a) activity = slope/3.67. see: Rosenberry and Sloggin, 1984.

(b) total amount of sample applied to affinity support in 6 separate batch extractions (i.e. 6 x 0.5 ml).

(c) total activity = (units x (1/dilution factor) x fraction volume)/ sample size (i.e. 10 ul).

Batch purification of AChE from 1 ml (3.1 mg) of *Naja naja atra* venom produced an enzyme with a specific activity of 2667 U/mg (Table 3) which represents a 338 fold purification in one step. The yield of 47 per cent (based on activity recovered) is comparable to the yield of AChE

TABLE 3.
BATCH AFFINTY PURIFICATION OF ACETYLCHOLINESTERASE FROM *Naja naja atra* VENOM

SAMPLE	FRACTION VOLUME(ul)	ACTIVITY U/ul(a)	TOTAL Units	PROTEIN (ug/ul)	SPECIFIC ACTIVITY
venom	1000	0.024	24.4	3.100	7.9
affinity AChEase	250	0.050	11.6	0.017	2667

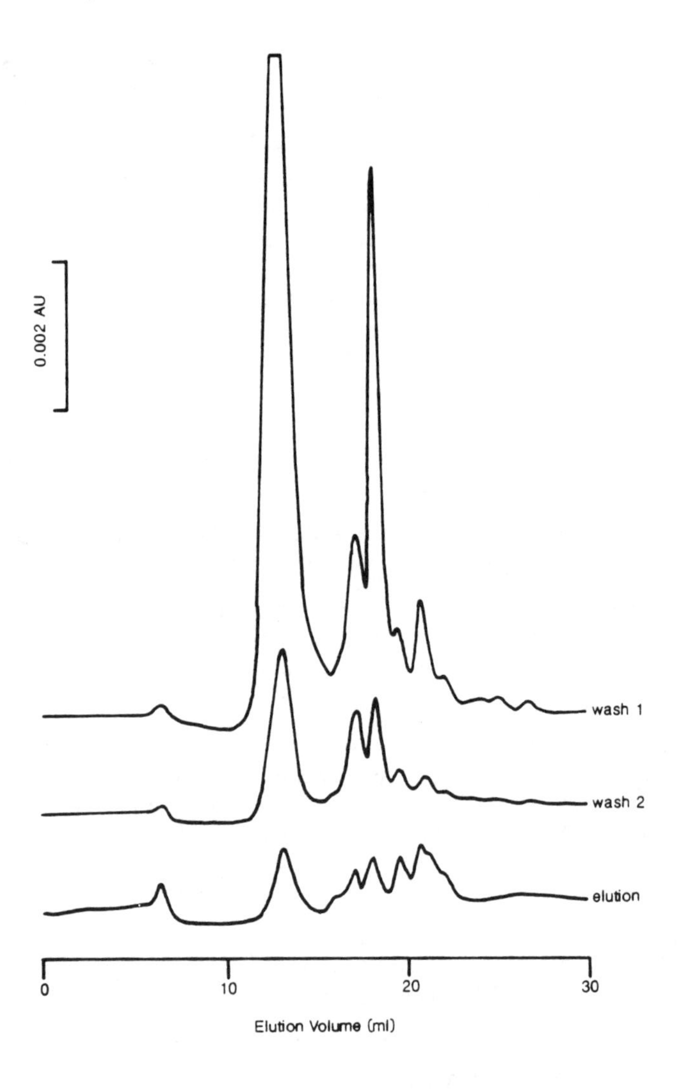

FIGURE 1. Size-exclusion chomatography of successive washes of the affinity support after application of approximately 350 units of <u>Naja naja kaouthia</u> acetylcholinesterase, and the eluted product after two washes.

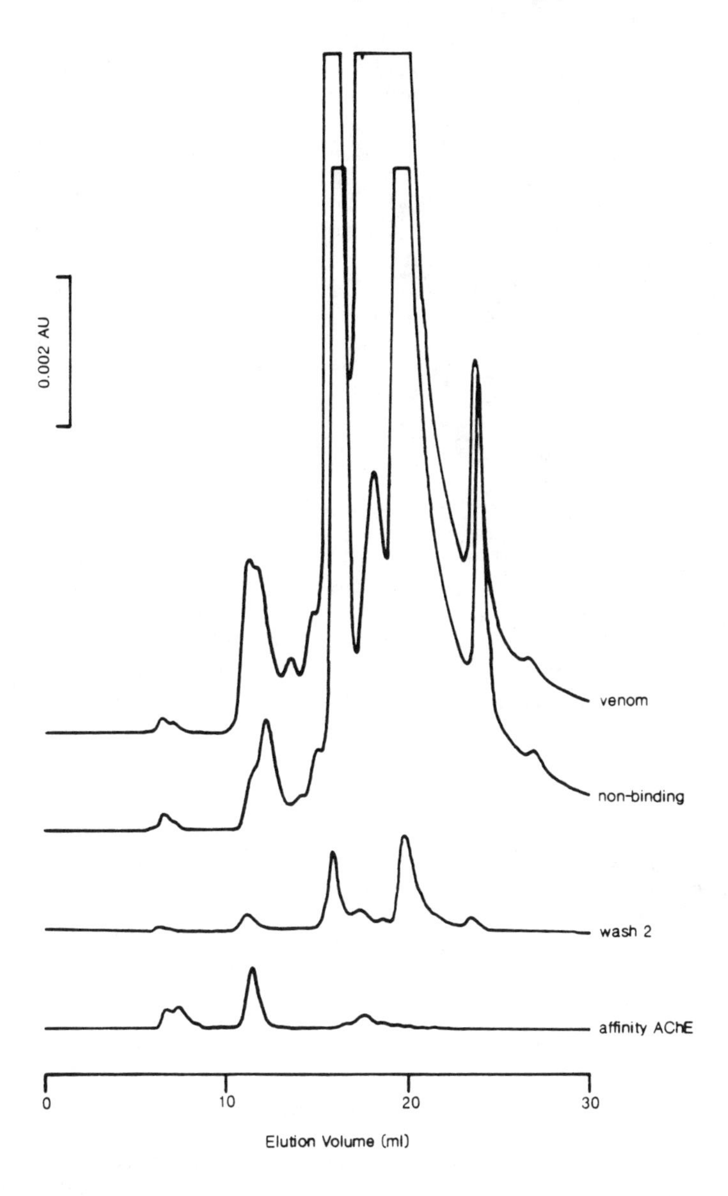

FIGURE 2. Size-exclusion chromatography tracing the purification of acetylcholinesterase from *Naja naja atra* venom from the starting material, through the non-binding fraction, the second wash, to the affinity purified enzyme.

from Naja naja kaouthia, and to the yield observed in prior studies (Sam Morris, unpublished). Based on the yield of cholinesterase activity and the specific activity of the isolated AChE it is estimated that Naja naja atra venom contains approximately 0.8 mg AChE per gram dried venom.

Characterization of the affinity purified Naja naja kaouthia AChE revealed that the sample was inhomogeneous by both size-exclusion and reversed phase HPLC. Further investigation of this phenomenon showned that a single wash of the affinity matrix prior to elution of the protein was not sufficient to remove all of the non-bound, or weakly bound, protein (Figure 1). As a result of this finding additional wash steps were added to the purification of Naja naja atra AChE.

The AChE from Naja naja atra shows some high molecular weight contaminants, i.e. 120-140 kDa (Figure 2), which appear to be enriched in the affinity fraction. These high molecular weigth components may be aggregation products of the AChE as AChE from other sources is known to aggregate in the absence of detergent (12). It should also be noted that the molecular weight of elapid venom AChE, i.e. 67 kDa, is the approximated molecular weight of the monomer from human erythrocyte AChE after removal of the hydrophobic domain (12). The high recovery of activity and speed of this method make it an excellent choice for the isolation and purfication of AChE.

ACKNOWLEDGEMENTS

This work was supported in part by a Fellowship from the Office of Naval Technology (ASEE/ONT program). The authors would like to thank Dr. Carl-Wilhelm Vogel and Gene Petrella of Georgetown University, School of Medicine, for the Naja naja kaouthia fractions used in this study, and Meosotis C. Curtis for her assistance in editing the manuscript.

REFERENCES

1. Rosenberry TL (1982). Acetylcholinesterase. The relationship of protein structure to cellular localization. In Martonosi AN (ed), "Membranes and Transport," Vol 2, New York: Plenum Publishing Corp, p 339.

2. Rosenberry TL (1975). Acetylcholinesterase. In Meister AJ (ed), "Advances in Enzymology," Vol 43, New York: Acedemic Press, p 103.
3. Stryer L (1975). "Biochemistry" (second edition). San Francisco: W.H. Freeman and Co., p 883.
4. Agbaji A, Gerassimidis K, and Hider R (1984). A comparison of eel electroplax and snake venom acetylcholinesterase. Comp Biochem Physiol 78C:211.
5. Tu AT (1977). "Venoms: Chemistry and molecular biology." New York: Wiley Interscience, 560 pp.
6. Raba R, Aaviksaar A, Raba M, and Siigur J (1979). Cobra venom acetylcholinesterase. Purification and molecular properties. Eur J Biochem 96:151.
7. Grossmann H and Lieflander M (1979). Affinitatschromatographische reinigung und eigenschaften der acetylcholinesterase aus kobragift (*Naga naga atra*). J Chrom 177:99.
8. Lee S, Latta JL, and Elliott WB (1977). Isoenzymes of elapid acetylcholinesterases. Comp Biochem Physiol 56C:193.
9. Kumar V and Elliott WB (1975). Acetylcholinesterase from *Bungarus fasciatus* venom -- I. Substrate specificity. Comp Biochem Physiol 51C:249.
10. Ellman GL, Courtney KD, Andres V, and Featherstone RM (1961). A new and rapid determination of acetylcholinesterase activity. Biochem Pharm 7:88.
11. Bradford MM (1972). A rapid and sensitive method for the quantitation of microgram quantities of protein utilizing the principle of protein-dye binding. Anal Biochem 72:248.
12. Dutta-Choudhury TA and Rosenberry TL (1984). Human erythrocyte acetylcholinesterase is an amphipathic protein whose short membrane-binding domain is removed by papain. J Biol Chem 259(9):5653.

Protein Purification: Micro to Macro, pages 217–223

HIGH YIELD NON-DENATURING PROCEDURE FOR THE PURIFICATION OF CARCINOEMBRYONIC ANTIGEN

Cristina Mottola, Raffaella Conti, Patrizia Lorenzoni, and Costante Ceccarini

Sclavo Research Center, Siena, Italy.

ABSTRACT A non-denaturing isolation procedure was developed for the carcinoembryonic antigen (CEA). This large glycoprotein was extracted from metastatic liver or cells in culture using sodium deoxycholate and subsequently purified by two steps of affinity chromatography, followed by separation in a HPLC system. Detergent treatment of metastases allowed a 5-fold higher recovery of CEA when compared to the generally adopted perchloric acid extraction. Our method prevented the appearance of CEA degradation products that were observed upon acid treatment.

INTRODUCTION

Several purification procedures have been described for the oncofetal antigen CEA (1-3). The microheterogeneity of CEA, a membrane-associated glycoprotein, and its immunological cross-reactivity with a number of probably related proteins (4) have led to sometimes contradictory observations. Here we present results obtained with a method that allows an improved recovery of CEA in its native molecular form.

METHODS

SDS-PAGE under reducing conditions was performed according to Laemmli (5) on 5-16% gradient slab gels. Immunoblots were obtained using peroxidase-conjugated secondary antibodies and 0.06% 4-chloro-1-naphtol as a revealing system. Specific polyclonal antisera were raised against purified CEA (Dr. J. Mendicino, UGA) in rabbit or goat and immunoglobulin (Ig) preparations were partially purified with the method of Garvey et al. (6).

Extraction protocol. Homogenized liver metastasis or packed cells were solubilized in 5-10 volumes of 50 mM Hepes, pH 7.4, 2% sodium deoxycholate (NaDOC) and anti-proteases for 4 h at 4°C under mixing. After 1:1 dilution with buffer without detergent, the suspension was clarified by centrifugation and dialyzed extensively.

Affinity chromatography. All the steps were routinely performed batchwise. The resin to solution ratio was 1:2 during the attachment and 1:1 during each elution cycle. The resin was abundantly washed between attachment and elution; complete desorption was carried out in six consecutive cycles. Chromatography on ConA-Sepharose was performed in 50 mM Na Acetate, pH 6.5, 0.1M NaCl, 1 mM $MgCl_2$, 1 mM $MnCl_2$, 1 mM $CaCl_2$; see "Results" for the elution procedure. Anti-CEA Ig linked to sepharose (4 mg/ml) was used for immunoaffinity chromatography in phosphate buffered saline (PBS); 0.5M NaCl was added during the washing step and the protein finally desorbed with 3M NH_4SCN, pH 7.4.

Final purification. Residual lipids and detergent were extracted in ice-cold acetone first, and then in a diethylether:ethanol (3:2) mixture. Molecular sieving separation was achieved by HPLC on an Ultropac TSK G3000 SW column, 7.5x600 mm. During the whole procedure, CEA recovery was measured immunologically (Abbott).

RESULTS AND DISCUSSION

Choice of the extracting agent. CEA was isolated from colon carcinoma-derived human liver metastases or human gastric carcinoma MKN-45 cells in culture. Extraction of CEA using the mild, dialyzable detergent NaDOC instead of perchloric acid resulted in two main advantages: higher recovery and prevention of degradation. In a representative test (metastases content of CEA is extremely variable), two pieces of equal weight from the same liver were extracted with each of the two methods. Perchloric acid-derived CEA amounted to 12% of the yield obtained in the presence of detergent. Moreover, while the harshness of acid treatment makes it unsuitable for cells in culture, our procedure enabled us to obtain CEA from cells as well as from tissue. We identified MKN-45 cells as active producers of the glycoprotein under study. NaDOC solubilization of this material yielded about 35 mg CEA/100 ml of packed cells, 90% of it being in the soluble

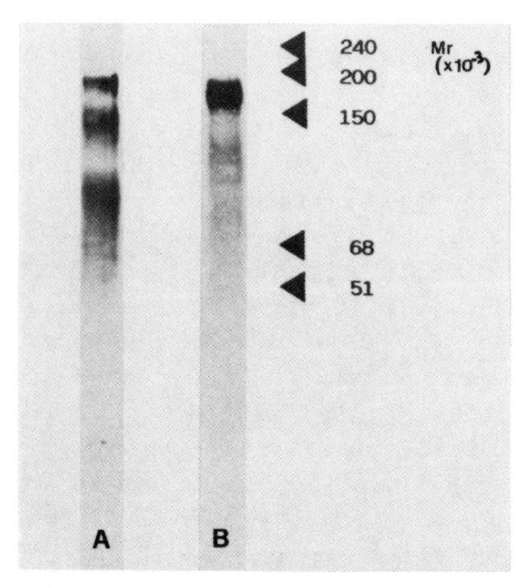

FIGURE 1. Comparative effect of extracting agents on CEA as seen on immunoblotting; 200 ng of CEA were loaded in each well.

extract. MKN-45 cells could be cultivated in monolayers, in suspension or on microcarriers and also be injected in nude mice to originate CEA-rich solid tumors.

Figure 1 shows the qualitative difference between CEA isolated with perchloric acid (lane A, Dr. J. Mendicino, UGA) or NaDOC (lane B). Exposure to acid produces degradation bands which are not very evident by staining of polyacrylamide gels but are clearly visible on immunoblotting. This effect is increased upon storage of the protein for long periods, while CEA prepared with detergent maintains its original molecular weight.

Chromatography on ConA-Sepharose. The glycoprotein fraction of the DOC extract was enriched on ConA-Sepharose; CEA binds with high affinity to the lectin and methyl-α-D-mannoside (α-mm) does not compete efficiently for this binding. We performed a first experiment on a column (15 ml bed volume), loading material extracted from MKN-45 cells. The resin was carefully washed and elution started with 0.1M α-mm, at a 0.5 ml/min flow rate. Fractions of 2.65 ml were collected and tested for their protein and CEA content; the elution profile obtained is reported in figure 2, panel A. We observed that CEA was released slowly and in a minor proportion. Subsequent experiments were carried out batchwise, always performing desorption of the proteins from the resin in six consecutive cycles. A relevant improvement of the yield was achieved by adding a chaotropic ion (SCN^-) in the elution buffer. This advantage is clearly shown by the data of figure 2, panel B. Three desorption conditions were tested in parallel, with three elution cycles each. A constant volume from each eluate was loaded on a gel and the recovered CEA was revealed by immunoblotting. For comparison, 193 µg of CEA were eluted using 0.1M α-mm (lanes a, b and c) while 3M NH_4SCN (lanes d, e and f) allowed a recovery of 331 µg of CEA. The best result was achieved adding the two chemicals simultaneously (lanes g, h and i) for a total recovery of 900 µg of CEA. This procedure was adopted henceforth; the average recovery of CEA at this step was 61% of the starting amount.

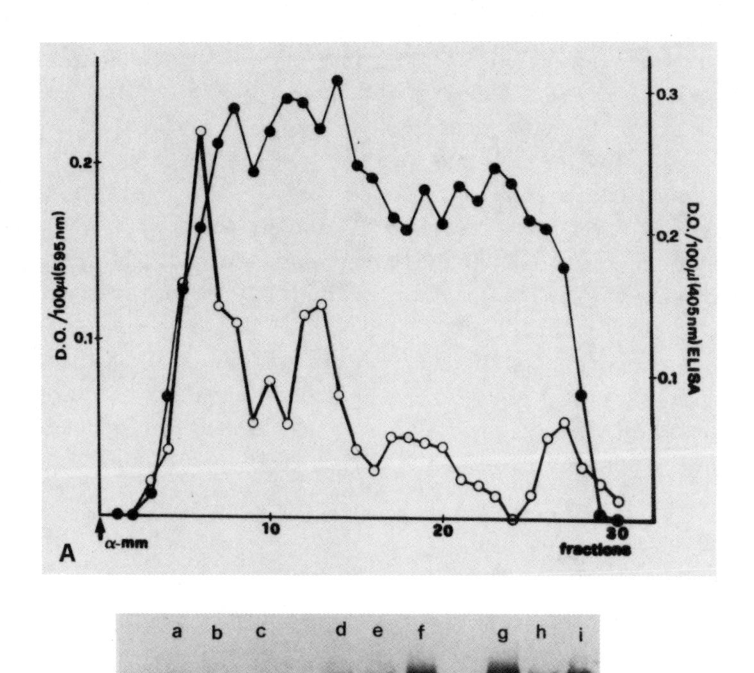

FIGURE 2. A. Elution pattern of CEA from ConA-Sepharose; ○ , protein content (BioRad); ● , CEA immunoreactivity (Abbott). B. Immunoblot showing the results of different elution conditions.

Immunoaffinity chromatography. CEA was further purified several fold by immunoadsorption on specific Ig bound to sepharose (refer to "Methods" for details). At this point, about 15% of the initial amount of tumor antigen was still present in our preparations; this value takes into account several dialysis and concentration steps and a certain loss of CEA immunoreactivity in time. The effective recovery from Ig-sepharose averaged at 45%.

Separation on HPLC. A few low molecular weight contaminants remained associated with CEA after immunoaffinity chromatography; these small proteins could not be separated from CEA on several gel filtration systems.

An improvement was achieved upon removal of residual lipids and detergent in the presence of ice-cold acetone followed, after sedimentation, by an ether-ethanol extraction. This procedure allowed the almost total recovery of CEA in the pellet and clarified the sample that could be finally loaded on a HPLC gel filtration column. Panel A of figure 3 shows the profile of this chromatography measured as absorbance at 280 nm. CEA immunoreactivity is overprinted: it is associated to a single peak eluting at 9.64 min. The fractions forming this peak were pooled and an aliquot was run again through the same system, showing a clean peak (inset). Analysis on SDS-PAGE of CEA at the end of the isolation procedure is shown in panel B and indicates the high purity of our preparation.

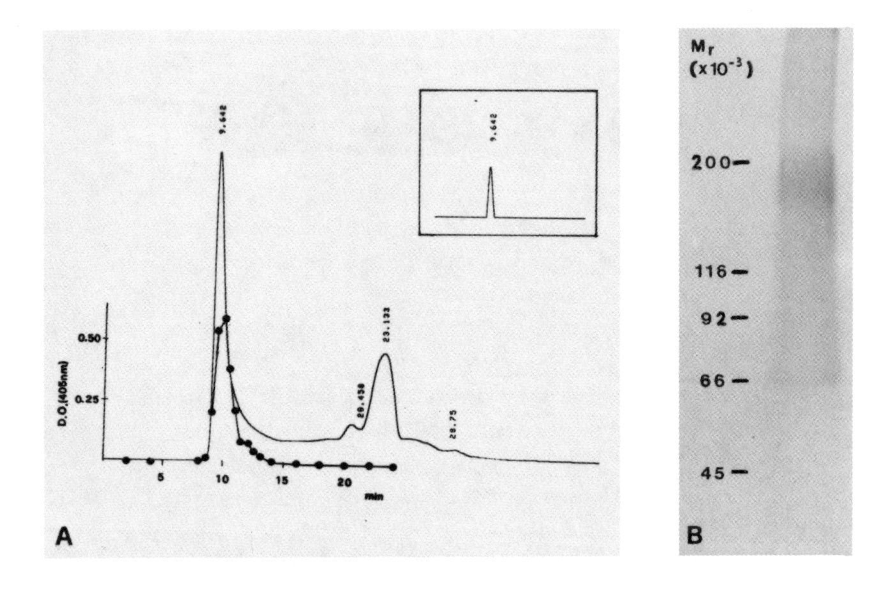

FIGURE 3. A. Gel filtration chromatography profile of CEA in a HPLC system. ●, CEA immunoreactivity. Inset: profile of the second separation. B. SDS-PAGE of purified CEA after silver staining (BioRad).

ACKNOWLEDGEMENTS

We thank Dr. Elvira Costantino-Ceccarini for helpful discussions. Nadia Bonelli and Salvatore Baldassano are acknowledged for their skilful technical help, and Giorgio Corsi for photographic work.

REFERENCES

1. Krupey J, Wilson T, Freedman SO, Gold P (1972). The preparation of purified carcinoembryonic antigen of the human digestive system from large quantities of tumor tissue. Immunochemistry 9: 617.
2. Carrico RJ, Usategui-Gomez M (1975). The isolation of carcinoembryonic antigen from tumor tissue at neutral pH. Cancer Res 35: 2928.
3. Chandrasekaran EV, Davila M, Nixon DW, Goldfarb M, Mendicino J (1983). Isolation and structures of the oligosaccharide units of carcinoembryonic antigen. J Biol Chem 258: 7213.
4. Neumaier M, Fenger U, Wagener C (1985). Monoclonal antibodies for carcinoembryonic antigen (CEA) as a model system: identification of two novel CEA-related antigens in meconium and colorectal carcinoma tissue by Western blots and differential immunoaffinity chromatography. J Immunol 135: 3604.
5. Laemmli UK (1970). Cleavage of structural protein during the assembly of the head of bacteriophage T4. Nature 227: 680.
6. Garvey JS, Cremer NE, Sussdorf DH (1977). Ammonium sulfate precipitation. In Reading MA (ed): "Methods in Immunology", third edition, WA Benjamin Inc., p 218.

Protein Purification: Micro to Macro, pages 225–238

AFFINITY IMMUNOELECTROPHORESIS AND CHROMATOGRAPHY FOR ISOLATION AND CHARACTERIZATION OF A PLACENTAL GRANULOCYTE ELASTASE INHIBITOR

Michael J. Sinosich, Michael D. Bonifacio[1]
and Gary D. Hodgen

Eastern Virginia Medical School, Norfolk, VA 23507
[1]Royal North Shore Hospital, St. Leonards
NSW 2065, Australia

ABSTRACT Because of the many physicochemical similarities between pregnancy-associated plasma protein-A (PAPP-A) and α2-macroglobubin (α2M), it has proved difficult to purify PAPP-A. Neither protein could be distinguished by lectin affinity immunoelectrophoresis or by chromatography on chelated divalent metal ions. However, after incubation with β-glucuronidase the immunoelectrophoretic properties of PAPP-A (but not α2M) were altered and thus suggested the presence of glucuronic acid (or chondroin sulphate) moieties. In the presence of heparin, an anionic glycosylaminoglycan, the electrophoretic mobility of PAPP-A was increased. The heparin-PAPP-A interaction was confirmed by chromatography on heparin-Sepharose matrix. Then, in conjunction with size fractionation and balanced negative immunoaffinity chromatography, positive affinity chromatography on heparin-Sepharose formed the initial specific concentration step for PAPP-A purification.

INTRODUCTION

Pregnancy-associated plasma protein-A (PAPPA),a large glycoprotein, was reported 14 years ago. Attempts to isolate PAPP-A from human pregnancy serum were based on classical physicochemical approaches and proved technically cumbersome (1). Subsequent purification schedules included affinity chromatography on concanavalin A-Sepharose (2) and positive immunoaffinity matrices. Although the purification factor increased, the resultant preparation was contaminated (3). Furthermore, chaotropic solvents required for dissociation of antigen-antibody complexes irreversibly denatured PAPP-A (4). None of these reports successfully isolated PAPP-A to purity and because of physicochemical similarities to α2-macroglobulin (α2M) and pregnancy zone protein (PZP) these latter glycoproteins were frequently identified as contaminants (5). Thus, to isolate PAPP-A to purity it was essential to separate this antigen from other macroglycoproteins (α2M, PZP). By affinity immunoelectrophoresis (AIE), interactions between these proteins and ligands were studied and evaluated as potential procedures for purification of PAPP-A.

MATERIALS AND METHODS

Affinity immunoelectrophoresis (AIE): PAPP-A (80 ug/ml) and α2M (100 ug/ml) were separately electrophoresed at 10 V/cm in 1% Litex HSA agarose (Denmark) containing one of 13 lectins (Table 1). After the bromophenol blue marker migrated 2.5 cm in a lectin free gel (control), the first dimensional gels were cut and electrophoresed into second dimensional gels containing 10 ul antibody/ml gel, respectively. Rabbit anti-PAPP-A and anti-α2M were purchased from Dakopatts (Bioscientific, Australia). LPA was prepared to 480 ug/ml gel, whereas the remaining lectins were made to 550 ug/ml gel (Table 1). LPA, BS-1 and BS-2 were purchased from Sigma (Bioscientific, Australia) and the remainder from LKB (Linbrook, Australia). Gels were pressed, dried and stained with Coomassie brilliant blue (6).

PAPP-A (11.6 ug) and α2M (9.3 ug) were separately incubated 1h at 37°C with 10 ul of B-glucuronidase (Sigma) or 50mM Tris-HCl,pH 7.4, containing 0.15M NaCl. Samples (10ul) were analysed by two dimensional immunoelectrophoresis (as above).

Separate pools of late pregnancy serum (LPS) and second trimester amniotic fluid (AF) were electrophoresed in 1% Indubiose A 37 agarose (L'Industrie Biologie Francaise, France) in absence and presence of 20U heparin/ml gel (C.S.L., Australia). When the bromophenol blue marker migrated 3 cm, gels were cut and electrophoresed into second dimensional gels containing respective antibodies (as above).

Metal chelate chromatography (MCC): Five columns (2ml) of iminodiacetic acid - Sepharose 6B (Pharmacia, Austriala) were respectively saturated with distilled water solutions of calcium chloride ($CaCl_2.2H_2O$; 3mg/ml), magnesium chloride ($MgCl_2.6H_2O$; 5mg/ml), manganese chloride (MnCl2; 5mg/ml), zinc chloride ($ZnCl_2$; 3mg/ml) and copper sulphate ($CuSO_4$; 3mg/ml) as per manufacturer's instructions. Samples of LPS (1ml), previously dialysed (4 x 50 volumes) against equilibration buffer (50mM sodium phosphate, pH7.0, containing 0.15M NaCl) were applied at 0.5ml/min. Bound proteins were eluted with 50mM EDTA and monitored at 280nm. Specific proteins were monitored by immunoassay.

Heparin-Sepharose chromatography: Pooled LPS (10ml), diluted 1:1 with 50mM Tris-HCl,pH7.8, containing 0.15M NaCl and 10mM sodium azide (TBS), was applied at 1ml/min on a 10 ml column of heparin-Sepharose (Pharmacia) preequilibrated with TBS. Elution of heparin-Sepharose bound proteins was initiated at fraction 40 with a linear (50ml) gradient of increasing concentration from 0.15 to 1.0M NaCl dissolved in 50mM Tris-HCl,pH7.8, containing 10mM sodium azide. Fractions (2 ml) were collected and total protein elution was monitored at 280 nm and by fused rocket immuno-electrophoresis (FRIE; 5). Another aliquot of pooled LPS (10ml) was diluted 1:1 with 50mM Tris-HCl,pH7.8, containing 0.45M NaCl and 10mM sodium azide (TBS-0.45M NaCl). The sample was chromatographed (as above) on a 10ml column of heparin-Sepharose equilibrated with 50mM Tris-HCl, pH7.8, containing 0.3M NaCl and 10mM sodium azide (TBS-0.3M NaCl). Protein desorption was achieved with a linearly increasing concentration gradient from 0.3 to 1.0M NaCl (TBS-1.0M NaCl). Protein elution was monitored at 280nm and by FRIE, and fractions (2ml) assayed for specific proteins by electroimmunoassay at antibody dilutions and sample volumes as listed in Table 2.

Negative immunoaffinity chromatography: Various preparations of anti-total human serum (ATHS) antisera (Dakopatts; Sigma; Silenus, Australia) were compared by two dimensional immunoelectrophoresis at 126ul of each antibody per ml gel, respectively, against heparin - binding proteins (2ul) isolated (as above) from pooled normal male serum (NMS). In addition, two NZ white male rabbits were immunized at two weekly intervals with 50ug heparin-binding NMS proteins. After the third dorsal subcutaneous injection, rabbits were bled from marginal ear vein and the antiserum compared against the commercial preparations.

TABLE 1. COMPARATIVE CARBOHYDRATE ANALYSIS OF PAPP-A AND α2-MACROGLOBULIN

	Abbreviation	Interaction[a]	
		PAPP-A	α2M
Lectin:			
Bandeiraea simplicifolia 1	BS-1	±	±
Bandeiraea simplicifolia 2	BS-2	-	-
Concanavalin A	ConA	+	+
Helix pomatia	HPA	+	+
Lens culinaris	LCA	+	+
Limulus polyphemus	LPA	+	+
Peanut agglutinin	PNA	-	-
Phytohaemagglutinin	PHA	+	+
Pokeweed mitogen	PWN	-	-
Ricinus communis	RCA	+	+
Soya bean	SBA	-	-
Ulex europaeus	UEA	-	-
Wheat germ	WGA	+	±
Glycolase:			
β-Glucuronidase	-	+	-

[a]+, elimination of immunoprecipitate; ± altered immunoprecipitate; -, no effect on immunoprecipitate

Purification schedule: Pooled LPS (100 ml) was diluted 1:1 with TBS-0.45M NaCl and applied on a 100 ml column of heparin-Sepharose equilibrated with TBS-0.3M NaCl. Column was developed at 1ml/min and 4 min fractions collected. Heparin-Sepharose bound proteins were eluted with stepwise increase in buffer ionic strength to TBS-1.0M NaCl. Fractions containing maximal PAPP-A concentration were pooled and applied at 11ml/h on a 480ml Pharmacia C26/100 column packed with Ultrogel AcA34 matrix (LKB) and developed with TBS. PAPP-A fractions were pooled and subjected to negative immunoaffinity chromatography on CNBr - Sepharose (Pharmacia) immobilized antibodies with specificities against heparin-binding NMS proteins. PAPP-A being unretained by the immobilized antibodies, eluted in the column void and was concentrated on 10ml column of heparin-Sepharose (6).

RESULTS

In the absence of ligands both PAPP-A and α2M have α2-electrophoretic mobility (Fig. 1). However, in the presence of lectins such as Con A, RCA, PHA and WGA, the immunoprecipitates of both proteins were eliminated (Table 1). Weaker but significant interactions were detected with LPA, LCA and BS-1, whereas the electrophoretic mobility and immunoprecipitate morphology of either PAPP-A or α2M was unchanged in presence of BS-2, PNA, PWM, SBA and UEA. Thus suggesting that both PAPP-A and α2M had undergone similar post-translational modifications since both proteins contained accessible sialic acid, glucose/mannose, N-acetyl-galactosamine and N-acetylglucosamine residues. However, immunoelectrophoretic analysis after incubation with β-glucuronidase virtually eliminated, qualitatively and quantitatively, the PAPP-A immunoprecipitate, whereas this glycolase had no effect on α2M electrophoretic mobility or immunoprecipitate morphology (Table 1).

The influence of heparin on the elctrophoretic migration of serum proteins is readily seen in Fig 1. The migration of six B-mobile proteins was increased in presence of heparin (Fig. 1 ATHS). With specific antibodies, significant increases in the anodic migration of PAPP-A (28%), antithrombin III (161%) and β-lipoprotein (47%) were demonstrated in presence of heparin. By contrast, heparin had no effect on the elctrophoretic mobility of PZP, α2M, and α1-antitrypsin (Fig. 1).

All of 20 proteins studied (albumin, alphafetoprotein, α1-antitrypsin, α2M, antithrombin III, β-lipoprotein, chorionic gonadotropin, complement factors C_4, C_3c, C_3d, factor B, fibronectin, immunoglobulin G, lactoferrin, placental lactogen, plasminogen, PAPP-A, pregnancy-specific β1-glycoprotein, PZP, transferrin) reversibly bound to chelated copper matrix. Only plasminogen interacted with chelated magnesium, and alphafetoprotein

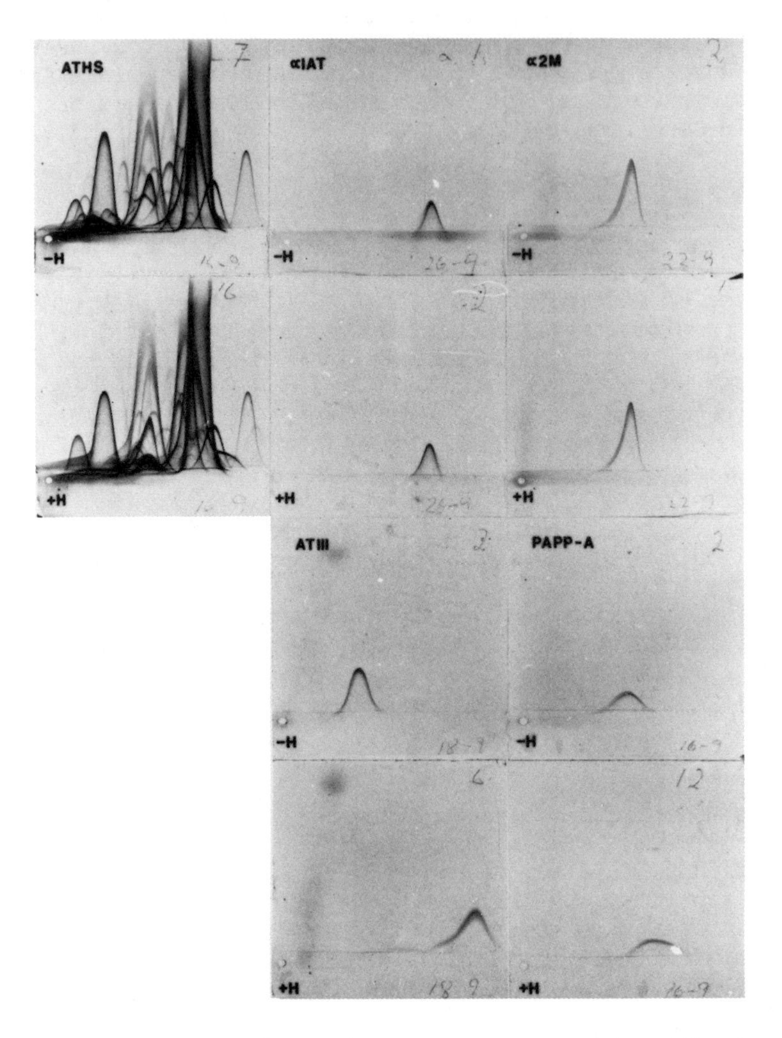

Figure 1: Two dimensional immunoelectrophoresis without (-H) and with (+H) heparin (20U/ml/gel) in the first dimensional gel. Antibodies used in second dimensional gels included anti-total human serum (ATHS), anti-α1-antitrypsin (α1AT), anti-α2 -macroglobulin (α2M), anti-antithrombin III (AT III) and anti-pregnancy-associated plasma protein-A (PAPP-A).

interacted with immobilized calcium and manganese ions. Weak interactions with zinc chelate matrix (α1-antitrypsin, β-lipoprotein, complement factor C_3d, placental lactogen) were readily eliminated by increasing the ionic strength of application buffer to PBS containing 0.8M NaCl. Under these conditions only 39% of the applied proteins bound to chelated zinc and these included PAPP-A, α2M, PZP, fibronectin, tranferrin and complement factors C_3c and C_4.

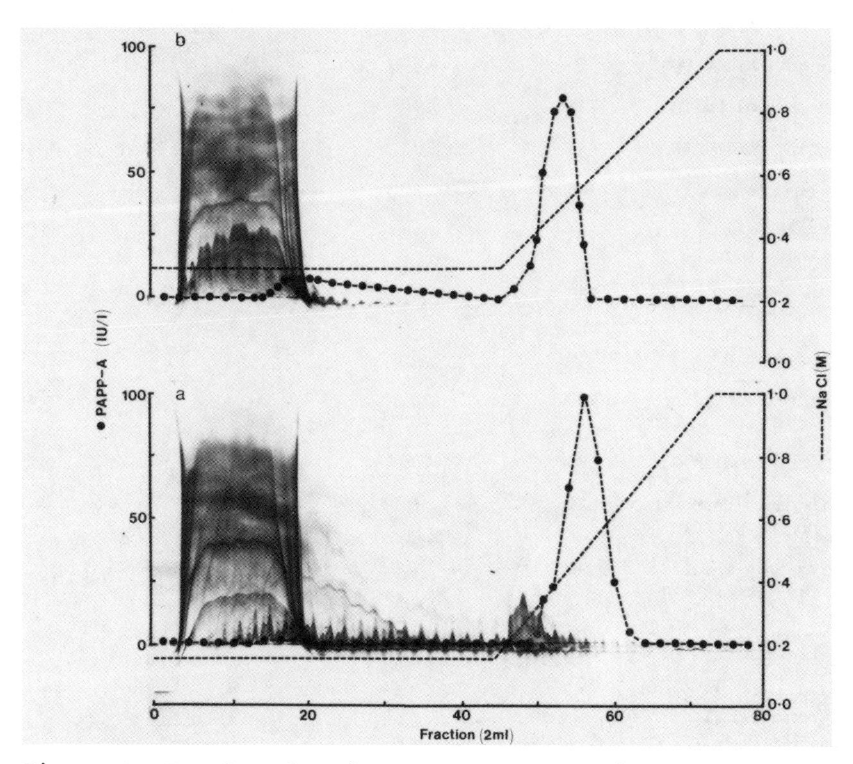

Figure 2: Fused rocket immunoelectrophoretic analysis of protein elution after chromatography of pooled late pregnancy serum on heparin-Sepharose, equilibrated in TBS-0.15M NaCl (a) or TBS-0.3M NaCl(b). Elution was achieved with a linearly increasing concentration gradient from equilibration buffer to TBS-1.0M NaCl and PAPP-A elution monitored by electroimmunosassay.

TABLE 2. INFLUENCE OF IONIC STRENGTH ON PROTEIN INTERACTIONS WITH HEPARIN-SEPHAROSE

Protein	Molecular Size (kd)	Antibody Dilution (ul/1ml gel)	Sample (ul)	Bound (%)	
				0.15M NaCl	0.3M NaCl
Albumin	66.3	20	2	<0.1[a]	-[b]
α1-antitrypsin	54	20	5	<0.1	-
α1-antichymotrypsin	68	10	5	-	-
α2-macroglobulin	725	10	5	<0.1	-
Antithrombin III	65	12.5	5	78.6	62.6
β-lipoprotein	2400	6.8	2	+[c]	+
Cl-esterase inhibitor	104	4	5	-	-
Complement factor 3	180	1.4	5	+	-
Fibronectin	350	5	5	+	+
Inter-α-trypsin inhibitor	160	5	5	5.5	-
Immunoglobulin G	165	5	5	-	-
Pregnancy-associated plasma protein-A	820	6.8	10	92.2	76.2
Pregnancy-specific β_1-glycoprotein	90	6.8	5	-	-
Pregnancy zone protein	360	6.8	5	<0.1	-
Total protein					
nonbound (%)				88.7	99.5
bound (%)				11.3	0.5

a, traces; b, undetectable; c, significant levels but not quantified.

FRIE analysis of pooled LPS fractionated on heparin-Sepharose is shown in Fig. 2. When sample was applied in TBS, 11.3% of the serum proteins bound to immobilized heparin, including PAPP-A (92.2%), antithrombin III (78.6%) and traces of α2M, PZP (Table 2). However, sample application in TBS-0.3M NaCl eliminated weak interactions (Fig. 2b) and only 0.5% of the serum proteins were retained by the matrix (Table 2). PAPP-A was completely recovered (100.3%) but only 76.2% bound to heparin-Sepharose and the remainder eluted with the non-interacting proteins (Fig. 2b). Under these conditions, α2M was not detected in the eluate. Other heparin-binding serum proteins included antithrombin III (free and protease complexed), β-lipoprotein and fibronectin (Table 2). In this single procedure PAPP-A was purified 153 fold, as compared to 5.5 fold purification when LPS was applied in TBS.

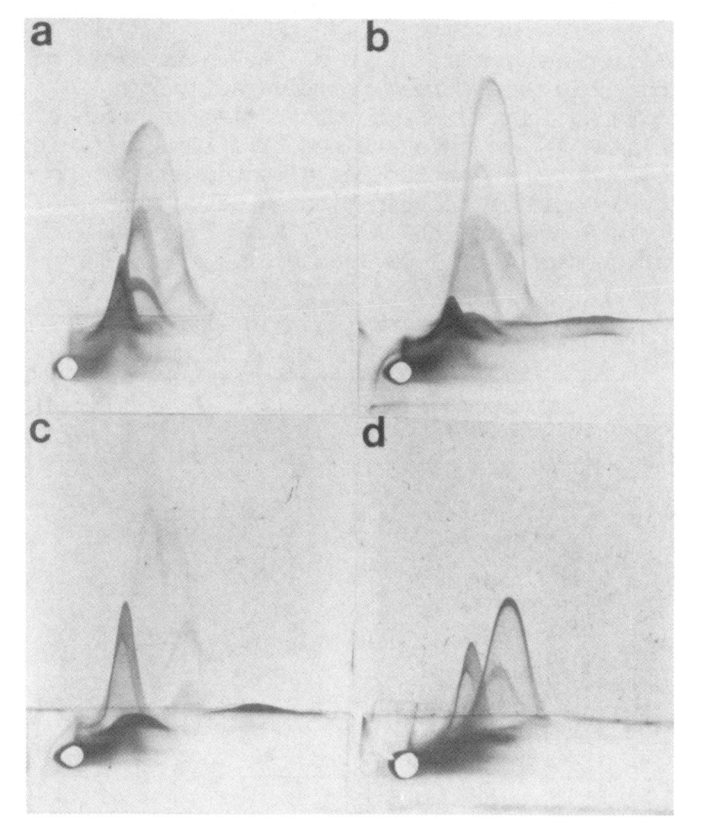

Figure 3: Two dimensional immunoelectrophoretic analysis of anti-total human serum antisera prepared, Dakopatts (b), Silenus (c) and Sigma (d). Rabbit anti-human male serum heparin-binding proteins was analyzed in (a). In all cases heparin-binding MNS proteins were used as antigen.

Of the commercial anti-human serum antiserum (ATHS), the Dakopatts preparation formed at least 11 immunoprecipitates when analyzed by two dimensional immunoelectrophoresis against NMS heparin binding proteins (Fig. 3). Fewer proteins were recognized by Silenus or Sigma preparations. Since additional immunoprecipitates were detected with our antiserum, then this preparation was immobilized for negative immunoaffinity chromatography. No crossreaction against PAPP-A was detected in any of these four antisera.

Isolation and purification of PAPP-A is based on three chromatographic procedures; positive affinity chromatography (heparin-Sepharose), size fractionation (Ultrogel AcA34) and balanced negative immunoaffinity chromatography (with antibodies directed against non-pregnancy-associated heparin-binding serum proteins. The overall yield of pure PAPP-A was 22.2% (1.1 mg) with purification factor of 1483 (Table 3).

TABLE 3. PAPP-A PURIFICATION SCHEDULE

	Fractionation mechanism	PAPP-A[a] (mg)	Recovery (%)	Yield (%)	Purification factor
Pooled pregnancy serum	starting material	4.85	100	100	1.0
Heparin-Sepharose	positive affinity	1.90	103.5	39.1	485.8
Ultrogel AcA34	size	1.84	98.8	37.9	826.9
Antibody matrix + Heparin-Sepharose	negative immunoaffinity positive affinity	1.08	58.7	22.2	1482.8

[a] 487 ug PAPP-A= 1 IU PAPP-A (WHO 78/610)

DISCUSSION

Unlike α2M, a broad spectrum protease inhibitor, PAPP-A is a pregnancy-associated specific inhibitor of leucocyte elastase (6). As these two protease inhibitors have many physicochemical similarities including size, homotetrameric conformation and electrophoretic mobility, fractionation by size and charge does not resolve these proteins (7). Hence, earlier attempts to purify PAPP-A by such techniques were unsuccessful (1). In addition, antiserum produced against such preparations were polyspecific and application of these antibodies for positive immunoaffinity isolation of PAPP-A yielded an impure product (3). Other specificities, in addition to PAPP-A, in such antisera (8) included anti-pregnancy-specific β1-glycoprotein, anti-PZP, anti- α2M, anti-antithrombin III and anti-β-lipoprotein (9).

When elution was effected with 50mM EDTA, chromatography on chelated divalent metal ions (Ca^{2+}, Cu^{2+}, Mg^{2+}, Mn^{2+}, Zn^{2+}) did not resolve PAPP-A from α2M (PZP and fibronectin). Gradient elution with linearly decreasing pH to 4.0 did resolve these proteins, but PAPP-A was immunologically and physicochemically denatured (10, 11). Although _in vitro_ function was retained (6) positive affinity chromatography on zinc chelate matrix was not applied for PAPP-A isolation.

Neither PAPP-A nor α2M were resolved by AIE with any of the 13 lectins, indicating the carbohydrate composition of both glycoproteins was qualitatively similar. Thus, despite different sites of production, both proteins may undergo similar post-translational glycosylation and further supporting the hypothesis that PAPP-A may be a placental or pregnancy-associated analog of α2M (6). However, after incubation with β-glucuronidase (and chondroctinase AC) the PAPP-A immunoprecipitate was virtually eliminated, whereas α2M immmunoactivity remained intact. Thus, α2M is a glycoprotein whereas PAPP-A is proteoglycan containing glucuronic acid in chondroitin sulphate moieties.

Of the broad spectrum of biological actions of proteoglycans, including inhibition of proteolytic activity, such molecules are considered as an adherence component of extracellular matrix. Although the mechanism of adherence remains to be defined, glycan moieties are highly charged and, indeed, PAPP-A reversibly interacted with anion exchange resin (1). Then, in the presence of heparin, a highly charged anionic glycosylaminoglycan, the electrophoretic mobility of PAPP-A was significantly increased. By contrast, the electrophoretic mobility of α2M and other pregnancy-associated antigens (PZP and pregnancy-specific β1-glycoprotein) remained unchanged, suggesting a specific heparin-PAPP-A interaction (12). Chromatography on immobilized heparin confirmed this interaction, and in addition to PAPP-A other proteins to interact with this anionic ligand included protease inhibitors (antithrombin III, inter-α-trypsin inhibitor), complement proteins (C_3), cell surface proteins (fibronectin), pregnancy-associated proteins (placental protein 5; 13) and β-lipoproteins.

The heparin-PAPP-A interaction was heterogeneous. Serum application in increased ionic strength buffers (TBS-0.3M NaCl) removed non- and weak heparin interacting proteins (99.5%), including low heparin affinity PAPP-A (24.8%). Since no α2M was detected in the high heparin affinity PAPP-A preparation, then for the first time PAPP-A was concentrated and quantitatively and qualitatively isolated from α2M, with complete retention of molecular, functional and immunological integrity.

Since the starting material was serum, it must be remembered that during coagulation, platelets undergo activation and aggregation, processes resulting in secretion of protease inhibitors, enzymes and platelet specific proteins (platelet factor 4, thromboglobulin, thrombospondin and platelet derived growth factor). With the exception of thrombospondin (Mr 450 kd) these proteins are smaller than albumin and readily separated from PAPP-A by gel filtration. Then because of the possibility of copurifying platelet derived factor(s) with PAPP-A, and the varied quality of commercial anti-human serum antisera, we produced a balanced antiserum for negative immunoaffinity chromatography.

Large scale purification of PAPP-A from placental extracts and pregnancy serum, could now be easily achieved by three mild chromatographic procedures: positive affinity chromatography on heparin-Sepharose, size fractionation by gel filtration and balanced negative immunoaffinity chromatography. Although this procedure isolates high heparin-affinity PAPP-A, no physicochemical, immunological or in vitro functional differences were detected between low and high heparin-affinity PAPP-A (6, 7, 15). Purity and homogeneity of isolated PAPP-A was confirmed by radioimmunoassay, isoelectric focusing, polyacrylamide gel electrophoresis, multidimensional immunoelectrophoresis and by immunizing rabbits which produced a monospecific anti-PAPP-A antiserum (not shown here). Thus, application of affinity immunoelectrophoresis is strongly recommended when developing a purification schedule.

ACKNOWLEDGEMENTS

The authors thank Ms. Cathy O'Brien, Ms. Lindy Benson, and Mr. Anthony Zakher for technical assistance and Ms. Dara Leary and Rose-Marie Bradley Jones for preparation of this manuscript.

REFERENCES

1. Lin TM, Halbert S., Keifer D. Spellacy WN, Gall S (1974). Characterization of four human pregnancy-associated plasma proteins. Am J Obstet Gynecol 118:223.

2. Biscof P (1979). Purfication and Characterization of pregnancy-associated plasma protein-A (PAPP-A). Arch Gynecol 27:315.

3. Folkersen J. Westergaard JG, Hinderson P, Teisner B (1979). Affinity chromatographic purification of a new high molecular weight pregnancy specific protein SP4. In Lehmann FG (ed): "Carcinoembryonic Proteins", Amsterdam: Elsevier-North Holland, p503.

4. Sutcliffe RG, Kukulska B, Nicholson LVB, Paterson WF (1979). The use of antibody affinity chromatography and other methods in the study of pregnancy-associated proteins. In Klopper A, Chard T (eds): "Placental Proteins", Heidelberg: Springer, p55.

5. Gore CH, Sutcliffe RG (1984). Pregnancy-associated plasma protein-A: purification under mild conditions, peptide mapping, and tests for possible interactions with trypsin, plasmin and complement. Placenta 5:293.

6. Sinosich MJ, Davey MW, Ghosh P, Grudzinskas JG (1982). Specific inhibition of human granulocyte elastase by

human pregnancy-associated plasma protein-A (PAPP-A). Biochem Internat 5:777.

7. Sinosich MJ (1985). Biological role of pregnancy-associated plasma protein-A in human reproduction. In Biscof P, Klopper A (eds): "Proteins of the Placenta", Basel: Karger, p158.

8. Biscof P, Haenggeli L, Sizonenko MT, Herrman W, Sizonenko PC (1981). Radioimmunoassay for the measurement of pregnancy-associated plasma protein-A (PAPP-A) in humans. Biol Reprod 24:1076.

9. Sinosich MJ (1986). Evolution of pregnancy-associated plasma protein-A. In Hau J (ed): "Pregnancy Proteins in Animals", Berlin: Walter de Gruyter, p 269.

10. Sinosich MJ, Davey MW, Teisner B, Grudzinskas JG (1983). Comparative studies of pregnancy-associated plasma protein-A and α2-macroglobulin using metal chelate chromatography. Biochem Internat 7:33.

11. Sinosich MJ, Davey MW, Grudzinskas JG (1984). Interaction of pregnancy-associates plasma protein-A (PAPP-A) and α2-macroglobulin (α2M) on metal chelate chromatography. Protides of Biol Fluids 32:19.

12. Sinosich MJ, Teisner B, Davey MW, Grudzinskas JG (1981). Pregnancy-associated plasma protein-A: interaction with heparin in crossed affinity immunoelectrophoresis. Aust N Z J Med 11:429.

13. Jones GRD, Davey MW, Sinosich MJ, Grudzinskas JG (1981). Specific interaction between placental protein 5 and heparin. Clin Chim Acta 110:65.

14. Holt JG, Niewiarowski S (1985). Biochemistry of granule proteins. Seminars in Hematology 22:151.

15. Davey MW, Teisner B, Sinosich MJ, Grudzinskas JG (1983). Interaction between heparin and human pregnancy-associated plasma protein-A (PAPP-A): a simple purification procedure. Anal Biochem 131:18.

Protein Purification: Micro to Macro, pages 239–245

CHOICE OF SALT AND FLOW RATE CAN AFFECT RECOVERY OF BIOMOLECULES IN ION EXCHANGE CHROMATOGRAPHY

David G. Maskalick, Marie A. Abbott
and Kelly J. Hoke

Eli Lilly and Company
Indianapolis, IN 46285

ABSTRACT Ion exchange chromatography of proteins usually involves losses due to either irreversible binding to the resin or inability to achieve the desired product purity and throughput without sacrificing yield. The salts present in the elution buffers may have an impact upon the product recovery analogous to the effect that different organics have upon performance in reversed phase chromatography. The utilization of alternate salts may lead to modified retention times and/or recoveries analogous to the use of propanol versus acetonitrile or the presence versus absence of triethylamine in a RP-HPLC buffer system. The relationship between binding kinetics and flow rates also may be exploited to effect better separations. Examples and explanations of these effects will be presented.
The impact of both the salt and the flow rate selection is primarily due to the ability of the biomolecules to interact with a single functional group in a multidentate fashion. The ramifications of this concept in the field of affinity chromatography is discussed.

INTRODUCTION

The challenges encountered when scaling-up a chromatographic purification process are: 1) purity, 2) yield, and 3) throughput. Purity levels which must be met have usually been dictated by work previously performed on a small lab scale where yield and throughput were not major concerns. On a large scale it is unacceptable to sacrifice product and yield using conservative main peak cuts to achieve the desired purity. Yield must be maximized and purity maintained. In order to maximize throughput the entire column operation must be as short as possible and/or the flow rates must be maximized. Different resins and/or different eluting salts will provide altered selectivity and thus should be used to maximize the resolution of the desired product from the remainder of the charge material before attempting to scale-up the process. Yield losses will also result from irreversible binding of the product to the resin. Alternate resins may afford better overall accountability but alternate salts may be a more cost effective means of achieving the same thing.

MATERIALS AND METHODS

The purification of EL-349 from solubilized granules was used to generate the data presented here. EL-349 is a globular protein, 22Kd, and was of recombinant DNA origin. The resin used is a commercially available DEAE anion exchange resin.

TRACE MULTIVALENT SALTS

Figure 1 indicates that trace amounts, lmM, of a multivalent salt can improve the overall recovery of the protein product from the resin by 33%. Elution in each case was accomplished using sodium chloride. The presence of the trace salt appears to have the greatest impact when the specific activity of the column charge is low (data not shown).

MULTIVALENT SALTS AND ELUTION

Figure 2 demonstrates that elution with multivalent salts may improve resolution. Elution using sodium chloride results in the largest main peak volume, that which contains the EL-349 (closed boxes), the EL-349 degradation products (open boxes), although not completely resolved, can be better separated when sodium tripolyphosphate (NATPP) and calcium chloride are used for elution. In general, the purity of the mainstream pool is greatest when using calcium chloride, slightly less using NATPP, and even less using sodium chloride for elution (data not shown). In this case

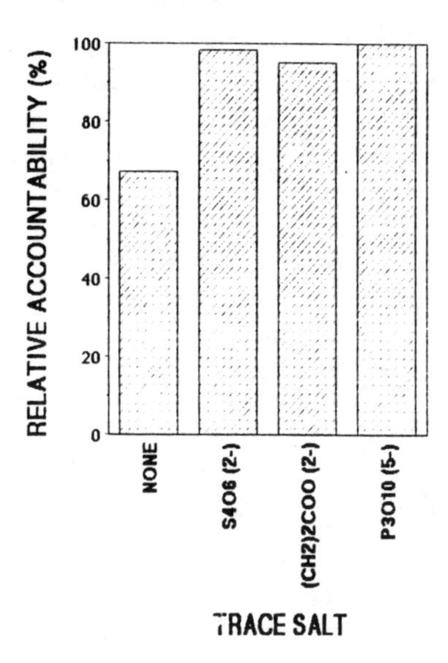

Figure 1: Accountability vs. trace salt

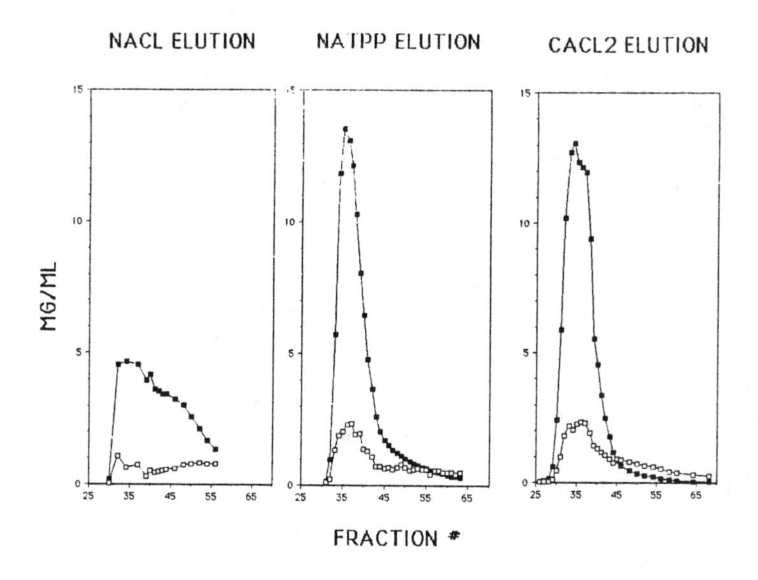

Figure 2: Eluting salt vs. elution characters

the resolution of the product from related substances, i.e. degradation products, is shown, however, the resolution of EL-349 from bacterial proteins is also improved using the multivalent salts.

CONTACT TIME & MULTIDENTATE BINDING

Figure 3A at first glance shows the obvious, i.e. at lower contact time with the resin less material is bound to the resin. What is now shown is that an equivalent quantity of EL-349 can be bound at the shortest and the longest contact time. This occurs even though EL-349 is one of the first components to elute from the column. In addition, Figure 3B indicates that some of the material which elutes last, i.e. in the high salt regeneration, does not bind as efficiently when the contact time is shorter. These results are consistent with a multidentate model as indicated in Figure 4A. The molecules bind to the resin using more than one functionality at a time. The molecules which bind to the resin most tenaciously are interacting with the resin with the most functionalities at the same time. The rate at which such a multidentate binding occurs places a lower limit on the contact time necessary for the tightest binding to occur. If the tight binding molecule is not allowed the time to form all of the required contacts it will not be bound to the resin. If a weak molecule can establish its binding to the resin very quickly then it will bind to the resin even at very short contact times. Thus, the binding rate limits the flow rate at which a column may be charged.

MULTIDENTATE BINDING & AFFINITY CHROMATOGRAPHY

Figure 4A explains why the multivalent salts used in Figure 1 improved recovery in addition to illustrating the multidentate binding of a protein to the resin. The multivalent salt may actually help displace the protein from the resin by binding to the resin in a multidentate fashion itself. The sodium chloride only helps reduce the charge-charge attractions by increasing the conductivity. The multivalent interaction of protein, resin, and salt should

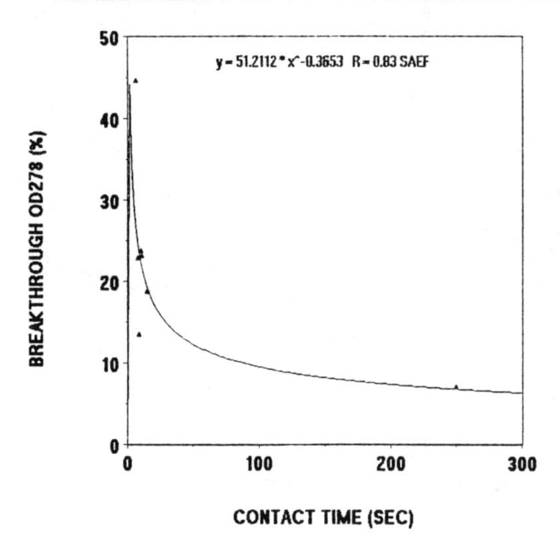

Figure 3A: Breakthrough OD vs. resin contact time

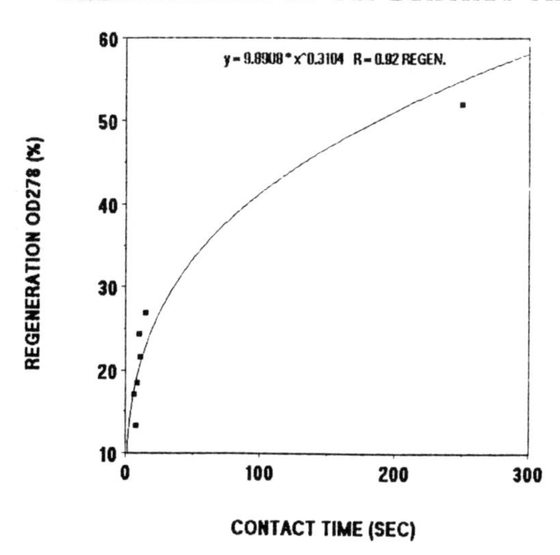

Figure 3B: Regeneration OD vs. resin contact time

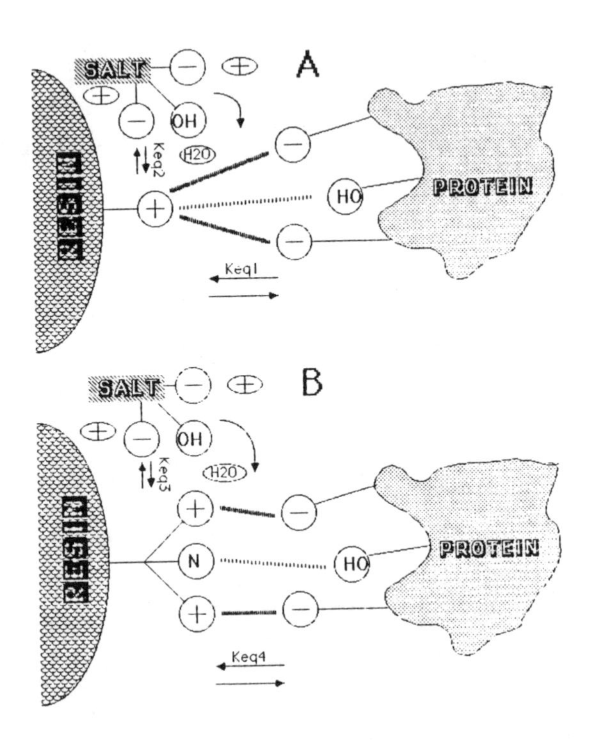

Figure 4: Multidentate binding in ion exchange and affinity chromatography

be extended to include the hydrophobic surface areas and the hydrogen bonding portions of the protein and the resin surfaces.

The situation developed in Figure 4A can be modified as indicated in Figure 4B so that specific affinity is created between protein and resin. The equilibrium K_{eq4} could be modified by both the static orientation of the functionalities on the resin and the flexibility, i.e. bond rotations and steric hinderances, given to the particular structure which is developed. The nature of the eluting salt will also play a key role in the performance of the affinity chromatography. The equilibria could be modified by adjustment of the pH, ionic strength, temperature, organic solvent, etc. without resorting to a change in the structure of the protein. Construction of the best equilibria through the design of the affinity ligand and eluting salt as well as the elution conditions will result in better recovery, better resolution from related substances, and better throughput.

DISCUSSION

Proteins bind to ion exchange columns in a multidentate fashion. Thus chromatographic purification may be optimized by considering various resins and/or multivalent salts. This optimization, although empirical in nature, may have significant impact upon the resolution and the yield. The resin and the charge conditions which are selected will also determine the rate at which the protein binds and thus the maximum flow rate that can be used during application of the charge to the column.

The multidentate binding concept can be applied to affinity chromatography as well. Small affinity ligands, more stable than monoclonal antibodies and the displacing salts could be designed either empirically or based on the protein structure if it is known. The work of Geysen, M.H., et al (1985), Immunology Today, 6 (12) 364-369 with mimotopes shows promise for the creation of small affinity ligands empirically. The selectivity and equilibria could then be modified in a scientific manner in order to maximize the chromatographic efficiency.

ACKNOWLEDGMENTS

Bibbie B. Cork is gratefully acknowledged for the typing and preparation of this paper.

Protein Purification: Micro to Macro, pages 247–253

MACROSORB KIESELGUHR-AGAROSE COMPOSITE ADSORBENTS

M.G.Bite, S.Berenzenko, F.J.S.Reed, and L.Derry

Sterling Organics R&D, Newcastle upon Tyne, NE3 3TT England

Incompressible composite adsorbents, whilst retaining all the desirable properties of traditional agarose based hydrogel media, overcome the operational limitations imposed by the use of soft hydrogels: they permit useful application of fast flow rates without restrictions on bed depth and they can be used in fluidised bed mode. These improvements are expected to have a significant impact on the realisation of economically viable downstream processing schemes.

The increasing number of protein isolations published each year is a combined reflection of the advances in the field of biotechnology and the power of the laboratory techniques which permit the recovery of single components from complex, heterogeneous mixtures. The isolation of proteins requires the application of methods which do not interfere with and damage their fragile structures. Consequently, chromatography in its various forms has been shown by experience to be the method of choice for the relatively simple and efficient recovery of proteins from dilute solutions. Appropriate application of ion-exchange and affinity adsorption methods can give extremely specific recovery of a particular material from a complex feedstock (1,2).

Although chromatographic techniques are capable of performing the desired separations under mild conditions, they have serious limitations when large scale operation is contemplated. Most of the difficulties arise from the physical properties of the soft hydrogels currently used in chromatographic practise, and the associated high cost of installing special plant to accommodate these gels in necessarily large-area / shallow bed configurations.

In many cases, chromatography is the only means whereby a biological product can be isolated in sufficient yield and purity, but it is only the production of low volume / high value materials which can sustain the capital overheads involved. The economic performance obtained using current large scale chromatography does not justify its use for the production of medium and lower value / higher volume type products, which require the use of larger scale equipment which is able to sustain the high volumetric throughputs.

There have been various attempts at the production of alternative media which would overcome the operational restrictions of currently used hydrogels, e.g.porous silicas with derivatised pellicular layers (3) and spirally wound radial flow cartridges (4,5). By contrast, Macrosorb kieselguhr-agarose composites (6) retain all the desireable features of an agarose hydrogel. This is achieved by entrapment of the gel within the macropores of a rigid support granule.

MACROSORB COMPOSITE ADSORBENTS

The support. The inorganic support skeleton which supports the gel in the Macrosorb composite adsorbent is Macrosorb-K (7), which is manufactured from specially purified kieselguhr (diatomaceous earth). The physical form of this material is spheroidal granules, typically with diameters in the range 100 - 300 microns, which possess an interconnected interior macroporous structure. The macropores have an average diameter of 15 microns; these are several orders of magnitude larger than the pores which are to be found in other available porous silicas, and offer a large interconnected pore volume which can be space-filled with hydrogels or other polymers.

The low surface area of Macrosorb-K (approx. 0.5 sq.m/g) ensures minimal non-specific adsorption by the matrix and makes the support effectively "transparent" to proteins. The low surface area also gives the support excellent stability in the presence of strong alkalis (which are commonly used to scrub and depyrogenate hydrogel beds).

The composite. Insertion of 6% strength agarose hydrogel in Macrosorb-K results in the formation of the composite material Macrosorb-K6A. After cross linking the gel to give Macrosorb-K6AX, the composite may be derivatised further to give a whole range of adsorbents. Composite materials based on 2% and 4% agarose gels are produced in a similar manner.

Characteristics of the composite. Macrosorb composite adsorbents contain about 0.45ml agarose hydrogel/ml adsorbent and are supplied in the nominal particle size range 100-300 microns. Packed beds of Macrosorb composites do not change volume when subjected to changes in pH and ionic strength. Since the gel itself does not contribute to the physical stability of the composite, the degree of cross linking employed is the minimum required to prevent disassociation of the gel structure due to disruptive forces exerted by mutually repelling ion-exchange groups. The low degree of cross linking is reflected in the greater hydrophilicity of the composite when compared to heavily cross linked gels. This translates in practise to a greater specificity of binding on the composites and significantly reduced levels of cumulative residue build up, which can have a great effect on product purity and column scrubbing requirements.

Preliminary lifetime studies show that subjecting Macrosorb-KAX.DEAE to 100 cycles of 25mM tris-HCl pH8 (10 volumes), 25mM tris-HCl pH8 containing 1M NaCl (5 volumes) and 0.2M NaOH (2 volumes) at a linear flow rate of 2.5cm/min has no effect on the capacity of the column. Similarly, continuous pumping of 0.2M NaOH for 80hr at 2.5cm/min was also shown to have no effect on column volume or capacity.

Macrosorb composites are manufactured in compliance with current good manufacturing practise (cGMP) on an FDA approved site, and drug master files (DMFs) are being submitted for approval by the appropriate authorities. This will ensure that processes which utilise Macrosorb adsorbents for the production of pharmaceutical products will comply with the current regulatory requirements for the quality assurance of chromatography media which are to be used for that purpose.

SCALED-UP PROCESSING : A CASE STUDY

There is a fundamental conceptual difference between analytical or laboratory scale chromatography and preparative chromatography (on whatever scale). In the former cases, the aim is to produce a small amount of pure product from a small sample, the result being achieved by the use of low flow rates to achieve maximum resolution. By contrast, in preparative chromatography, the primary objective is to maximise throughput at the expense of optimising resolution. This is of particular concern where economic production is contemplated.

The separation of albumin from desalted bovine serum. This has been studied in depth (8) to determine the comparative performance of Macrosorb-KAX.DEAE when compared to a compressible adsorbent: DEAE-Sepharose, which is normally used in the production process. In this separation, albumin is adsorbed to the DEAE column while the rest of the serum components of the sample pass through. The column is then washed before selective desorption of albumin is performed. Finally, residual proteins are desorbed by applying a high ionic strength wash and the column is re-equilibriated before being used for the next process cycle.

Column capacity. The capacity of a column is an important factor which determines throughput. In a Macrosorb composite only about 50% of the volume of the adsorbent is agarose, and so the equilibrium capacity of the adsorbent is only about half that of a purely agarose-based matrix derivatised to the same degree. When, as in this case, sample is loaded to the point just before which albumin breakthrough occurs (to minimise waste of feedstock), one has to offset the benefits gained by utilising a fast flow rate against the capacity achieved at that flow rate and arrive at optimised conditions which will maximise throughput (performing several fast cycles which each produce less material may be more efficient than running one cycle under optimum adsorption conditions).

In small scale columns (30 x 1.6 cm) it was determined that Macrosorb-KAX.DEAE achieved albumin adsorption and subsequent elution at a flow rate of 5cm/min (compared to 1.25 cm/min when using DEAE-Sephrose) to give albumin of the same quality. Re-equilibration on the composite adsorbent could be performed at 10cm/min.

Large scale operation. In a scaled-up comparative study simulating production conditions, both media were operated in a 16L Pharmacia KS370 "Stack" column. For this particular separation, both gels were loaded to 30% of their specific capacity for albumin under the operating conditions to be used. In this case, the capacity of the DEAE-Sepharose for albumin was a factor of 2.66 higher than that of the Macrosorb-KAX.DEAE when the gels were operated at 0.4L/min and 5L/min respectively, the quality of the albumin produced being the same in both cases. The disparity in the flow rates used is a reflection of the back pressures generated when using soft adsorbents. The pressure drops exerted by Macrosorb adsorbents in columns of this size are negligible.

For the DEAE-Sepharose column, 0.5kg albumin was applied as a 25L sample of desalted serum. The total cycle, including re-equilibration, was 612 min and yielded 0.45kg albumin in 23L of eluate. By contrast, the Macrosorb column was loaded with 0.19kg albumin in 9.4L of sample, and in a 51 min cycle yielded 0.17kg albumin in 25L of eluate. Time for time, the throughput of albumin in the Macrosorb column was ca. 2kg albumin in 300L (using 12 cycles).

Concentration of eluate does not add significantly to the economics of the process, and it can be shown that the Macrosorb process produces albumin for approximately one third of the cost at four times the throughput.

Alternative sample preparation. When using the Macrosorb column, serum can be diluted to the correct ionic strength as an alternative to desalting. This requires the application of a sample which has three times the volume but only increases the Macrosorb process cycle time by 11%. Thus, desalting can effectively be eliminated from the process scheme.

Scrubbing requirements. One consequence of the low degree of cross linking in this composite adsorbent is low cumulative build up of residually bound proteins. When the process is run using DEAE-Sepharose, the column requires a 0.2M NaOH scrub/re-equilibration treatment every three process cycles to recover column capacity. On the composite column it has been shown that after 12 cycles there is no loss of capacity and only a very low level of residually bound protein on the column. The down-time factor for scrubbing would have a significant effect on the economics of the separation.

Process optimisation. The albumin separation has been successfully performed using an adjustable bed-depth Amicon G300x500 glass column containing 21L of Macrosorb-KAX.DEAE, with loading and elution flow rates of 4cm/min and the re-equilibrations being run at 10cm/min. Operation on this column showed an increase in specific throughput (per L of column packing) of 30% over that achieved in the Macrosorb column described previously).

The process is not yet fully optimised in terms of economic operation. Small scale experiments have shown that it is possible to load the Macrosorb-KAX.DEAE column to 60% albumin capacity (compared to 30% capacity as described earlier), whilst suffering only a 10% breakthrough loss of albumin (which could be recovered by appropriate recycling). This increases throughput by another factor of two.

THE USE OF COMPOSITE ADSORBENTS IN FLUIDISED BEDS

There is a growing interest in the use of adsorbents in fluidised beds for the extraction of proteinaceous materials from culture broths. It is often required to process feedstocks as rapidly as possible in order to maximise product recovery. With highly diluted or unstable feedstocks, any process which permits the rapid processing of large volumes is useful. Fluidised bed configurations are able to accommodate these requirements, and are particularly appropriate when unclarified feedstocks are to be treated, the advantage over column operation being that blinding does not occur (9).

Fluidised bed configurations for adsorption processes have not hitherto been exploited mainly because of the lack of suitable adsorbents, and also because of the lack of fundamental understanding of the liquid fluidisation process when used for this type of application.

Pilot studies in which Macrosorb composite ion-exchangers have been evaluated in fluidised bed mode show them to operate successfully on both clarified and unclarified feedstocks. Macrosorb-KAX.CM has been used to extract asparaginase from clarified alkaline cell lysate and Macrosorb-KAX.DEAE has been used to recover aryl acyl amidase from unclarified Pseudomonas putida culture broth (10).

Researchers at the Oak Ridge National Laboratory are developing a process in which Macrosorb-KAX.DEAE is used to recover cellulase both from aqueous process liquors and undigested solid residues to which cellulase is still adsorbed after the hydrolysis step is complete. This work is of significance because in this particular process, the cost of the cellulase represents approximately 60% of the total production cost if the enzyme is used on a zero recovery basis. Cellulase is recovered by selective adsorption and removal from suspended microparticulate residues by exploiting the higher density of the composite adsorbent, thus achieving an easy separation by fluidisation or decantation (11).

CONCLUSION

Composite adsorbents enable the downstream process designer to scale up from laboratory scale to large volume processing by eliminating the compressibility problem, but without losing the otherwise highly desirable characteristics of agarose-based adsorbents. Thus, Macrosorb composite adsorbents enable the use of fast flow rates in low pressure column equipment. This permits the use of dilution of sample to the correct ionic strength as an alternative to desalting. In appropriate cases, fluidised bed operation enables extractions directly from unclarified broths.

ACKNOWLEDGEMENTS

We would like to thank staff at Advanced Protein Products Ltd. (Birmingham UK) and at C.A.M.R. (Porton Down UK) at whose establishments much of the work described in this paper was performed under contract.

REFERENCES

1. Kanekanian ADA, Lewis MJ (1986) Dev Food Proteins 4:135.
2. Low DKR (1986) in Stowell JD et al (eds): "Bioactive microbial products 3: Downstream processing" Academic Press (ISBN 0 12 672960 3) pp 121-145.
3. Cueille R, Tayot J-L (1985) in "World Biotech Report 1985 Vol. 1" p 141.
4. Hou KC, Cogswell G Eur Pat Appl 180,766
5. McGregor WC, Szesko DP, Mandaro RM, Rai VR (1986) Bio/Technology 4:526.
6. Bite MG (1986) in Conf Procs "Eurochem '86: Process chemistry today" p 137.
7. Miles BJ Br Pat 1,586,364.
8. Shakespeare M, Terry M (1987) Sterling Organics UK contract research report APP/6.
9. Biscans B, Sanchez V, Couderc JP (1985) Entropie no. 125/126:17.
10. Scawen M (1986) Sterling Organics UK contract research report CAMR/3,4.
11. Anon (1986) Bioprocess Technol 8:1.

Protein Purification: Micro to Macro, pages 255–261

PURIFICATION AND STRUCTURAL CHARACTERIZATION OF INSULIN FROM THE HOLOCEPHALAN FISHES: THE RATFISH AND RABBIT FISH[1]

J.Michael Conlon and Lars Thim

Clinical Research Group for Gastrointestinal Endocrinology of the Max-Planck-Gesellschaft at the University of Göttingen, FRG and Novo Research Institute, Bagsvaerd, Denmark

ABSTRACT A general scheme for the purification of novel insulins is illustrated by the isolation of the hormone from the pancreas of two Holocephalan fishes: _Hydrolagus colliei_ (Pacific ratfish) and _Chimaera monstrosa_ (rabbit fish). The purification procedure involves (1) extraction of tissue with ethanol/0.7 M HCl (3:1 vol/vol) at 4°C (2) concentration of proteins/peptides on Sep-pak C18 cartridges (3) gel filtration on Sephadex G-50 and (4) successive reverse-phase HPLC separations using octadecyldimethylsilylsilica (C18) and diphenylmethylsilylsilica columns. After reduction with dithiothreitol and derivatization of the cysteine residues with 4-vinylpyridine, the A- and B-chains of insulin are separated on a C18 reverse-phase HPLC column. Amino acid sequence analysis of the intact A- and B-chains and of peptide fragments produced by digestion with _Armillaria mellea_ protease is accomplished by automated gas-phase Edman degradation. The insulins from the ratfish and the rabbit fish are probably identical and contain 21 amino acid residues in the A-chain and 37 residues in the B-chain.

INTRODUCTION

The primary structure of insulin has been determined (or may be deduced from the nucleotide sequence of the gene) for at least 30 different vertebrate species (Dayhoff protein sequence data base, National Biomedical Research

[1] This work was supported by the Stiftung Volkswagenwerk

Foundation, Georgetown, Washington DC). Examples of insulins from primitive species include the Atlantic hagfish (Cyclostome)(1), Torpedo marmorata (Elasmobranch)(2) and the spiny dogfish (Elasmobranch)(3). The isolation of a prothoracicotropic hormone with 40% homology to human insulin from the silkworm, Bombyx mori (4) suggests that genes encoding the insulin family of polypeptides may have arisen before the evolution of the vertebrates. The Holocephalan fishes are phylogenetically related to the Elasmobranchian fishes but diverged from the line of evolution leading to contemporary sharks, rays and skates at least 250 million years ago. In the present day, the Holocephalan fishes are limited to three families: Hydrolagus (ratfishes), Chimaera (rabbit fishes) and Callorhynchus (elephant fishes). This report describes the chromatographic procedures used to purify insulins from two Holocephalan fishes, Hydrolagus colliei and Chimaera monstrosa, and the strategy employed to elucidate the primary structures of the peptides.

METHODS AND RESULTS

Ratfish were collected at Bamfield Marine Biology Station, Vancouver Island, Canada and rabbit fish at Bergen Marine Station, Norway. The approach used to purify insulins from these species is outlined in Figure 1. A full account of the extraction of ratfish pancreas and the isolation of ratfish insulin has been provided (5). The gel filtration step was included to remove high and low molecular weight components from the extract so as not to overload the HPLC columns. The preferred column packing material for the first HPLC separation was wide-pore (300 Å) Vydac-218 TPB (5μ) supplied by The Separations Group, Hesperia, CA. This material gave good resolution, peak shape and a high recovery of insulin. A Supelcosil LC-18-DB column (deactivated octadecyldimethylsilylsilica, 5μ), supplied by Supelco Inc. Bellefonte, PA, also provided good resolving power but slightly lower recoveries of insulin. The insulins were purified to homogeneity using a Supelcosil LC-3DP (diphenylmethylsilylsilica, 5μ) column. Contaminants that co-eluted with insulin during reverse-phase HPLC on a C18 column were generally separated on a phenyl column. All HPLC separations were carried out using gentle linear gradients formed from 0.1%(vol/vol) trifluoroacetic acid and acetonitrile. The use of volatile solvents ensured

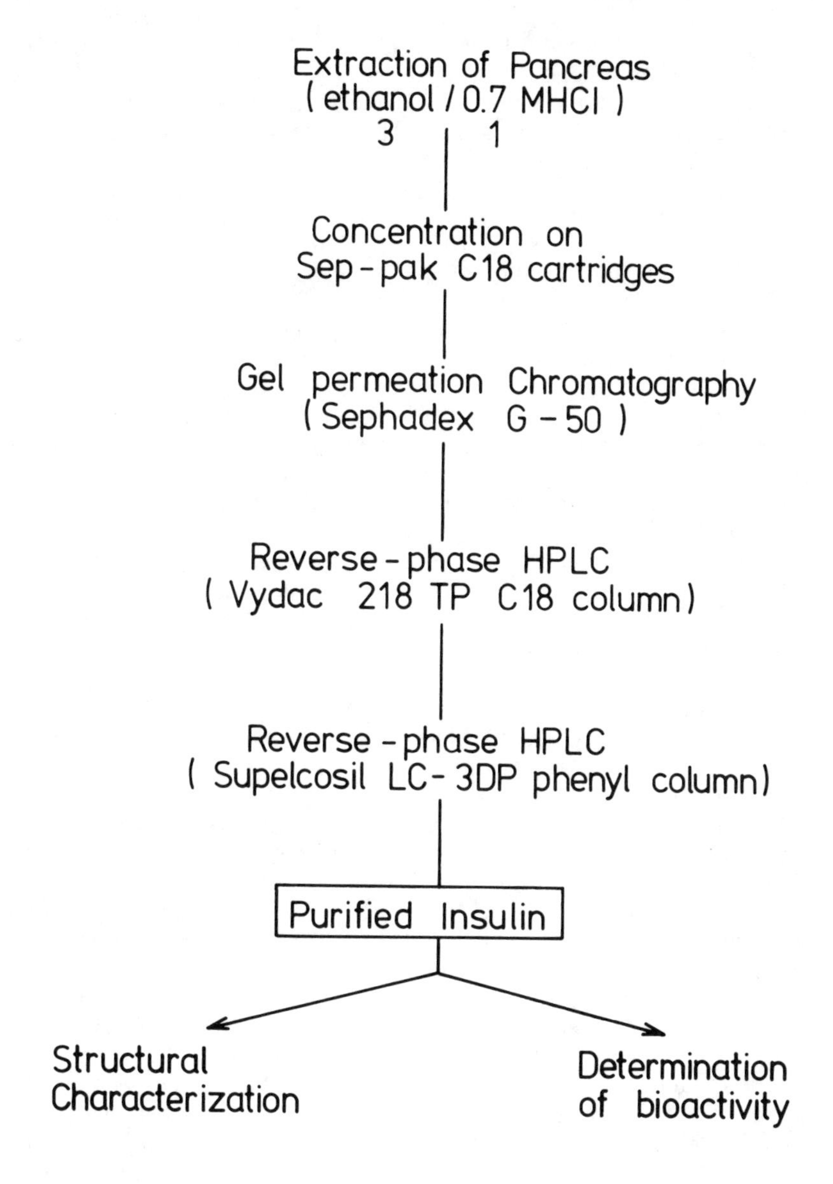

FIGURE 1. General scheme for the purification of novel insulins

that the final product was salt-free. Ratfish and rabbit fish insulins showed negligible immunoreactivity in a radioimmunoassay using an antiserum raised against porcine insulin but identification was facilitated by the fact that their retention times and the ratio of absorbance at 214nm and 280nm did not differ markedly from a porcine insulin calibration standard. The yield of ratfish insulin was approximately 60nmol from 100g of pancreas and the yield of rabbit fish insulin was 6nmol from 7g pancreas.

Structural characterization of ratfish and rabbit fish insulins.

The general strategy for determination of the primary structures of the insulins is outlined in Figure 2. Three methods were investigated for the preparation of the free A- and B-chains of insulin: (1) oxidation with performic acid (2) reduction with dithiothreitol and derivatization of the cysteine residues with iodoacetamide and (3) reduction with dithiothreitol and derivatization with 4-vinylpyridine. The reactions in (3) proceeded in quantitative yield and the pyridylethylated A- and B-chains were readily separated by chromatography on a C18 column. The amino acid compositions of the A- and B-chains of ratfish and rabbit fish insulins indicated that the structure of the hormone was probably the same in both species with 21 residues in the A-chain and 37 residues in the B-chain.

Automated Edman degradation of ratfish A-chain (10nmol) ratfish B-chain (6nmol), rabbit fish A-chain (4nmol) and rabbit fish B-chain (4nmol) was carried out using an Applied Biosystems model 470 A gas phase sequencer. The average repetitive yield during the determinations ranged from 90 to 93% and the detection limit for phenylthiohydantoin amino acids was 0.5pmol (6). The full sequence of the A-chain was obtained but unambigous assignation of only 34 residues of the B-chain was possible (Figure 3). The amino acid sequences of the peptides from ratfish and rabbit fish were identical and the discrepancy between the composition and sequence analysis data indicated that the B-chain of both insulins was further extended by (Glx Pro Leu).Treatment of the derivatized ratfish B-chain with *Armillaria mellea* protease (7), an enzyme that cleaves at the NH_2-terminal side of lysyl residues, generated a nonapeptide which terminated in Glu Pro Leu.

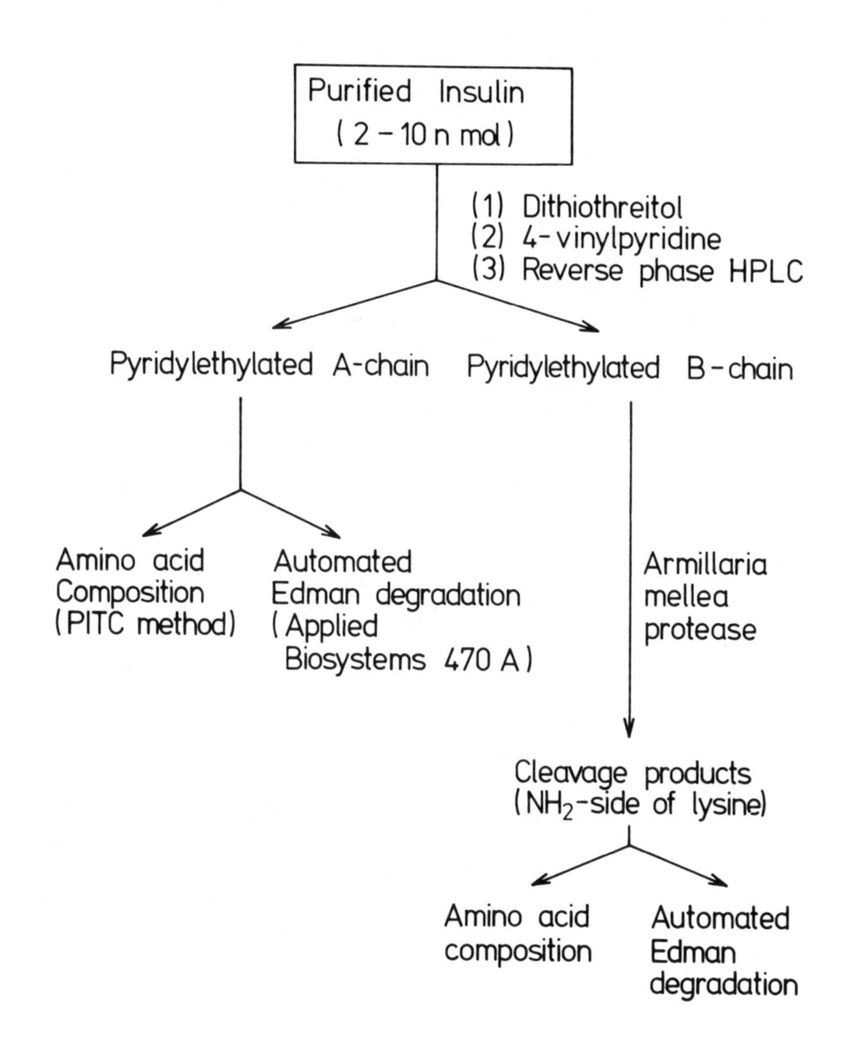

FIGURE 2. General scheme for the structural characterization of novel insulins

```
A-chain
                    5         10         15         20
  G I V E Q C C H N T C S L A N L E G Y C N
  → → → → → → → → → → → → → → → → → → → → →

B-chain
                    5         10         15         20         25
  V P T Q R L C G S H L V D A L Y F V C G E R G F F Y S P K
  → → → → → → → → → → → → → → → → → → → → → → → → → → → → →
                                                            ⇀
 30        35
  P I R E L E P L
  → → → → →
  ⇀ ⇀ ⇀ ⇀ ⇀ ⇀ ⇀ ⇀
```

FIGURE 3. Primary structures of the A-chain and B-chain of insulins from the ratfish and rabbit fish. The structures are the same in both species. The sequences were deduced from Edman degradation of the intact peptides (→) and an *A. mellea* protease fragment (⇀).

DISCUSSION

A general strategy has been outlined for the isolation and characterization of novel insulins that has been applied to purification and elucidation of the primary structure of insulin from the ratfish and rabbit fish. The strategy has also been used, with minor modifications, to characterize insulins from the Elasmobranchian fish, *Torpedo marmorata* (2), from the Teleostean fish, *Platichthys flesus* (Flounder) (8) and from human and rat tumours (unpublished data). The A-chain of the Holocephalan insulins shows strong homology (86%) to the A-chain of Torpedo insulin and, in common with all mammalian insulins yet studied except that of the coypu, has 21 amino acid residues. The primary structure of the B-chain suggests an unusual mode of proinsulin processing. In all species yet studied, proinsulin is converted to insulin and the C-peptide of proinsulin by specific proteolytic cleavages at sites of pairs of basic amino acid residues. The COOH-terminal region of ratfish B-chain (Ile^{31} Arg^{32} Glu^{33} Leu^{34} Glu^{35} Pro^{36} Leu^{37}) shows homology to the NH_2-terminal region of human C-peptide (Arg Arg Glu^{1} Ala^{2} Glu^{3} Asp^{4} Leu^{5}). The substitution of arginine in mammalian insulin by isoleucine in ratfish at B31 requires only a single base change in the corresponding region of the DNA (AUA for AGA). It is proposed, therefore, that the increased length of the Holocephalan insulin B-chain is a consequence of a mutation at the processing site linking the B-chain to the C-peptide.

The processing enzyme is unable to cleave at the monobasic site and so an alternative cleavage takes place within the C-peptide region of proinsulin. Preliminary experiments indicate that the extension to the B-chain does not markedly reduce the binding affinity of ratfish insulin for receptor sites on rat fat cells compared with human insulin.

ACKNOWLEDGEMENTS

The authors thank Dr. E.Dafgård, Dr. U.Askensken and Professor S.Falkmer, Karolinska Institute, Stockholm, Sweden for providing the ratfish and rabbit fish pancreata.

REFERENCES

1. Chan SJ, Emdin SO, Kwok SCM, Kramer JM, Falkmer S, Steiner DF (1981). Messenger RNA sequence and primary structure of preproinsulin in a primitive vertebrate, The Atlantic hagfish. J Biol Chem 256:7595.
2. Conlon JM, Thim L (1986). Primary structure of insulin and a truncated C-peptide from an Elasmobranchian Fish, Torpedo marmorata. Gen Comp Endocrinol 64:199.
3. Bajaj M B undell TL, Pitts JE, Wood SP, Tatnell MA, Falkmer S, Emdin SO, Gowan LK, Crow H, Schwabe C, Wollmer A, Strassburger W (1983). Dogfish insulin: primary structure, conformation and biological properties of an elasmobranchian insulin. Eur J Biochem 135: 535.
4. Nagasawa H, Kataoka H, Isogai A, Tamura S, Suzuki A, Mizoguchi A, Fujiwara Y, Suzuki A, Takabashi SY, Ishizaki H (1986). Amino acid sequence of a prothoracicotropic hormone of the silkworm Bombyx mori. Proc Natl Acad Sci USA 83:5840.
5. Conlon JM, Dafgård E, Falkmer S, Thim L (1986). The primary structure of ratfish insulin reveals an unusual mode of proinsulin processing. FEBS Lett 208:445.
6. Thim L, Hansen MT, Sørensen AR (1987). Secretion of human insulin by a transformed yeast cell. FEBS Lett 212:307.
7. Lewis G, Basford JM, Walton PL (1978). Specificity and inhibition studies of Armillaria mellea protease. Biochim Biophys Acta 522:551.
8. Conlon JM, Davis MS, Thim L (1987). Primary structure of insulin and glucagon from the flounder (Platichthys flesus). Gen Comp Endocrinol 66:in press.

Protein Purification: Micro to Macro, pages 263–269

Preparative Purification of Viral Polypeptides from Mouse Mammary Tumor Virus

Joseph K.K. Li[1,2],
Thomas Mercolino[2] and Jeffrey Bruton[2].

[1]Department of Biology, Utah State University, Logan, Utah 84322; [2]Becton Dickinson Research Center, Research Triangle Park, North Carolina 27709.

Abstract

Gradient-purified Mouse Mammary Tumor Virus (MMTV), a B-type retrovirus, was disrupted with 1% Triton X-100 in the presence of a high salt (0.4 M KCl) and high pH (9.2) lysis buffer. All of the seven viral structural proteins were then purified to near homogeneity and with good yield by sequential affinity and ion-exchange chromatography. Unlike the traditional approach, the smaller molecular weight viral proteins, P14, P12 and P10 were first purified by oligo-d(T) column chromatography, a technique which significantly reduced the loss of these minor proteins. All seven MMTV polypeptides, GP52, GP36, P28, P23, P14, P12 and P10 were isolated to near homogeneity as revealed by silver staining and autoradiography of SDS-PAGE. High titer monospecific polyclonal antibodies were produced against all of the isolated MMTV proteins except P12 and P10. These antisera show little or no cross-reactivity among all the isolated MMTV proteins in radioimmunoassays.

Introduction

The congenitally transmitted B-type retrovirus, mouse mammary tumor virus (MMTV), is the known causative agent for mammary carcinomas in mice (1). Many biological properties characteristic of MMTV have also been found in those particles present in human breast cancer tissues (2). These reports suggest that both mice and humans might harbor provirus, which is clearly under strict control and tight regulation. The recent reports on AIDS viruses further indicate that humans may occasionally become infected with horizontally transmitted exogenous retroviruses (3). Intensive efforts have been made to isolate and identify these retroviruses and their viral polypeptides in order to determine their structural and functional relationships, to assess their oncogenic potentials and their possible contribution to cell transformation, malignancy and metastasis.

The virion of MMTV consists of a nucleoprotein core, a core shell and a lipid-containing membrane envelope whose outer surface is covered with projections. Seven structural polypeptides have been identified and are designated as GP52, GP36, P28, P23, P14, P12 and P10 based on the relative molecular weights from gel electrophoresis (1). Several studies have reported partial success in the purification of GP52, P28 and P12. However, only low titer antiserum against GP52 and P28 were produced (4,5,6,7). This report concerns the development of a different approach to isolate to near homogeneity all the seven MMTV structural viral polypeptides successfully from the same batch of MMTV. The large yield of native viral proteins has led to the production of high titer antisera against all MMTV proteins except P12 and P10.

Methods

Viral Protein Purification: MMTV grown in C3H mice and isopycnically banded in a sucrose gradient was collected, pooled and pelleted by centrifugation. The virus pellet (50 mg) was incubated in a high salt and high pH extraction buffer (5 mM Tris-HCl, pH 9.2, 0.4 M KCl, 1 mM EDTA and 1% Triton X-100) at 37C for 10 min. The extract was centrifuged for 10 min in a microfuge and the supernatant was dialyzed against 10 mM Tris-HCl (pH 7.5), 1 mM EDTA, 50 mM KCl and 0.1% Triton X-100 before used for viral protein purification. The dialysate was applied to an oligo-d(T) cellulose column pre-equilibrated with the dialysis buffer minus 50 mM KCl. The flow-through fractions were collected and saved before the absorbed viral proteins were eluted with a 0.1 - 1.0 M KCl gradient in the same buffer.

The flow-through from the oligo-d(T) column was dialyzed against phosphate-buffered saline (PBS, pH 7.2) and then applied to a lentil lectin-sepharose 4B column. The flow-through was saved. The column-bound viral glycoproteins were eluted with PBS containing Ø.2 M alpha-methylmannose and 1 M NaCl.

The flow-through from lentil lectin column was dialyzed against 1Ø mM BES (pH 6.5) and 1 mM EDTA. The dialysate was then applied to a phosphocellulose (PC) column. Viral proteins were eluted with a Ø.1 - 1.Ø M KCl gradient in the same buffer. The fractions containing the co-purified GP36, P28 and P23 from the PC column were pooled and dialyzed against 1Ø mM Tris-HCl (pH 7.1) and 1 mM EDTA. They were then separated by a heparin column using a Ø.1 - 1.Ø M NaCl gradient. The co-purified P28 and P23 were rechromatographed through another heparin column using a shallower gradient of Ø.1 - Ø.4 M NaCl.

Antisera production and characterization: Animals (goats and rabbits) were immunized with purified MMTV viral proteins that had been emulsified with adjuvants as described by Strand and August (8). Animals were boostered and antiserum were collected biweekly. The titers of the monospecific polyclonal antisera were determined by radioimmunoprecipitation (RIP) using radioiodinated viral polypeptides as substrate prepared according to Li and Fox (9).

Results

A high salt (Ø.4 M KCl) and high pH (9.2) lysis buffer facilitated the dissocation of the viral proteins of MMTV as well as ten other retroviruses we have tested (unpublished data). Fewer protein complexes were formed, as indicated by the reduced cross-contamination of viral proteins in the fractions of different columns used for their purification. All the previously published methods (4-7) failed to isolate the minor MMTV structural proteins P14, P12 and P1Ø. GP36 and P23 had also never been isolated and purified. To minimize loss of these minor structural proteins, we isolated them with an oligo-d(T) column first before purifying the major core proteins and glycoproteins. All of the major core (P28 and P23) and glycoproteins (GP52 and GP36) did not bind to the oligo-d(T) column and were found in the flow-through or wash fractions. However, the three minor components, P14, P12 and P1Ø showed different affinities and were eluted using a Ø.1 - 1.Ø M KCl gradient in a buffer containing 2Ø mM Tris-HCl (pH 7.5), 1ØØ mM NaCl and 1 mM EDTA. P12, P1Ø and P14 were eluted at Ø.1, Ø.3 and Ø.5 M KCl, respectively.

When the flow-through fractions from the oligo-d(T) column were applied to the lentil lectin column, GP52, but not GP36, was selectively absorbed and was eluted with PBS containing Ø.2 M alpha-methylmannose and 1 M NaCl. The flow-through fractions were chromatographed through a PC column. GP36, P28 and P23 were always co-eluted between Ø.4 - Ø.55 M KCl of the Ø.1 - 1.Ø M KCl gradient. However, a new peak of P12 was obtained between Ø.55 - Ø.65 M KCl. After the co-purified GP36, P28 and P23 were dialyzed, they were applied to a heparin column. GP36 was not absorbed and could be recovered in the flow-through fractions in pure form. Using a Ø - Ø.4 M NaCl gradient, P28 and P23 were selectively eluted at Ø.1 M and Ø.18 M NaCl, respectively.

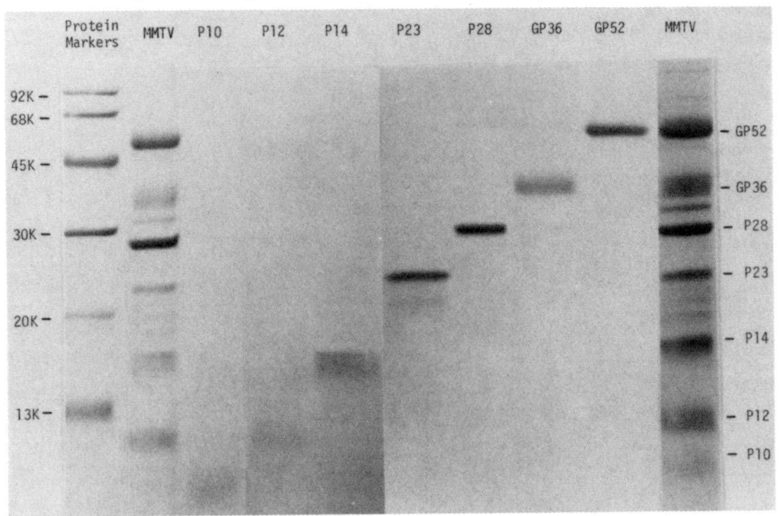

Fig. 1. SDS-PAGE Analysis of Isolated MMTV Polypeptides.

The purity of the individually isolated MMTV viral structural proteins was analyzed by SDS-PAGE (9) and by silver staining of the gel (10). The results indicated that all the seven MMTV structural proteins have been purified to a homogeneity exceeding 95% (Figure 1).

Antisera against these seven MMTV proteins were analyzed and characterized by RIP using radioiodinated polypeptides as substrates. Animals immunized with GP52, GP36 and P28 produced very high titer antisera. P23 and P14 were slightly less immunogenic and their titers were approximately 4-10X less than that of GP36. However, rabbits produced no detectable antisera against P12 and P10 even after receiving 10-12 boosters in a period of 6 months. Table 1 summarizes the antiserum titers and monospecificities of the serum produced against five of the seven MMTV structural proteins.

Table 1. Summary of radioimmunoassays of monospecific antibody against five isolated proteins Mouse Mammary Tumor Virus

M M T V Proteins	Antiserum Titer*				
	Anti-GP52	Anti-GP36	Anti-P28	Anti-P23	Anti-P14
P14	50	50	50	50	12,000
P23	50	50	50	5,000	50
P28	50	50	18,000	200	50
GP36	50	96,000	400	50	50
GP52	48,000	50	50	50	50

*Titer represents antiserum dilutions which can precipitate 50% of 50 ng of radioiodinated viral proteins. The titer denotes average of triplicates.

Discussion

The new method described here shows that all of the seven MMTV structural proteins can be isolated and purified to greater than 95% homogeneity as determined by silver staining of SDS-PAGE (Fig. 1). This is also the first time that GP36, P23, P14 and P1Ø have been isolated and purified. The large yield of native proteins has permitted us to produce very high titer monospecific antisera. Radioimmunoassays and RIP revealed little or no cross-reactivity among these MMTV viral proteins (Table 1).

The inability to dissociate the MMTV virions and the uncontrollable loss of the minor viral components are the major reasons why previously published methods (4-7) failed to obtain any GP36, P23, P14, P12 and P1Ø. We attribute our success to both the use of lysis buffer containing high salt and high pH and the initial use of oligo-d(T) column. It has been known that smaller molecular weight proteins are usually and comparatively less antigenic than larger molecular weight ones. This might explain why we were unable to produce antisera against P12 and P1Ø. However, our successful isolation of all these seven MMTV proteins will permit us to study some of their structural and functional relationships, to determine both their amino acid composition and sequences and to investigate their oncogenic potentials to cell transformation and malignancy.

Acknowledgement

We are grateful to Drs. Wolfgang Joklik, Dani Bolognesi, Tom Vanaman, Ralph Smith, Mitte Strand for timely discussions and suggestions. James Schram, Robert Campbell, Teri McClusky, Lilly Lou and Shawn Millinder have provided excellent technical assistance. Most of the MMTV was provided by John Cole III of National Cancer Institute (NCI). This work was supported in part by Grant CP-91ØØ4 (NCI) and Utah AES Projects 537 and 538.

References

1. Moore DH, Long CA, Vaidya AB, Sheffield JB, Dion AS, Lasfargues EY (1979) Mammary tumor viruses. Adv Cancer Res 29:347
2. Ohno T, Spiegelman S (1977) Purification and characterization of the DNA polymerase of human breast cancer particles. Proc Natl Acad Sci USA 74:764
3. Gallo R (1987) The Aids virus. Scientific American Jan:47
4. Park WP, Howk RS, Scolnick EM, Oroszlan S, Gilden RV (1974) Immunochemical characterization of two major polypeptides from murine mammary tumor virus. J Virol 13:1200
5. Sarkar NH, Dion AS (1975) Polypeptides of the mouse mammary tumor virus. I. Characterization of two group-specific antigens. Virology 64:471
6. Ritzi E, Baldi A, Spiegelman S (1976) The purification of a gs antigen of the murine mammary tumor virus and its quantitation by radioimmunoassay. Virology 75:188
7. Westenbrink F, Koornstra W, Bentvelzen P (1977) The major polypeptides of the murine mammary tumor virus isolated by plant-lectin affinity chromatography Eur J Biochem 76:85
8. Strand M and August T (1974) Structural proteins of ribonucleic acid tumor viruses. J Biol Chem 251:4859
9. Li JK-K and Fox CF (1975) Radioiodination of the envelope proteins of Newcastle disease virus. J Supramol Structure 3:51
10. Merril CR, Goldman D, Sedman SA, Ebert MH (1981) Ultrasensitive stain for P proteins in polyacrylamide gels shows regional variation in cerebrospinal fluid proteins. Science 227:1437

Protein Purification: Micro to Macro, pages 271–278

BACTERIOCINS FROM HALOBACTERIA[1]

Ursula Rdest and Margarete Sturm

Institut für Genetik und Mikrobiologie,
Universität Würzburg, D-8700 Würzburg, West-Germany

ABSTRACT Some halobacterial strains secrete bacteriocinogenic agents (halocins) into the medium. Hal R1, the halocin from strain Halobacterium spp. GN101, is strictly specific for halobacteria (not for eubacteria or eukaryotes) and exhibits a bacteriostatic effect on growing cells. Hal R1 is sensitive to proteinases but relatively resistant to heat (only 50 % inactivation by boiling for 1 h). Analysis on gel filtration columns (sephacryl S200 or superose) shows some of the halocin activity in the void volume and some corresponding to the relative molecular mass (Mr) of 6.5 kDalton (kD). This seems to be due to formation of large aggregates since in chromatography on silica thin-layer in a basic eluent halocin activity migrates shortly below the front. The data indicate that Hal R1 is a peptide of Mr = 6.2 kD possibly complexed with a larger protein.

INTRODUCTION

Bacteriocins are bactericidal substances which are distinguished from other antimicrobial agents by the presence of an essential protein moiety and a narrow inhibitory spectrum for closely related organisms. A variety of examples of antagonism between particular bacterial strains are described for eubacteria (1) with a wide diversity in the

[1]This work was supported by a grant from the Deutsche Forschungsgemeinschaft (Goe 168/12-3, Rd 1/4-4).

mode of action.

On the basis of several characteristics halobacteria have been recently included in the primary kingdom archaebacteria whose phylogenetic relationship with eubacteria and eukaryotes is still poorly defined. In certain aspects they resemble eubacteria whereas in others they resemble eukaryotes (2). Meseguer et al. (3) first described the production of bacteriocins by halobacteria, termed them halocins and characterized a protein (halocin H4) of 28 kD from strain Halobacterium mediterranei which exhibited a bacteriolytic effect on other halobacteria.

The present report describes the purification and characterization of halocin produced by strain Halobacterium spp. GN101. The bacteriocin, designated Hal R1 is not inactivated in low-salt, has an inhibitory effect for a broad range of halobacteria, is of proteinaceous nature and exhibits a bacteriostatic effect on sensitive cells. A second halocin Hal R2 was characterized from Halobacterium spp. TuA4. It closely resembles Hal R1 with respect to its size and its biological properties, although the two strains are not closely related and were isolated from different locations (GN101 from Mexico, TuA4 from Tunesia).

METHODS

Bacterial strains and growth conditions. The halobacteria used in this study were isolated from natural sources which were described before (4). All halobacteria strains were grown in salt medium, consisting of 4 M NaCl, 0.12 M $MgSO_4$, 0,03 M KCl, 0.01 M trisodium citrate, 1 % peptone (Oxoid) pH 7.2 with shaking and illumination at 37°C for 7 days. Agar plates contained salt medium and 1.5 % agar.

Bacteriocin activity was determined by a modified critical dilution method (5). 1 l salt medium agar was supplemented with 8 ml stationary phase H. halobium cells before pouring the plates. The bacteriocin preparation was serially diluted in 0.01 M sodium phosphate buffer pH 7.0 and 10 ul of each dilution were spotted onto the agar plates containing H. halobium. The lysis zones could be seen after 2 days of incubation at 37°C. Bacteriocin activity is expressed as arbitrary units per milliliter (AU/ml) where an arbitrary unit is the reciprocal of the highest dilution of bacteriocin that yields a visible zone of inhibition of cell growth.

Isolation and purification of Hal R1. 10 l of culture supernatant were concentrated 100fold by a Diaflow Hollow

Fiber Cartridge HP10-20 from Amicon. NaCl was removed by dialysis overnight, dialyzed material was freeze-dried and resuspended in 0.01 M sodium phosphate buffer pH 7.0. Final concentration was 2000fold.

Materials. Sephacryl S200 and the prepacked FPLC Column Superose 12HR10/3 0 were obtained from Pharmacia. Column fractions were concentrated by centricon-10 microconcentrators, 10 000 Mr cutoff from Amicon.

Polyacrylamide gel electrophoresis was performed by the method of Laemmli (6). A kit for molecular weight standards (MW-SDS 17) from Sigma was used.

RESULTS AND DISCUSSION

Detection of halocins synthesized by halobactia.

Fourteen different halobacteria isolates were surveyed for excretion of and sensitivity to halocins. The strains were previously defined by genetic and physiological criteria (4). The most potent halocins were obtained from *Halobacterium* spp. GN101, spp. TuA4 and spp. LaPaz which caused mutual growth inhibition. The red colour of *H. halobium* (hal^-) made this strain a good indicator for halocin. The bacteriocinogenic effect was restricted to halobacteria. The most active halocin was that of strain GN101. This strain in culture was affected by its own halocin. This could be due to incomplete immunity of this strain to its own bacteriocin. Alternatively the production of halocin may be a lethal event for the producing cell as was shown for certain colicins (7). All further investigations presented here were carried out with this halocin (Hal R1) from strain GN 101 and a closely related one termed Hal R2 from strain TuA4.

Stability of Hal R1.

Since bacteriocin activity was mainly found in the supernatant, Hal R1 was isolated from culture medium. When culture medium was dialyzed against 10 mM sodium phosphate buffer, pH 7.0 and assayed by the twofold critical dilution method 80 % loss of halocin activity was observed. No activity was found in the dialysis buffer which is probably due to dilution effect. A film of the dialysis bag placed on the agar surface of an indicator plate allowed active material applied on top to pass through the membrane and to inhibit growth of the indicator strain. This was not the case when a Spectraphor dialysis membrane with a cutoff of 3.5 kD was used.

Digestion of halocin with proteinase K, pronase P and elastase destroyed the Hal R1 activity, whereas more specific proteases like papain, trypsin or thermolysin as well as DNase or RNase did not affect its activity.

Isolated halocin showed no decrease of activity over the pH range of 2-12. Incubation of halocin at 25°C or 37°C for up to 24 h did not affect the activity and boiling for 1 h reduced activity only by 50 %. Organic solvents like n-propanol, acetonitrile, trifluoracetic acid or methanol did not inhibit halocin activity when added to dialysed material (equal volumes, 1 h mixing at RT). Meseguer et al. (8) described the adsorbtion of halocin H4 to the target cells with a concomittant inactivation of H4. This is not observed for Hal R1. Incubation of crude preparation of Hal R1 (12 800 AU/ml) together with 4×10^8 cells/ml for 24 h at 37°C under shaking did not decrease Hal R1 activity.

Effect of Hal R1 on sensitive halobacterial cells.

To determine whether halocin had a bacteriostatic, bactericidal or bacteriolytic effect on halobacteria, viability and lysis of the indicator strain were monitored in the presence of halocin. Stationary phase cells did not show a decrease of optical densitiy (A_{600}) after treatment with Hal R1 (1280 AU/ml) over a period of seven days. Morphological changes of the treated halobacterial cells were also not observed. Addition of Hal R1 to a fully grown lawn of _H. halobium_ on solid medium did not lyse the cells. Hal R1 affected only growing cells (Fig. 1). The addition of increasing amounts of Hal R1 resulted in proportional inhibition of growth. A concentration of 2560 AU/ml inhibited growth of H. halobium completely. After diluting the culture containing Hal R1 1:100 with medium cell growth resumed after a lag phase of four days. No Hal R1 resistant mutants could be selected from these Hal R1-treated cultures. Thus Hal R1 seems to exhibit a bacteriostatic effect on growing sensitive cells but no irreversible killing although lysis of a certain amount of these cells cannot be excluded.

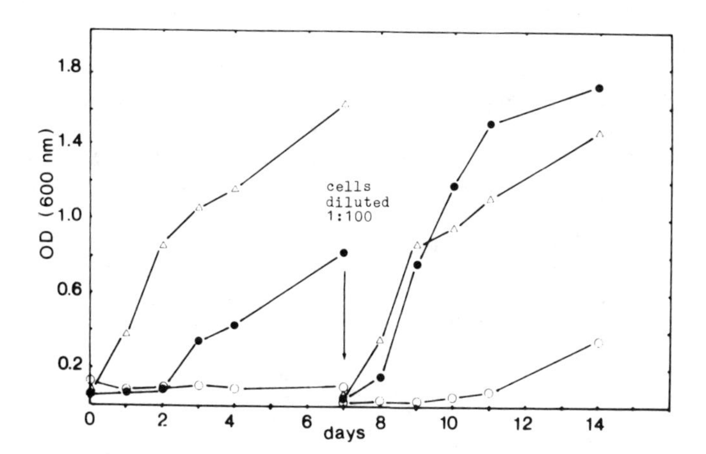

Figure 1. Growth curve of Halobacterium halobium beginning with 4 x 10^6 cells/ml. Cell growth is represented by the optical density at 600 nm. In the absence of halocin (△), in the presence of 25 AU halocin/ml (●) and of 2560 AU halocin/ml cells (○). On day seven the cultures were diluted 1:100 with medium.

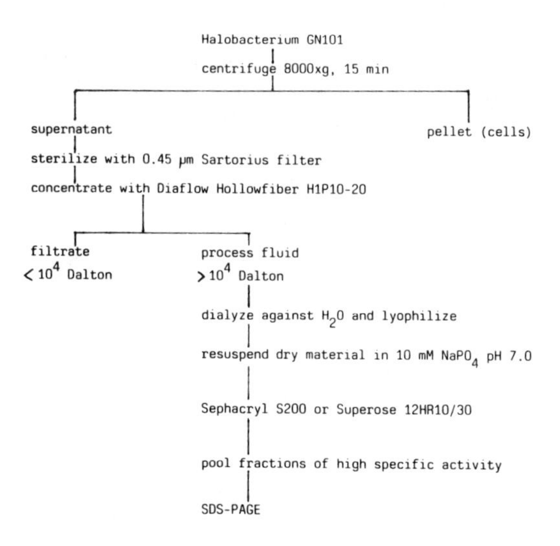

Figure 2. Purification of halocin Hal R1 from Halobacterium spp. GN101.

Purification of Hal R1.

The procedure for purification of Hal R1 is summarized in Fig. 2. The activity of concentrated material eluted mostly in the void volume V_0 (pool 1) from a sephacryl S200 (size exclusion Mr = 200kD) or a superose 12HR10/30 (FPLC) column and to a lesser extend (pool 2) at a position corresponding to a protein of low molecular mass (Mr = 6.5 kD) (Fig. 3). The large complex structure may be due to specific or unspecific association of Hal R1 with the

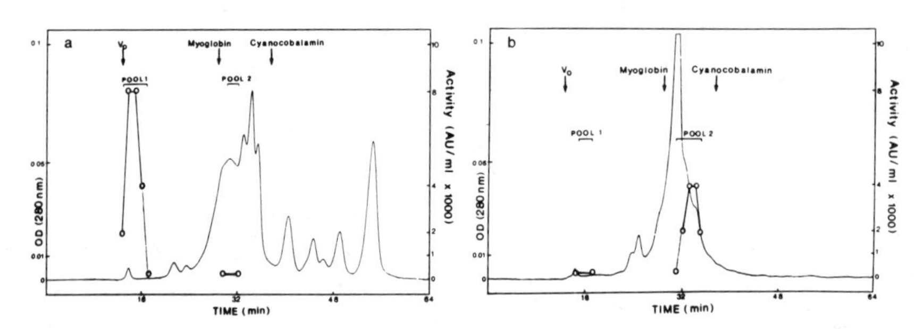

Figure 3. Gel filtration of concentrated material from (a) GN101 (Hal R1) and (b) TuA4 (Hal R2) on a superose 12 column. A 4 ul sample containing 20 000 AU (Hal R1), 10 000 AU (Hal R2) was injected onto the column and eluted in 0.5 ml fractions at a flow rate of 0.5 ml/min with 50 mM $NaPO_4$-0.15 M NaCl pH 7.2. Activity was recovered in pool 1 and pool 2. Marker proteins are myoglobin (17 500) and cyanocobalamin (1355).

membrane or membrane components. Bacterioruberin, a carotinoid found in most halobacterial membranes is present in all column fractions corresponding to pool 1.

The formation of larger aggregates has been described for the bacteriocin from *Bacteroides fragilis* (9). This bacteriocin has a Mr of 5 kD in its monomeric active form and was recovered from culture medium as a complex larger than $2x10^7$ D.

When the halocin-active fraction from pool 1 was analysed by 10 % SDS-PAGE only some faint protein bands were detected (Fig. 4). In 18 % SDS-PAGE however a major band corresponding to a protein of 6.2 kD and 4-5 minor bands were

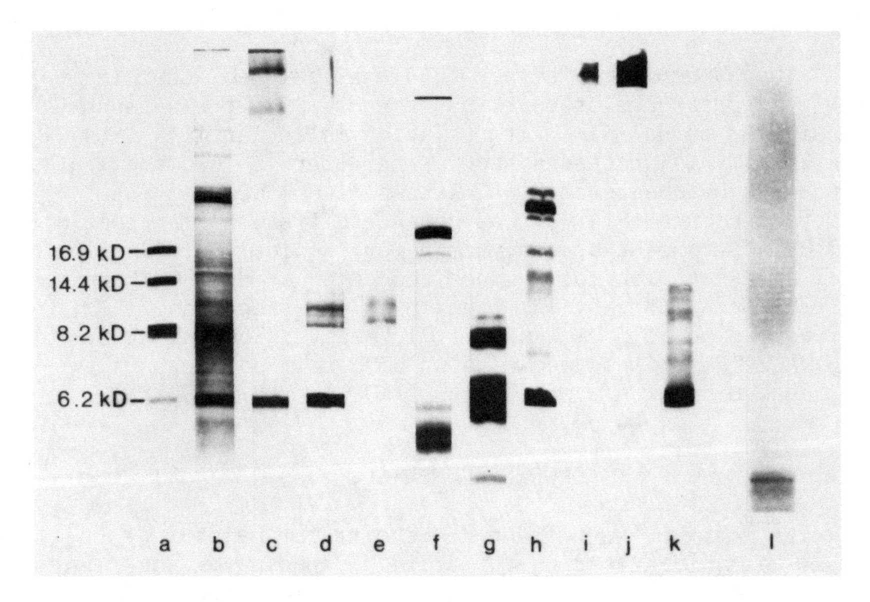

Figure 4. 18 % SDS-PAGE stained with silver stain. Mr markers (a), extract from GN101 supernatant (b-e): crude Hal R1 (b), extract from thin-layer plate (c), pool 1 (d) and pool 2 (e) from superose column. Extract from TuA4 supernatant (f-g): pool 1 (f) and pool 2 (g) from superose column. Protease treated Hal R1 (h-k): trypsin (h), pronase P (i), proteinase K (j), untreated control (k). 10 % SDS-PAGE, crude extract from GN101 supernatant (l).

visible. The same major band is obtained from Hal R1 active material extracted from thin-layer plates (a basic eluent was used, $CHCl_3:CH_3OH:NH_3 = 2:2:1$). From pool 2 this protein band could not be visualized unless the halocin preparation was treated with 6M guanidinium hydrochloride prior to sephacryl column chromatography (data not shown). By cutting the 18 % gel horizontally into 10 mm slices and eluting the proteins by diffusion, halocin activity could be detected only in the slice corresponding to the band of Mr = 6.2 kD. From a 10 % gel the activity was only extracted in the protein front.

The same proteases which inactivate Hal R1 (proteinase K, pronase P) destroyed also the 6.2 kD protein, whereas treatment of the 6.2 kD protein with papain or trypsin which did not inactivate Hal R1 did also not change the size of the 6.2 kD protein (Fig. 4).

Halocin Hal R2 from H. spp. TuA4

The halocin from strain TuA4 was isolated according to the same procedure as described for Hal R1. It was found to be similar to Hal R1 in its relative molecular mass, its sensitivity to proteases and in its effect on growing sensitive halobacteria. The elution of Hal R2 from the superose column (Fig. 3) was however different from that of Hal R1. The amount of active monomeric halocin eluting in pool 2 was considerably higher than that of Hal R1 whereas less halocin activity is detected in the fractions corresponding to pool 1 of Hal R1. Analysis of Hal R2 by 18 % SDS-PAGE (Fig. 4) showed a major band from pool 2 corresponding to a protein of 6.2 kD.

ACKNOWLEDGEMENTS

We thank W. Goebel and W. Schmitt for helpful discussion, S. Kathariou for critical reading of the manuscript and E. Appel for her help in its preparation.

REFERENCES

1. Konisky J (1982). Ann. Rev. Microbiol. 36:125.
2. Woese CR (1982). Zentralbl. Bakt Hyg I Abt. Orig C3:1.
3. Meseguer I, Rodriguez-Valera F (1985).FEMS Microbiol Letters 28:177.
4. Ebert K, Goebel W, Moritz A, Rdest U, Surek B (1986). System. Appl. Microbiol. 7:30.
5. Tagg JR, Dajani AS, Wannamaker LW, Gray D (1973). J. Exp. Med. 138:1168.
6. Laemmli UK (1970). Nature 227:680.
7. Ozeki H, DeMargerie H (1959). Nature 184:337.
8. Meseguer I, Rodriguez-Valera F (1986). J. Gen. Microbiol. 132:3061.
9. Hayes TJ, Cundy KR, Fernandes PB, Hober K (1983). J. Bact. 155:1171.

Protein Purification: Micro to Macro, pages 279–305

PURIFICATION OF PROTEINS IN THE DENATURED STATE

Mark W. Knuth and Richard R. Burgess

McArdle Laboratory for Cancer Research and U.W. Biotechnology Center, University of Wisconsin-Madison Madison, Wisconsin 53706

ABSTRACT

Most proteins must be properly folded into a specific three-dimensional conformation in order to express their normal biological activities. Anfinsen demonstrated that the information required for the protein to attain this folded state was encoded in the primary sequence. For a long time prior to this, it was thought that denatured proteins could not be renatured, once unfolded. Oddly, this dogma tended to persist even after Anfinsen's revelation, thereby restricting the types of techniques attempted for the preparative purification of proteins.

Recent literature strongly suggests that once persons interested in purifying proteins can free themselves from the bias that proteins should never be purposely denatured, great numbers of potentially powerful separation techniques become available. In fact, many proteins can be purified in the denatured state and subsequently renatured. This paper attempts to present an overview of some of the cases in which this has been done, and to encourage the reader to try some of the techniques.

INTRODUCTION

Most proteins must be folded into a specific three-dimensional, usually globular, conformation in order to express their biological activities. This conformation is called the "native" state, and is one of the thermodynamically more stable forms of the protein. When in the "native" state, the protein does not exist rigidly in a single form, but interconverts among a variety of native-like states. Thus, the "native" conformation and activity of a protein solution refers to an average of the population of protein molecules at a given time.

When agents destabilize the structure of a protein, it may adopt a different conformation (denature), lose biological activity, or precipitate. People formerly thought that proteins, once denatured, could not be restored to their active state. Anfinsen's work in the 1960's showed that to the contrary, the primary amino acid sequence of many proteins completely encoded the information for the protein to adopt its native structure. In his work, he denatured several enzymes and showed that complete enzymatic activity was recovered by slow removal of denaturant (1,2). It has since turned out that many simple proteins can be renatured in such a fashion. In other cases where covalent modification of the protein occurs, chemical intervention is needed. Other times, disaggregants such as mild detergents can be helpful. Finally, in cases where irreversible chemical modification of amino acid side chains or cleavage of the peptide backbone takes place, the protein is said to be irreversibly denatured (3).

Persons interested in protein purification were long held captive to a degree by the notion that denatured proteins were difficult to work with and nearly impossible to renature. In recent years, experimenters have become more willing to accept that it may be relatively easy to remove denaturants and renature proteins at the end of a purification. A number of innovative experimenters have extended the concept, allowing or causing the protein to be denatured during the course of the purification with the intent of recovering the activity later. Many found that denaturing the protein gave them access to very powerful purification techniques. This strategy of purifying proteins in the denatured state is the main topic of this paper.

TYPES OF USEFUL DENATURANTS

Temperature and Pressure

Proteins appear to be stabilized to a large degree by the "hydrophobic effect", wherein the protein seeks to internalize amino acid side chains that prefer to be in a non-aqueous environment (4). Since heat and pressure both tend to decrease the stabilization energy associated with hydrophobic effects, they are capable of denaturing proteins. The pressures required to denature proteins are prohibitively high, but many proteins denature at temperatures as low as 45-60 degrees Centigrade (4). Interestingly, cold can also denature proteins through entropic effects, although this has not been used preparatively (5). Since higher temperature also enhances chemical reaction rates, it is not surprising that some thermal denaturations of proteins involve modifications of amino acid side chains and are irreversible (6). There are reports that thermally-induced irreversible denaturation follows a faster reversible denaturation step, so rapid heat treatment may be safer in many cases (3).

Chaotropes, Organic Solvents

Water is highly ordered, and in protein solutions adopts a different structure in the vicinity of the protein than in the bulk solution. Agents which increase the disorder of the bulk water (chaotropic agents) weaken the aversion of hydrophobic protein side chains to water, thus destabilizing the compact "native" structure. In general, ions with low charge density appear to be chaotropic, whereas ions with high charge density may have an anti-chaotropic effect (4). The relative potency of various chaotropic ions have been ranked and are commonly known as the Hofmeister series. For anions, $CBr_3COO^- > CCl_3COO^- > SCN^- > I^- > Br^- > CF_3COO^- >> CH_3COO^-$ > citrate > $PO_4^- > SO_4^-$. For cations, guanidinium > $Ca^{2+} > Li^+ > Na^+ > K^+$. Thus, guanidine thiocyanate will unfold proteins at lower salt concentrations than will sodium iodide.

Anti-chaotropic salts like ammonium sulfate will precipitate proteins, but they do so by causing the water to order itself so as to exclude protein entirely, and the protein, though precipitated, will be in a compact form. Thus, ammonium-sulfate is known as a "structure-forming" precipitant, although it must be kept in mind that the structures that it forms and stabilizes may also be non-native (7). Aqueous solutions of organic solvents can also be denaturants, since they decrease the aversion of the amino acid side chains to the bulk solvent (8).

Detergents

Detergents may denature a protein, depending on the concentration and type of detergent being used. Detergents typically consist of a hydrophobic hydrocarbon chain with a polar group at one end. At low detergent concentrations, the hydrocarbon chain of the detergent interacts with hydrophobic amino acid side chains of the protein, and the detergent polar group extends out into the bulk water. The detergent thus masks certain hydrophobic patches on the protein surface and may enhance solubility. At higher concentrations, detergent may insert into the hydrophobic protein core and complex with newly externalized hydrophobic amino acids, denaturing the protein. Detergents may also form micelles, with protein binding either on the inside or on the surface of the micelle (9,10,11).

Surfaces

Boundaries of two phases may denature proteins. Probably everyone who has purified a protein has been told not to allow the solution to foam. The reason is that the air-water interface is a boundary between a hydrophobic and a hydrophilic phase. As the protein seeks to reorient itself to achieve maximum stability in this new environment, it may become denatured (12). Similarly, protein conformation can be altered at the surface of detergent micelles (13), or upon interaction with chromatographic packings (14).

Charge Effects

As the pH of a protein solution is changed, various amino acid side chains gain or lose charge. If they gain charge, they will tend to be externalized, and if they lose charge, they become hydrophobic and internalize. If pH changes cause the native structure to be destabilized, the protein will unfold. Chemical modifications which alter charge can also change protein conformation and/or solubility (15). Sometimes these effects can be used to advantage, as seen below.

Ligands and Substrates

Ligand binding can affect protein conformation. An example would be allosteric enzymes such as hemoglobin which change conformation upon binding one or more molecules of ligand. As discussed later, some authors add ligand to a protein refolding solution to provide a folding template. Keyes has carried this principle to an extreme by refolding and crosslinking proteins in the presence of an ester substrate. Using these techniques, he could alter the specificity of esterases, or more strikingly could impart esterase activity to serum albumin, a normally nonenzymatic protein (16).

Combinations of Denaturants

Finally, it should be remembered that the effects of denaturants are often additive or synergistic. That is, addition of chaotropic agents tend to lower the temperature or change the pH at which a protein denatures (17,18). The effects of chaotropes in solution are also additive: addition of antichaotropic ammonium sulfate can reverse the denaturing effect of sodium chlorate (7).

USES OF DENATURANTS TO ENHANCE PROTEIN SEPARATIONS

Sodium Dodecyl Sulfate - Polyacrylamide Gel Electrophoresis (SDS-PAGE)

SDS is used in this technique to dissociate oligomers, to give each protein an equal negative charge density, and to denature proteins to a rod-like conformation so that the electrophoretic mobility is proportional to the molecular weight. This technique is popular analytically because of the superior resolution that is attainable, but has not been commonly used until recently in a preparative mode. This was partly because of loading limitations, but often because of a persistent dogma that SDS-treated proteins are difficult or impossible to renature.

In fact, a large number of proteins of various types have been recovered in active form from SDS-PAGE gels after elution from the gel and treatment to remove the detergent (see below). Among these have been the sigma subunit of RNA polymerase (19), DNA topoisomerase (19), the alpha subunit of the *Torpedo Californica* acetylcholine receptor (20), lactate dehydrogenase (21), 1,25-dihydroxyvitamin D3 receptor (22), estrogen receptor (23), chymotrypsin (24), urease (25), fructosyltransferase (26), and ß-lactamase (27). Some authors report better results after removing residual acrylamide (28). An article in this book written by R. Burgess treats the subject of renaturation from gels in more detail.

In a variant of the normal protocols, proteins were electroblotted from an SDS-PAGE gel through a second gel containing the non-ionic detergent NP40. The normally-cytosolic proteins passed through the second gel, but many integral membrane proteins did not, and were recovered in purified form from the second gel (29).

Reversed Phase Liquid Chromatography (RPLC)

RPLC is excellent for high-resolution protein separations. Many proteins denature upon prolonged contact with the hydrophobic column packing, however (14). Strategies for avoiding this include minimizing on-column dwell time (30), lowering the temperature (31), or using more mildly denaturing solvent systems such as triethylammonium phosphate (30). Others have resorted to the yet milder conditions of hydrophobic interaction chromatography (32). Many proteins can probably be renatured after RPLC by dialysis or GuHCl denaturation/renaturation (see below). RPLC is popular in the purification of proteins that have already been denatured in prior steps. In some cases, use of denaturants in the mobile phase can improve recovery of very hydrophobic proteins from RPLC columns (33).

Size Exclusion Chromatography (SEC)

SEC of proteins is often most effective when done in denaturing solutions, since aggregation effects and protein-packing interactions in nondenaturing solutions often skew elution patterns severely (34). Typically, urea, guanidine hydrochloride, or SDS are used (35), although some have used organic solvents (36). The rationale for using these agents in the running buffer is much the same as for using SDS in gel electrophoresis. Since the Stokes radius of proteins in denaturants is greater than for proteins in the native state (37,38), columns run using denaturants should be calibrated using protein molecular weight standards run in the same denaturant. It should also be noted that protein-SDS complexes of low Stokes radii (<70 Å) may elute anomalously on some size exclusion resins (39). SDS will alter column packing properties, so an SDS-treated column should not later be used for other purposes.

Ion Exchange Chromatography (IEX)

Denaturing solvents are sometimes used in combination with high performance IEX. In these cases, denaturants are typically used to preserve the solubility of hydrophobic proteins, or to dissociate the subunits of a multimeric protein for individual isolation. Examples of these applications would include the purification of viral coat proteins from Sendai virus (40), and subunits of alpha crystallin (41) and tubulin (42).

Detergent Extraction of Proteins

This technique has become popular for the purification of membrane proteins from a wide variety of sources, and good review articles exist for these techniques. (43,44,45). Many of these proteins can be reconstituted into lipid bilayers in a functional form by detergent exchange (26).

Inclusion of Denaturants in Initial Cell Extracts

Denaturants can be very useful for preventing proteolysis in crude tissue extractions. The aim of this strategy is that the proteases are also denatured by the treatment and are removed before protein is renatured (46,47). This approach may avoid the use of protease inhibitors, many of which are toxic and not completely effective. Usually, this strategy is used after tests on small quantities of previously purified protein have established renaturability. It should also be noted that quite a number of proteases are highly stabilized by disulfide bridges, and can maintain activity in fairly high concentrations of denaturants, even 0.1% SDS (48). Thus, unless rather high concentrations of denaturant are used, the denaturation-resistant proteases will degrade the partially denatured proteins of interest.

Denaturation to Confer Unusual Properties to Proteins

In some cases, denaturation of a protein imparts an aberrant behavior which can be exploited in a purification scheme. In an unusual purification, the authors noted that while many proteins precipitate more easily when in the reduced form, alpha-1-protease inhibitor did not. Thus, reduction of the mixture allowed precipitation of other disulfide-containing proteins, leaving the desired protein in free solution (49). Another purification scheme used denaturing detergent solutions to purify myelin proteolipids. Addition of detergent caused myelin basic protein to bind irreversibly to gel filtration media, while the proteolipid flowed through (50). A final example is a reported purification of human antihemophilic factor in which the authors wished to remove contaminants which lacked cysteine. They selectively reduced certain disulfides in the factor, and used thiol-disulfide interchange chromatography to purify it (51).

Denaturation to Confer Predictable Properties to Proteins

Even if the primary amino acid sequence of a protein is known, it is very difficult to use this knowledge to predict the chromatographic or precipitation behavior of the native protein from the primary sequence. For instance, bovine growth hormone has high sequence homology with growth hormone from other species, yet it is the only one of the family which is extremely insoluble in water after reduction of the disulfide bonds (52). Another example is from the work of Kopaciewicz (53), which suggested that the cause of deviations from idealized retention behavior of proteins on ion exchangers near their isoelectric points was the existence of local surface charge on the protein which was independent of the net charge.

It is important to recognize that chromatographic retention properties of many proteins are mediated through surface groups more than global properties, and that many proteins may have asymmetric surface characteristics. Well known examples are DNA-binding proteins which often are acidic proteins, yet have a region of basic surface charge which enables them to bind the phosphate groups of DNA and columns of phosphocellulose (54).

Theoretically, the inclusion of a denaturant in the chromatographic mixture should eliminate the effects of charged patches of these kind, resulting in a more predictable separation. Work by Parente and Wetlaufer tends to support this prediction, in that they show strong effects of urea, an uncharged yet denaturing solute, on cation-exchange chromatography of lysozyme (55). Another use of denaturants to impart predictable chromatographic properties is the inclusion of denaturants in SEC solvents to ensure that determined molecular weights are accurate (34,35,36,37).

Chemical Modification

There are some rather startling papers coming out in which authors have considerably modified the protein by reaction with chemical agents while still preserving biological activity. In some cases, this chemical modification was minor, such as the report regarding antihemophilic factor. In this paper, the authors could S-alkylate a particular disulfide bond without affecting activity (51). In more drastic examples, various persons have covalently added PEG (56), or other hydrophobic agents (57) to proteins for the purpose of extracting them into organic solvents. Others have added charged groups to proteins to increase solubility in aqueous solvents (58). In several cases, the proteins were active, although their conformation and in some cases their activity was altered (59,60).

Extraction Into Organic Solvents

Sulfolane (61), acidic ethanol (62), hexane (63), and glacial acetic acid (64) have all been used for the initial extraction of proteins from cells, with subsequent recovery of full biological activity. In most cases, testing of previously prepared samples of the protein gave predictions of the success of these techniques. In some cases, the enzymes were kept in the organic solvents, and exhibited biological activity (56,63,65,66). A review article on the behavior of proteins in organic solvents has recently appeared (67). Interesting extractions with reversed micelles in organic solvents have been reported (68), and are discussed by A. Hatton elsewhere in this book.

rDNA Inclusion Bodies

Many rDNA-modified E.coli cells express eukaryotic proteins in a fully-reduced form that is often precipitated in hard-to-dissolve inclusion bodies. This was originally thought to be harmful, and people tried with some success to have the protein produced in a secreted or properly folded form. In recent years, however, some have come to regard the formation of inclusion bodies as desirable, for they can simplify purification of the protein. In a typical process, the cells are lysed by various means, and the majority of the contaminants are rinsed away by a solution containing a weak detergent or denaturant of insufficient strength to solubilize the inclusion bodies. The washed protein at this point can be 85% pure or better. The inclusion bodies are then dissolved in a denaturing buffer, and subsequently renatured using a variety of protocols (69). Two recent reviews on this topic have appeared (70,71).

In one particularly unusual separation, the washed inclusion bodies were dissolved in 7 M guanidine HCl, and RPLC (C-18) packing material was added to absorb the protein. The pelleted packing material was poured into a loose bed, and the desired protein was eluted from the packing in high yield and purity (72).

WAYS TO RENATURE PROTEINS

From Detergent Solutions

It was long thought that detergents and proteins should never mix. One of the reasons for this was that the detergents used were not pure enough for protein separations. Many commercial preparations of detergent contain oxidizing impurities that can react with proteins. Detergents notorious for this are those in the Brij, Tween, Lubrol, and Triton series (73). These impurities can be removed, or detergent lots lacking them can be ordered.

Another complication results from the fact that many commercial detergents are mixtures of many related compounds of identical protomer composition but different molecular weight. For instance, medium-molecular weight fractions of Triton X-100 were more easily removed from Sindbis virus envelope proteins than were high or low molecular weight fractions of the same detergent series (74). Another report indicates that traces of C-14 and C-16 analogs of SDS are more difficult to remove from proteins than SDS itself and result in poorer renaturation yields. The level of the analogs varied between lots of detergent tested (75).

A wide variety of methods have been used to remove detergent from proteins. In some of them, renaturation is concurrent with the removal of detergent, and in others, it takes place in a later step. Ionic detergents can be removed from proteins by IEX, or detergents with a long hydrophobic sidechain can be removed by selective absorption onto a hydrophobic resin (76). Gel filtration can be used in those cases where the protein can be pre-dissociated (often by acidic conditions) (77). Dialysis or ultrafiltration can be employed, but the speed can be limited by the rate at which the nondialyzable detergent micelles convert to dialyzable monomers (78).

In some cases, strong detergents like SDS can be competed off the protein by excess quantities of weaker detergents such as Triton (79). The weaker detergent is then either removed by conventional means, or left in the solution if it doesn't inhibit activity. In the case of integral membrane proteins, this weaker detergent is often competitively replaced by phospholipid micelles (78). Detergent can also be removed by solvent extraction (80).

Finally, the detergent can often be removed by precipitating the protein with acetone, ethanol, or acid (19,81). In such cases, a secondary renaturation step may be required (see below). In some cases, renaturation of the protein may be enhanced by inclusion of the natural substrate as a template in the refolding buffer (27). Such procedures have been used to detect activity in-situ in SDS gels (27,82).

Slow Dialysis/Dilution in Physiologic Buffer

Many proteins are designed to function in near-physiologic solutions. For this reason, many proteins may revert to their native form if gradually shifted from denaturing conditions to those reflecting their normal environment. The procedure often works best when done gradually; this allows a protein that misfolds time to try out other conformations before the potential energy barrier of interconversion is too high. This gradual dialysis is often referred to as allowing the protein to "shuffle" its way to the native state (1,2,83).

It is very important to attempt renaturation using various rates of dilution. Many experimenters have diluted their protein from 6M to 1M GuHCl over the course of a few hours, have been disappointed by the resultant precipitate, and never pursued the matter further. This ignores the fact that some protein folding steps are quite slow. As an example, the transition of glucagon in solution from helix to pleated sheet has been shown to take more than 5 days (84), as did complete refolding of ligninase H8, as reported in this book by K. Javaharian.

Formation of aggregated protein also competes kinetically with monomeric refolding. Brems has shown an aggregative folding intermediate in bovine growth hormone involves an amphipathic alpha helix which forms early in the folding pathway (85,86). If the protein does not get a chance to fold properly around this helical section, the amphipathic helix can then associate with a homologous segment of another molecule of growth hormone, thus initiating the aggregative pathway (87). It is thus intuitive that rates of dilution of denaturant might influence the choice of pathway for this and other proteins that fold by similar mechanisms. Similar competitions between refolding and aggregation pathways have been reported for rhodanese (88,89), phosphorylase b (90), and antithrombin (91). A recent review exists on irreversible folding problems (92).

Guanidine/Urea Denaturation/Renaturation (GDR)

One popular strategy for the renaturation of proteins that have been denatured by the action of detergents, precipitants, or heat is to fully denature the protein in a solution of a chaotropic salt, and then to renature by the slow dialysis or dilution method. The rationale for this approach is that high concentrations of chaotropic salts are thought to denature proteins nearly to random coils, thus undoing the effects of prior denaturation steps. Because of ease and versatility, most reported protein renaturations are performed using some variant of this technique. Since concentrated urea solutions generate protein-reactive cyanate ions (2), guanidine salts are becoming favored.

Burgess et.al. successfully renatured proteins that had been purified by SDS-PAGE by eluting the protein into SDS-containing buffer by diffusion, precipitating the eluate with acetone, and renaturing the precipitate by GDR (19). This procedure has been successfully performed with dozens of other proteins. Other investigators have used GDR to recover activity from heat-denatured proteins (93). GDR in some form is involved in many of the large-scale preparative methods for recombinant proteins such as interferon (47), chymosin (94), interleukin-2 (95), E.coli sigma factor (19),urokinase, prorennin, growth hormone, and tissue plasminogen activator (69,96). In some cases, pretreatment of denatured proteins with detergents or inclusion of detergents in the folding buffer has favorable effects on the renaturation of the protein by counteracting aggregation (97,98).

Renaturation of Proteins Containing Disulfides

Most renaturation protocols for proteins containing reduced disulfide bonds use the basic guidelines established by Anfinsen. A typical protocol would involve slow oxidation of the protein during dialysis from a urea or guanidine-containing solution. The theory is that each protein molecule will be allowed to try out a large number of disulfide bond pairings by thiol-disulfide interchange (disulfide shuffling). Usually, if the protein concentration is kept low to prevent polymer formation, the majority of the protein will find its way to the thermodynamically most stable and usually native form. Some have found it convenient to react thiols with protecting groups so that premature oxidation can be prevented. Air can be used as an oxidant, or thiol reagents such as glutathione or thiol-disulfide interchange enzymes are added to enhance shuffling (99,100,101). In the past, the enzymes had to be isolated from a natural source. Recently, a commercial preparation of thioredoxin has been introduced which reportedly works well for this purpose (102).

Some researchers have devised means for oxidizing proteins to the correct disulfide pairings at high protein concentrations in the mg/ml range. Such techniques are unfortunately being kept as trade secrets by many companies. One of the more interesting publicly-divulged methods is that of Creighton, who immobilizes the protein of interest on ion-exchange resins prior to oxidation. By thus physically preventing protein-protein interactions, he can prevent polymer formation (103). In another novel approach, Light managed to refold reduced proteins by first reacting the protein with citraconic anhydride, which converted basic lysine residues to a derivative carrying a net negative charge. This modification increased the solubility of the enzyme and increased the amount of reoxidized protein in monomeric form, presumably because the negative charge density of the derivatized protein causes repulsion of individual monomers. After oxidation, the blocking groups were completely removed by treatment with dilute acid (58).

SITUATIONS IN WHICH RENATURATION IS LIKELY TO BE DIFFICULT, BUT NOT IMPOSSIBLE

Many proteins will refold well using the general techniques presented above. There are specific cases, however, where the refolding problem is likely to be more difficult. These would include proteins which are normally glycosylated but are being produced in prokaryotes, proteins which are normally found in multisubunit assemblies or which are enzymatically processed from a prefolded proform, or proteins which contain cofactors. Since proteins from all of these categories have been successfully refolded, however, the reader should be cautioned but not discouraged. A nice example of the renaturability of complex proteins is presented elsewhere in this book by K. Javaharian, who was successful in refolding ligninase H8, a glycosylated protein 344 amino acids long, containing hemin, four disulfides, and 10% proline residues.

USES OF DENATURED PROTEIN

If one gets to the end of a purification and finds that the pure protein is irreversibly denatured, all is not lost. One can still perform amino acid analysis or sequencing (104); with tryptic peptide mapping, one might even obtain the entire amino acid sequence. This information may then be used to design a better purification, possibly involving immunoaffinity chromatography using monoclonal antibodies to a peptide fragment. Alternatively, one may be able to raise useful antibodies to the denatured protein itself. In one account, the antiserum raised to denatured protein was more effective than antiserum raised to native protein for the purpose of screening a library of *E.coli* for the expression of a eukaryotic protein, possibly because the cells were producing the protein in a denatured form (105). In another interesting use of denatured protein, the authors immobilized the denatured protein on a resin. They then used the immobilized enzyme to purify more enzyme by subunit exchange chromatography. The immobilized denatured protein proved more effective for this purpose than did the native enzyme (106).

CAVEATS AND CAUTIONS

Although the above accounts are intended to encourage the reader to consider trying purification of certain proteins in the denatured state, certain cautions must apply. First, for every success noted above, there has likely been a failure. It is very difficult to try to guess what the failure to success ratio is for these sorts of techniques, since negative results of purifications are rarely reported in the literature. Therefore, it is prudent to test samples of previously purified material for lability and renaturability in the contemplated systems. Second, although proteolysis is inhibited by denaturants, other chemical reactions involving amino acid side chains are not inhibited by denaturants. Examples of these would be deamidation of glutamine and asparagine side chains in either alkaline or acidic solutions (107), disulfide shuffling being catalyzed by alkaline conditions (1,2), and selective cleavages of the protein backbone at asp-pro residues in acidic conditions (108). Finally, as noted above, some reagents are not manufactured with protein applications in mind, and may contain impurities.

SUMMARY

Once persons interested in purifying proteins can free themselves from the bias that proteins should never be purposely denatured, great numbers of potentially powerful separations techniques are available. In fact, many proteins can be purified in the denatured state and subsequently renatured, once conditions are found that allow them to reattain the three-dimensional conformation encoded in their primary sequence. This paper has given an overview of some of these techniques, and perhaps will encourage the reader to try some of them. It is likely that work in this relatively new area will produce many more useful techniques in the near future.

REFERENCES

1. Epstein CJ, Goldberger RF, Young DM, Anfinsen CB (1962). A study of the factors influencing the rate and extent of enzymic reactivation duringreoxidation of reduced ribonuclease. Arch Biochem Biophys Suppl 1:223.
2. Haber E, Anfinsen CB (1962). Side-chain interactions governing the pairing of half-cystine residues in ribonuclease. J Biol Chem 237:1839.
3. Zale SE, Klibanov AM (1983). On the role of reversible denaturation (unfolding) in the irreversible thermal inactivation of enzymes. Biotechnology and Bioengineering 25:2221.
4. Tanford,C (1973). "The Hydrophobic Effect", New York: Wiley Interscience.
5. Privalov PL, Griko YV, Venyaminov SY (1986). Cold denaturation of myoglobin. J Mol Biol 190:487.
6. Ahern TJ, Klibanov AM (1985). The mechanism of irreversible enzyme inactivation at 100 degrees centigrade. Science 228:1280.
7. Hatefi H, Hanstein WG (1974). Destabilization of membranes with chaotropic ions. Meth Enzymol 31:770.
8. Sadler AJ, Micanovic R, Katzenstein GE, Lewis RV, Middaugh CR (1984). Protein conformation and reversed-phase high-performance liquid chromatography. J Chromatogr 317, 93-101.
9. Miyake J, Takagi T (1981). A low-angle laser light scattering study of the association behavior of a major membrane protein of *Rhodospirillium Rubrum* chromatophore at various concentrations of sodium dodecyl sulfate where polypeptides derived from water-soluble globular proteins are solubilized monomerically. Biochim Biophys Acta 668:290.
10. Chang PL, Ameen M, Lafferty KI, Varey PA, Davidson AR, Davidson RG (1985). Action of surface-active agents on arylsulfatase-C of human cultured fibroblasts. Anal. Biochem. 144:362.

11. Makino S, Woolford JL, Tanford C, Webster RE (1975). Interaction of deoxycholate and of detergents with the coat protein of bacteriophage f1. J Biol Chem 250:4327.
12. Henson AF, Mitchell JR, Mussellwhite PR (1970). The surface coagulation of proteins during shaking. J Colloid Interface Sci 32:162.
13. Wilson ML, Dahlquist FW (1985). Membrane protein conformational change dependent on the hydrophobic environment. Biochemistry 24:1920.
14. Katzenstein GE, Vrona SA, Wechsler RJ, Steadman BL, Lewis RV, Middaugh CR (1986). Role of conformational changes in the elution of proteins from reversed-phase HPLC columns. Proc Natl Acad Sci USA 83:4268.
15. Hollecker M, Creighton TE (1982). Effect on protein stability of reversing the charge on amino groups. Biochim Biophys Acta 701:395.
16. Keyes, MS (1982). Process for preparing semisynthetic enzymes. Australian Patent #525924.
17. Sureshchandra BR, Rao AGA, Rao MSV (1986). Denaturation of glycinin by urea and guanidine hydrochloride. Int J Pept Protein Res 27:493.
18. Von Hippel PH, Schleich T (1969). Ion effects on the solution structure of biological macromolecules. Acc Chem Res 2:257.
19. Hager DA, Burgess RR (1980). Elution of proteins from sodium dodecylsulfate-polyacrylamide gels, removal of sodium dodecyl sulfate, and renaturation of enzymatic activity: results with sigma subunit of *Escherichia coli* RNA polymerase, wheat germ DNA topoisomerase, and other enzymes. Anal. Biochem 109:76.
20. Haggerty JG, Froehmer SC (1981). Restoration of the alpha-bungarotoxin binding activity to the alpha subunit of *Torpedo* acetylcholine receptor isolated by gel electrophoresis in sodium dodecyl sulfate. J Biol Chem 256:8294.
21. Clarke S (1981). Direct renaturation of the dodecyl sulfate complexes of protein with Triton X-100. Biochim Biophys Acta 670:195.

22. Dame MC, Pierce EA, DeLuca H (1985). Identification of the porcine intestinal 1,25-dihydroxyvitamin D3 receptor on sodium dodecyl sulfate/polyacrylamide gels by renaturation and immunoblotting. Proc Natl Acad Sci USA 82:7825.
23. Sakai D, Gorski J (1984). Reversible denaturation of the estrogen receptor and estimation of polypeptide chain molecular weight. Endocrinology 115:2379.
24. Bieger W, Scheele G (1980). Two-dimensional isoelectric focusing/sodium dodecyl sulfate gel electrophoresis of protein mixtures containing active or potentially active proteases: analysis of human exocrine pancreatic proteins. Anal Biochem 109:222.
25. Shaik-M MB, Guy AL, Pancholi SK (1980). An improved method for the detection and preservation of urease activity in polyacrylamide gels. Anal Biochem 103:140.
26. Russell RRB (1979). Use of Triton X-100 to overcome the inhibition of fructosyltransferase by SDS. Anal Biochem 97:173.
27. Tai PC, Zyk N, Citri N (1985). *In Situ* detection of ß-lactamase activity in sodium dodecylsulfate-polyacrylamide gels. Anal Biochem 144:199.
28. Brooks KP, Sander EG (1980). Preparative polyacrylamide gel electrophoresis: removal of polyacrylate from proteins. Anal Biochem 107:182.
29. Ito K, Akiyama Y (1985). Protein blotting through a detergent layer, a simple method for detecting integral membrane proteins separated by SDS-polyacrylamide gel electrophoresis. Biochem Biophys Res Commun 133:214.
30. Benedek K, Dong S, Karger BL (1984). Kinetics of unfolding of proteins on hydrophobic surfaces in reversed-phase liquid chromatography. J Chromatogr 317:227.
31. Wu S, Benedek K, Karger BL (1986). Thermal behavior of proteins in high-performance hydrophobic interaction chromatography: on-line spectroscopic and chromatographic characterization. J Chromatogr 359:3.

32. Fausnaugh JL, Kennedy LA, Regnier FE (1984). comparison of hydrophobic-interaction and reversed-phase liquid chromatography of proteins. J Chromatogr 317:141.
33. Sharifi BG, Bascom CC, Khurana VK, Johnson TC (1985). Use of a urea and guanidine-HCl-propanol solvent system to purify a growth inhibitory glycopeptide by high-performance liquid chromatography. J Chromatogr 324:173.
34. Montelaro RC, West M, Issel CJ (1981). High-performance gel permeation chromatography of proteins in denaturing solvents and its application to the analysis of enveloped virus polypeptides. Anal Biochem 114:398.
35. Konishi K (1985). Gel permeation of proteins by high-performance gel chromatography in denaturing solvents. Progress in HPLC 1:43.
36. Swergold GD, Rubin CS (1983). High Performance Gel-Filtration Chromatography of polypeptides in a volatile solvent: rapid resolution and molecular weight estimations of proteins and peptides on a column of TSK-G3000-PW. Anal Biochem 131:295.
37. Burgess RR (1969). Separation and characterization of the subunits of ribonucleic acid polymerase. J Biol Chem 244:6168.
38. Fish WW, Reynolds JA, Tanford C (1970). Gel chromatography of proteins in denaturing solvents. Comparison between sodium dodecyl sulfate and guanidine hydrochloride as denaturants. J Biol Chem 245:5166.
39. Wong P, Barbeau A, Roses AD (1985). Gel chromatography on a Sepharose 4B column: Earlier elution of protein-sodium dodecyl sulfate complexes of low Stokes radii. Anal Biochem 146:191.
40. Welling GN, Groen G, Welling-Wester S (1983). Isolation of Sendai virus F protein by anion-exchange high-performance liquid chromatography in the presence of Triton X-100. J Chromatogr 266:629.
41. Bloemendale H, Groenewoud G (1981). One-step separation of the subunits of alpha-crystallin by chromatofocusing in 6 M urea. Anal Biochem 117:327.

42. Lu RC, Elzinga M (1977). Chromatographic resolution of the subunits of calf brain tubulin. Anal Biochem 77:243.
43. Hjelmeland LM, Chrambach A (1984). Solubilization of membrane proteins. Meth Enzymol 104:305.
44. VanRenswoude J, Kempf C (1984). Purification of integral membrane proteins. Meth Enzymol 104:329.
45. Klausner RD, VanRenswoude J, Rivnay B (1984). Reconstitution of membrane proteins. Meth Enzymol 104:340.
46. Kung HF (1984) US Patent US 4,476,049.
47. Honda S, (1984) European Patent EP 138,087.
48. Weber K, Osborn M (1975). Proteins and sodium dodecyl sulfate: Molecular weight determination on polyacrylamide gels and related procedures. In Neurath H, Hill RL (eds): "The Proteins" 3rd Edition, New York: Academic Press, p 217.
49. Glaser CB, Chamorro M, Crowley R, Karic L, Childs A, Calderon M (1982). The isolation of alpha-1-protease inhibitor by a unique procedure designed for industrial application. Anal Biochem 124:364.
50. Riccio P, DeSantisA, Quagliarello E (1985). On the binding of brain myelin basic protein to chromatographic resin. Biochem Biophys Res Commun 126:233.
51. Harris RB, Johnson AJ, Hodgins LT (1981). Partial purification of biologically active, low molecular weight human antihemophilic factor free of Von Willebrand factor II. Further purification with thiol-disulfide interchange chromatography and additional evidence for disulfide bonds susceptible to limited reduction. Biochim Biophys Acta 668:471.
52. Santome JA, Dellacha JM, Paladini AC (1976). Chemistry of growth hormone. Pharmac Ther B 2:571.
53. Kopaciewicz W, Rounds MA, Fausnaugh J, Regnier FE (1983). Retention model for high-performance ion-exchange chromatography. J Chromatogr 266:3.

54. Burgess RR (1976). Purification and physical properties of E.coli RNA polymerase. In Losick R, Chamberlin M (eds): "RNA Polymerase," Cold Spring Harbor: Cold Spring Harbor Laboratory, p 71.
55. Parente ES, Wetlaufer DB (1984). Influence of urea on the high-performance cation-exchange chromatography of hen egg lysozyme. J. Chromatogr 288:389.
56. Inada Y, Takahashi K, Yoshimoto T, Ajima A, Matsushima M, Saito Y (1986). Application of polyethylene glycol-modified enzymes in biotechnological processes: organic solvent-soluble enzymes. Trends Biotechnol 4:190.
57. Criado M, Aguilar JS, De Robertis E (1980). The use of p-toluene sulfonate to dissolve synaptosomal membrane proteins into organic solvents. Anal Biochem 103:289.
58. Light A (1985). Protein solubility, protein modifications and protein folding. BioTechniques 3:298.
59. Sadana A, Henley JP (1986). Effect of chemical modification on enzymatic activities and stabilities. Biotechnol Bioeng 28:256.
60. Inada Y, Yoshimoto T, Matsushima A, Saito Y (1986).Engineering physicochemical and biological properties of proteins by chemical modification. Trends Biotechnol 4:68.
61. Vecchio G, Righetti PG, Zanoni M, Artoni G, Gianazza E (1984). Fractionation techniques in a hydro-organic environment I. sulfolane as a solvent for hydrophobic proteins. Anal Biochem 137:410.
62. Erickson JS, Paucker K (1979). Purification of acid ethanol-extracted human lymphoid interferons by blue sepharose chromatography. Anal Biochem 98:214.
63. Ayala G, Nascimento A, Gamez PA, Danszon A (1985). Extraction of mitochondrial membrane protein into organic solvents in a functional state. Biochim Biophys Acta 810:115.
64. Knuth MW, Friedman AR (1983). unpublished observations.

65. Bowers LD, Johnson PR (1981). Characterization of immobilized ß-glucuronidase in aqueous and mixed solvent systems. Biochim Biophys Acta 661:100.
66. Martinek K, Semenov AN (1981). Enzymatic synthesis in biphasic aqueous-organic systems II. Shift of ionic equilibria. Biochim Biophys Acta 658:90.
67. Waks M (1986). Proteins and peptides in water-restricted environments. Proteins: Struct Funct Genetics 1:4.
68. Kadam KL (1986). Reverse micelles as a bioseparation tool. Enzyme Microb Technol 8:266.
69. Olson KC (1985). Purification and activity assurance of precipitated heterologous proteins. US Patent US 4,518,526.
70. Marston FAO (1986). The purification of eukaryotic polypeptides synthesized in _Escherichia coli_. Biochem J 240:1.
71. Sharma SK (1986). On the recovery of genetically engineered proteins from _Escherichia coli_. Sep Sci Technol 21:701.
72. Wolfe RA, Case J, Familletti PC, Stein S (1984). Isolation of proteins from crude mixtures with silica and silica-based adsorbents. J Chromatogr 296:277.
73. Ashani Y, Catravas GN (1980). Highly reactive impurities in Triton X-100 and Brij-35: partial characterization and removal. Anal Biochem 109:55.
74. Scheule RK, Gaffney BJ (1981). Reconstitution of membranes with fractions of Triton X-100 which are easily removed. Anal Biochem 117:61.
75. Lacks SA, Springhorn SS, Rosenthal AL (1979). Effect of the composition of sodium dodecyl sulfate preparations on the renaturation of enzymes after polyacrylamide gel electrophoresis. Anal Biochem 100:357.
76. Furth AJ (1980). Removing unbound detergent from hydrophobic proteins. Anal Biochem 109:207.

77. Amons R, Schrier PI (1981). Removal of sodium dodecyl sulfate from proteins and peptides by gel filtration. Anal Biochem 116:439.
78. Furth AJ, Bolton H, Potter J, Priddle JD (1984).Separating detergents from proteins. Meth Enzymol 104:318.
79. Clarke S (1981). Direct renaturation of the dodecyl sulfate complexes of proteins with Triton X-100. Biochim Biophys Acta 670:195.
80. Horikawa S, Ogawara H (1979). A simple and rapid procedure for removal of Triton X-100 from protein solutions. Anal Biochem 97:116.
81. Henderson LE, Oroszlan S, Konigsberg W (1979). A micromethod for complete removal of dodecyl sulfate from proteins by ion-pair extraction. Anal Biochem 93:153.
82. Dottin RP, Manrow RE, Fishel BR, Ankerman SL, Culleton JL (1979). Localization of enzymes in denaturing polyacrylamide gels. Meth Enzymol 68:513.
83. Acharya AS, Taniuchi H (1982). Implication of the structure and stability of disulfide intermediates of lysozyme on the mechanism of renaturation. Mol Cell Endocrinol 44:129.
84. Moran EC, Chou PY, Fasman GD (1977). Conformationaltransitions of glucagon in solution: the alpha-beta transition. Biochem Biophys Res Commun 77:1300.
85. Brems DN, Plaisted SM, Havel HA, Kaufmann EW, Stodola JD, Eaton LC, White RD (1985). Equilibrium denaturation of pituitary- and recombinant-derived bovine growth hormone. Biochemistry 24:7662.
86. Brems DN, Plaisted SM, Kaufmann EW, Havel HA (1986). Characterization of an associated equilibrium folding intermediate of bovine growth hormone. Biochemistry 25:6539.
87. Brems DN, Plaisted SM, Dougherty JJ, Holzman TF (1987). The kinetics of bovine growth hormone folding are consistent with a framework model. J Biol Chem 262:2590.
88. Horowitz P, Criscimagna NL (1986). Low concentrations of guanidinium hydrochloride expose apolar surfaces and cause differential perturbation in catalytic intermediates of rhodanese. J Biol Chem 261:15652.

89. Horowitz P, Bowman S (1987). Reversible thermal denaturation of immobilized rhodanese. J Biol Chem 262:5587.
90. Price NC, Stevens E (1983). The denaturation of rabbit muscle phosphorylase b by guanidinium chloride. Biochem J 213:595.
91. Fish WW, Danielsson A, Nordling K, Miller SH, Lam CF, Bjork I (1985). Denaturation behavior of antithrombin in guanidinium chloride. Irreversibility of unfolding caused by aggregation. Biochemistry 24:1510.
92. King J (1986). Genetic analysis of protein folding pathways. BioTechnology 4:297.
93. Martinek K, Mozhaev VV, Berezin IV (1980). Reactivation of enzymes irreversibly denatured at elevated temperature: trypsin and alpha-chymotrypsin covalently immobilized on sepharose 4B and in polyacrylamide gel. Biochim Biophys Acta 615:426.
94. McConnor MJ, King JF (1985) World Patent WO 85-05,377.
95. Kung HF, Yamazaki S (1985) European Patent EP 147,819.
96. Olson KC (1985). Purification and activity assurance of precipitated heterologous proteins. US Patent US 4,512,922.
97. Stewart WE, DeSommer P, DeClerq E (1974). Renaturation of inactivated interferons by "defensive reversible denaturation". Prep Biochem 4:383.
98. Reynolds J, Tanford C (1970). The gross conformation of protein-sodium dodecyl sulfate complexes. J Biol Chem 245:5161.
99. Lambert N, Freedman RB (1983). Structural properties of homogeneous proteins disulphide-isomerase from bovine liver purified by a rapid high-yielding procedure. Biochem J 213:225.
100. Swaisgood HE (1980). Sulphydryl oxidase: properties and applications. Enzyme Microb Technol 2:265.

101. Saxena VP, Wetlaufer DB (1970). Formation of three-dimensional structure in proteins. I. Rapid nonenzymatic reactivation of reduced lysozyme. Biochemistry 9:5015.
102. Pigiet V (1986). The role of thioredoxin in refolding disulfide proteins. In November issue, Chemalog HiLites (Chemical Dynamics Corp) p 3.
103. Creighton T (1985) personal communication.
104. Vandekerckhove J, Bauw G, Puype M, Van Damme J, VanMontagu M (1985). Protein-blotting through polybrene-coated glass-fiber sheets. Eur J Biochem 152:9.
105. Timmins JG, Petrovskis EA, Marchioli CC, Post LE (1985). A method for efficient gene isolation from phage lambda gt11 libraries: use of antisera to denatured, acetone-precipitated proteins. Gene 39:89.
106. Carrea G, Pasta P (1983). Purification of chymotrypsin by subunit exchange chromatography on the denatured protein. Biotechnol Bioeng 25:1331.
107. Lewis UJ, Cheever EV, Hopkins WC (1970). Kinetic study of the deamidation of growth hormone and prolactin. Biochim Biophys Acta 214:498.
108. Rittenhouse J, Marcus F (1984). Peptide mapping by polyacrylamide gel electrophoresis after cleavage at aspartyl-prolyl peptide bonds in sodium dodecyl sulfate-containing buffers. Anal Biochem 138:442.

Protein Purification: Micro to Macro, pages 307–314

SUPERCRITICAL GELS FOR PROTEIN CONCENTRATION

E. L. Cussler

Department of Chemical Engineering
and Materials Science
University of Minnesota
Minneapolis, Minnesota 55455

ABSTRACT Modified polyacrylamides can be used to concentrate dilute beers and protein solutions. The gels absorb small solutes like water but not large solutes like proteins. The gels can be collapsed and reused by slight warming, a consequence of the proximity of a lower critical solution temperature. As a result, the gels are promising alternatives to rotary vacuum filtration or ultrafiltration.

INTRODUCTION

In this paper I have two goals which don't fit together. The first is to suggest a template by which bioseparations can be generally approached. I mean this template to be a rough strategy for thought, and not a definitive or exhaustive list about how to proceed. I've always found that a template is useful in thinking about any given problem.

My second goal in the paper is to describe one new way for concentrating proteins. This way has the potential for being less expensive yet faster than ultrafiltration. It's very much of a novelty, and will need more development before it is routinely useful.

I'll discuss each of these goals sequentially.

An Overall Template for Bioseparations

A chief characteristic of biotechnology is the tremendous variety of products which are produced. A typical chemical petrochemical company makes about ten products, but a drug company may make more than two hundred. To be sure, all two hundred products may not be exclusively made microbiologically, but many will be partially biologically converted.

Because of this diversity of products, bioseparations involve a tremendous spectrum of separation methods. One way to see this is to simply list the separation methods suggested in books like C. Judson King's Separation Processes. Of the forty two processes which King lists, I know of thirty five which are used in bioseparations. All modes of operation are common: steady and unsteady state, batch and continuous, co-current and counter-current contacting. The scale of operations varies widely from micrograms to tons. This variation of scale is exacerbated for drugs, where the New Drug Applications often require inefficient but quick syntheses of significant amounts of material for clinical trials. This initial inefficient production often becomes the basis for later large scale production.

In the face of such diversity in bioseparations, how can we hope to succeed?

The tremendous diversity of these products obscures an overall similarity between many of the processes used. This similarity is by no means exact and has numerous exceptions. Still, the similarity provides a template for thinking, just as the periodic table emphasizes parallels between elements.

The template involves four sequential steps.

1) Removal of Insoluables - Most bioseparations begin with this step, often accomplished by filtration and centrifugation. Relatively little concentration of the product or improvement of its quality occurs here.
2) Isolation of Product - These steps are relatively non-specific, and remove materials of widely divergent properties compared to the desired product. The resulting concentrate can be effectively processed by more routine chemical means. Examples are extraction and adsorption.
3) Purification - Once an impure concentrate is obtained, we purify the product. This is the point where bioseparations involve the greatest diversity. Chromatography is probably the most common example.
4) Polishing - Finally, most bioseparations involve an additional purification to produce the final commercial product. If possible, this fourth step

depends on crystallization. It almost always includes drying.

We remember these four steps as a RIPP sequence: (R)emoval of Insoluables; (I)solation of Products; (P)urification; and (P)olishing.

To me, the most important aspect of this template is the split between isolation and purification. Isolation primarily tries to concentrate the product. If some purification also occurs, great; if it doesn't, we will handle it later. I find that microbiologists implicitly understand this distinction; but engineers, trapped in the equations of unit operations, do not appreciate it.

Super Critical Gels

I now turn from my first goal of this overall template to my second goal, the description of a method of protein isolation using gels.

Our gels are an alternative to ultrafiltration. In ultrafiltration, a feed solution of protein is forced through a membrane. The membrane retains the protein solution. This retention can cause the significant concentration polarization which is commonly minimized by high crossflow. Even with this crossflow, concentration polararization can dramatically compromise ultrafiltration flux.

We would greatly prefer a method of ultrafiltration which retains water molecules and passes the protein molecules as a concentrate. We can achieve this using a crosslinked polymer gel. The gel absorbs water but not protein. When the resulting suspension is filtered, the gel containing the water is retained, but the protein passes through the membrane.

The success of this process depends on having a gel which can be regenerated and reused. After all, silica gel would work; it would absorb water but not protein. However, to be reused, silica gel would have to be dried under high heat, a step which would make the process expensive.

Instead of silica gel, we plan to use gels like that shown in Fig. 1. In this figure, there are three test tubes each containing the same mass of gel. The test tube at the left contains dry gel, but the two test tubes at the right contain gel immersed in water. You can see the meniscus across the tubes. The only other difference between these three test tubes is the temperature. The

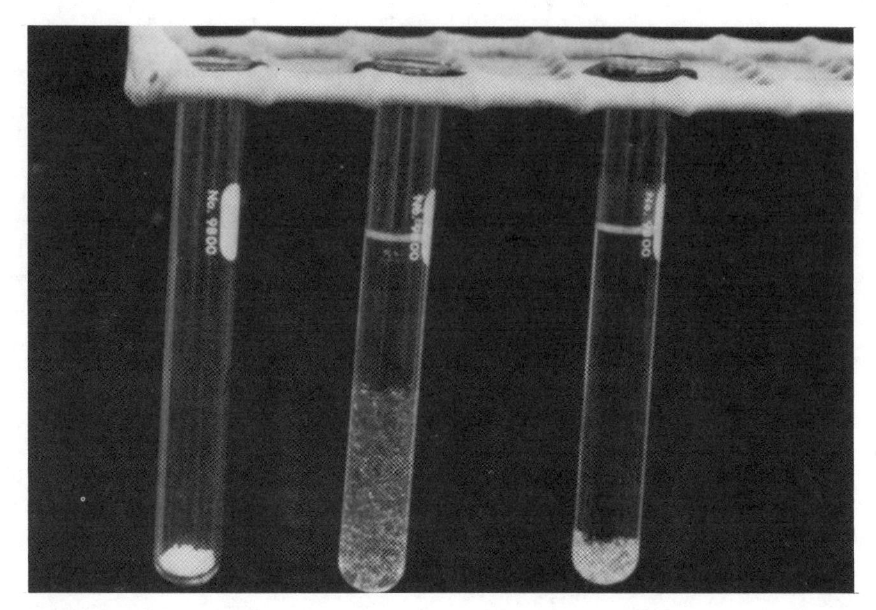

FIGURE 1. Gel Volume vs. Temperature - Each tube contains the same mass of polyisopropylacrylamide, one of the gels used here. That in the left hand tube is dry and at 33°C; that in the center tube is wet and at 33°C; that in the right hand tube is wet and at 34°C. (Permission granted by Pergamon Journals Ltd.)

test tube at the left is at 33°C. The test tube in the center is also at 33°C. You can see how much the gel is swollen, to about 20 times its dry volume. The test tube on the right is at 34°C and yet has a volume 7 times less than that in the center. Thus we can regenerate a gel like this simply by warming it slightly.

The way in which this gel can be used to concentrate proteins is shown in Fig. 2. In this figure, a collapsed gel is initially added to a cool protein solution. The gel swells. Because of its crosslinked polymer nature, it absorbs water but excludes macromolecular solutes like proteins. It then is filtered out of the solution, leaving behind a protein concentrate. The gel is then warmed slightly with waste heat to release most of the absorbed water. The gel is again removed by filtration, cooled, and reused. We have shown that the gel can be reused without significant degradation for at least one hundred cycles. Moreover, the gel particles are not at all sticky and so are easily filtered.

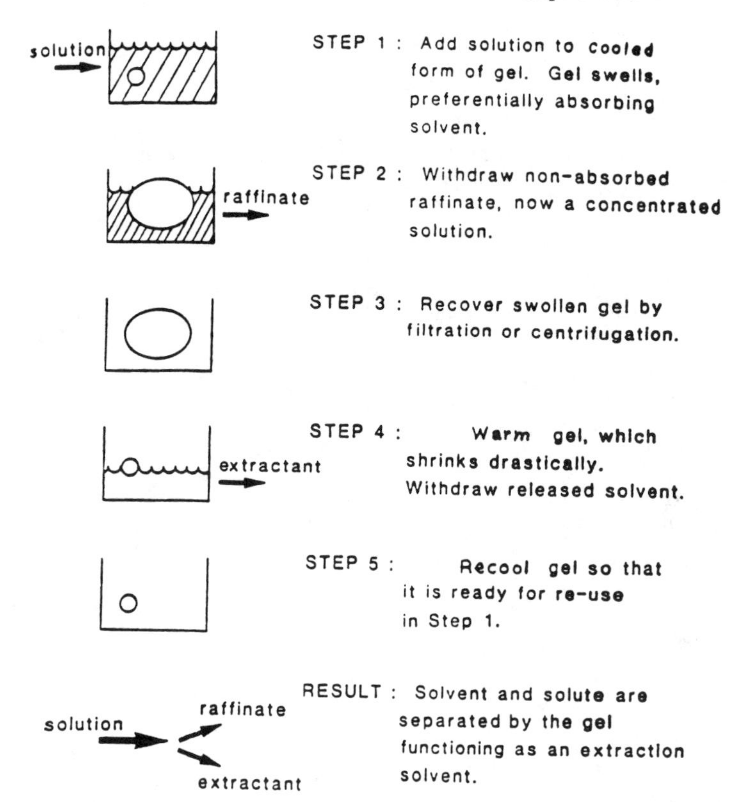

FIGURE 2. The Gel Process - This scheme can cheaply concentrate proteins and microorganisms from dilute solution. As such, it is part of the isolation step described in the text.

I want to now move to two aspects of this gel which are important. First, what will it separate? Second, why does the gel show this unusual behavior?

Gel Selectivity

There are three aspects of gel selectivity which are important to this paper. The first and the most important is that the gel will separate largely by size, as shown in Table 1. In this table, the left hand column gives the

TABLE 1

GELS SEPARATE LARGE SOLUTES

(results for polyisopropylacrylamide)

Solute	Molecular Weight	Separation Efficiency
latex	70,000,000	95%
blue dextran	2,000,000	97
gelatin	500,000	98
albumin	45,000	97
vitamin B-12	1,355	32
urea	60	2

solute and the center column gives an approximate molecular weight, an indication of solute size. The right hand column gives the efficiency of separation. This efficiency is very much like a percent rejection. An efficiency of 100% would mean that the solute is completely excluded from the swollen gel. An efficiency of 0% would mean that the solute can move freely from solution into the gel and back again.

The data in Table I show that the gel separates largely on the basis of solute size. Large solutes like proteins are excluded from the gel but small solutes like sugar and antibiotics penetrate the gel. Other experiments show that gel selectivity is also a function of solute shape, that compact solutes penetrate the gel more easily than extended solutes.

Gel selectivity can also be altered by changing the crosslinking in the gel. In a variety of experiments detailed elsewhere, we've shown that the logarithm of this selectivity depends approximately of the density of crosslinking. However, these adjustments in selectivity are not precise: we have difficulty separating between two proteins of molecular weights 30,000 and 50,000. We can coat the gels with reverse osmosis and ultrafiltration membranes to sharpen selectivity, but this makes the separations slower. In the same sense, most of the gels which we've studied are relatively uneffected by solute charge, although in some cases, Donnan effects are important.

I don't think this situation is going to change. I think that our gels will remain a simple, crude method and not have the exquisite selectivity promised by, for example, affinity adsorption.

The Causes of Gel Behavior

The success of these separations leads to the second question: why do the gels show these unusual properties? In particular, why does their swelling change so violently as a function of temperature?

The gels which we've used are largely based on crosslinked, substituted acrylamides. As such, they are cousins of the gels used in gel permeation chromatography. In chromatography, however, one seeks gels whose properties are as constant as possible. The last thing we want in chromatography is gels whose swelling changes with temperature or pressure or pH. In contrast, the gels which we've used here can have tremendous changes with temperature or pH.

These tremendous changes occur because the gels are close to critical points or phase transitions as implied by the title of this paper. As such, they are rough parallels to a van der Waals gas, as suggested in Fig. 3. In the left of this figure, we've drawn the usual picture of pressure vs. volume for a van der Waals gas. Such a gas has a two phase dome. The top of this two phase dome is of course the critical point. Isotherms make a characteristic S shaped curve above this critical point and show an even more abrupt change in volume with the phase transition from liquid to vapor.

The gels are analogous: their phase diagram has a similar shape. However, to get this same shape we must plot temperature vs. volume instead of pressure vs. volume. We still have a two phase dome, a collapsed gel phase to the left of the dome, and a swollen gel phase to the right of the dome. We still have a critical point. We still have S shaped lines, although these are now isobars - lines of constant pressure - rather than isotherms - lines of constant temperature. Still, the effect is parallel: small changes of temperature can

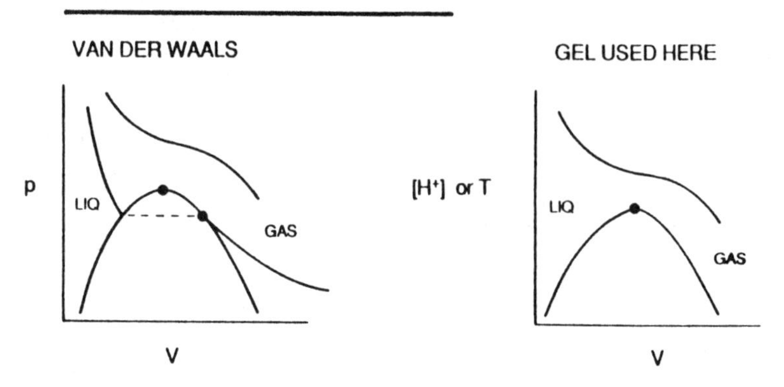

FIGURE 3. Gels vs. a van der Waals Gas - The gels show violent swelling changes because they are near a critical point. These changes can be analyzed in parallel to the pressure induced changes of a van der Waals gas.

produce large changes in gel volume. Parenthetically, Toyoshi Tanaka at M.I.T. has studied this critical point explicitly and has found scaling laws and critical opalescence, just like other critical points. As a result, these gels really are a supercritical solvent which you can hold in the palm of your hand.

I should caution you that the thermodynamics of these gels are substantially more elaborate than this superficial analogy with the van der Waals gases has suggested. In particular, the phases in our system separate on warming and the critical point which occurs is that special kind called a lower consolute solution temperature. It is a critical point associated not with enthalpic effects, but with non-ideal entropy, and as such, retains some mystery compared with more common critical phenomena.

Conclusions

In this paper I've had two goals. I've talked first about a way which biological separations can be designed in terms of a four step template. I've then described one new technique, supercritical gel extraction. The supercritical gel extraction provides an easy and gentle way to concentrate proteins. While it is experimental, it appears to be a promising alternative to ultrafiltration.

Protein Purification: Micro to Macro, pages 315–328

RECYCLING INSTRUMENTATION FOR PREPARATIVE SCALE ELECTROPHORESIS

Richard A. Mosher, Ned B. Egen and Milan Bier

Center for Separation Science, University of Arizona, Tucson, AZ 85721

ABSTRACT A recycling instrument is described for high resolution, free fluid separation of proteins by isoelectric focusing. This device employs a modular design which provides an easily scaled process volume. Proteins which differ by 0.1 pH units can be routinely resolved. In addition to isoelectric focusing in a batch mode the instrument can operate, with only slight modification, in a continuous zone electrophoretic mode called feed and bleed.

INTRODUCTION

The ability of the biotechnology industry to produce specific proteins in very large quantities has led to a search for improved methods of protein fractionation. Chromatographic techniques are routinely used when a high resolution step is required. While these methods provide good resolution in the laboratory, there is often a loss in performance with scale-up. In addition, they are batch procedures which require expensive supports having finite lifetimes and often requiring cleaning and/or regeneration after each use. These shortcomings point out the need for additional high resolution techniques for protein purification.

Electrophoretic methods are unsurpassed for the resolution of complex proteinaceous mixtures on an analytical scale. They are most often performed in gels, which provide high resolution due to their very high degree of fluid stability. While some scale-up of these systems is possible, the use of a solid support limits the physical dimensions of the separation space because of the need to dissipate Joule heat. The matrix also limits the flow rate of a continuous process. Additionally, it must be removed from the recovered samples and it contributes to adsorptive losses of the product. For these reasons, successfully scaled-up electrophoretic processes have eliminated any solid support and perform the separation in free fluid. One of the first concepts for the performance of free fluid electrophoresis was developed by Hannig (1). The separation chamber was a rectangular narrow gap between 2 flat plates, through which buffer flowed. The electric field was applied transverse to the flow. This device was designed for continuous operation in the zone electrophoretic (ZE) mode. Sample was introduced to the buffer as it entered the chamber and the electric field caused the constituents to be deflected to a greater or lesser degree. An array of tubes at the chamber outlet collected the separated fractions.

Although ZE is a high resolution method for protein fractionation, isoelectric focusing (IEF) provides an even greater resolving power. In ZE the buffer is uniformly distributed, and the buffer and sample components are continuously transported by the electric field. In contrast, IEF is characterized by the presence of a pH gradient which monotonically increases in the direction of the cathode. This pH gradient is usually formed by the action of the electric field on a complex mixture of amphoteric buffering components. The mode is unique in that all components present, both buffer and sample, produce a final steady state distribution which is approached asymptotically at constant voltage. For this reason it is difficult to perform an IEF separation in the continuous-flow Hannig type devices. The formation of the background pH gradient and the focusing of the sample components within the gradient is a lengthy process. Even if the cooling capacity of the instrument will allow these slow flow rates, the throughput advantage offered by continuous flow is lost.

The stationary steady state achieved in IEF makes this process amenable to operation in a recycling mode. Two devices which utilize this mode have been developed at the Center for Separation Science. Each employs a flow through separation chamber with an electric field perpendicular to the flow. The fluid passing through this chamber exits into a tubing array which ultimately carries it back to the entrance and reinjects it at the same axial position. With each pass through these devices the buffers and sample components undergo a partial migration toward their isoelectric points. Several passages through the focusing cell are required to produce the steady state distribution. The 2 instruments differ most fundamentally in the manner in which fluid is stabilized against convective mixing within the separation cell. One utilizes a Hannig type cell with a very narrow gap (0.75mm) between the plates. In the other device the rectangular separation cell has membrane defined subcompartments. The first is called RF3 for recycling free flow focusing and the second RIEF for recycling isoelectric focusing. Protein fractionations achieved with this second device will be described in this paper.

THE RIEF INSTRUMENT

A schematic of the device is presented in Fig. 1. Its 3 essential components include a multichannel, peristaltic pump which recycles the process solution between a heat exchange reservoir and a focusing cell. The heat exchange reservoir holds the bulk of the process fluid and comprises a plexiglass box through which pass 12 glass tubes, ten for the sample channels and two for the electrolytes. The focusing cell consists of ten plexiglass plates with a center portion removed, separated from each other by nylon screens with 10 micron pore size. The cavities in the plexiglass plates, and the nylon screens, define the sample subcompartments. A port at the top and bottom of each subcompartment allows fluid ingress and egress. The screens prevent bulk fluid flow between sample chambers but allow unhindered electromigration of proteins and buffers. Each plexiglass endplate has a recessed cavity for the electrolyte and a platinum wire electrode. The electrolyte chambers are separated from the focusing compartment by ion-exchange membranes. The distance between these membranes is 3 cm and the total volume of sample in the focusing cell is 25 ml. In normal operation the heat exchange reservoir is positioned above the focusing cell. Fluid is gravity fed from the reservoir into the top of the focusing cell. The pump then removes the fluid from the bottom of the cell and returns it to the top of the heat exchange reservoir. The design allows the process volume to be easily changed by inserting a different reservoir.

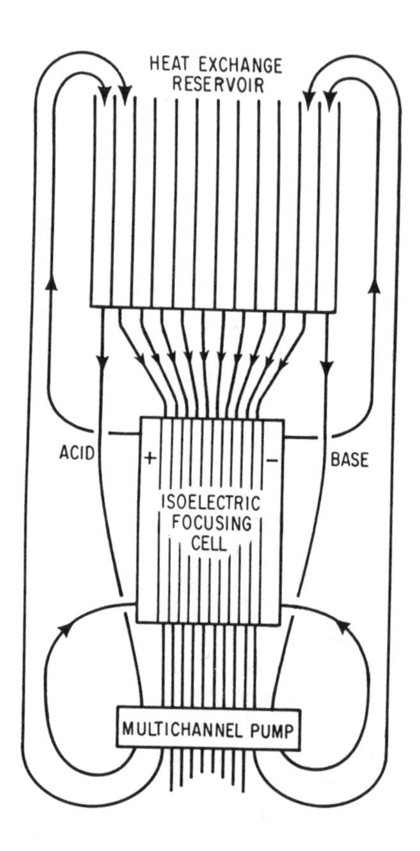

Fig. 1. Schematic of the fluid flows in the RIEF apparatus in the IEF mode.

Operating Conditions

The maximum power which can be applied to the device is primarily dependent on the characteristics of the pump which was constructed in our lab. It can maintain a flow rate of 30 ml/min/channel with less than 3% difference between channels. This allows 100 watts to be applied, corresponding to 400 V/cm and 80 mAmp. Greater flow rates result in increased flow rate differences between the channels and the pressure gradients produced in the focusing cell cause mixing between the chambers.

The maximum throughput of the instrument is about 1 g/hr, depending on the nature of the sample and the width of the pH gradient. Proteins which differ in pI by 0.1 pH units can be routinely separated, with the best resolution being 0.05 pH difference. Synthetic carrier ampholytes such as Ampholine are normally used to create the pH gradient. In some cases when the protein of interest tightly binds these buffers, it is possible to create a pH gradient using simple, well defined buffers such as amino acids (2).

One of the inherent limitations in IEF is the tendency of proteins to precipitate during the process. This is due to the low ionic strengths exhibited by systems of focused ampholytes and because proteins are minimally soluble at their isoelectric points. If precipitation occurs, the run is frequently aborted because of occlusion of the nylon screens and/or enhanced electroosmosis (EEO). Two approaches are useful to avoid these problems. Since most complex protein solutions contain constituents which will precipitate at low ionic strengths, all samples are routinely desalted before focusing. Any precipitate is removed by filtration or centrifugation. This step generally eliminates precipitation during the run. If the protein of interest is removed by these treatments, a variety of additives can be used to promote solubility. These include glycine, glucose, glycerol and non-ionic detergents such as Chaps, Chapso, NP-40 or Brij. An additional effect of salt removal is maximization of the number of fractions in the linear portion of the pH gradient. If allowed to remain, these strong electrolytes migrate to the end channels where they cause extremes of pH.

RESULTS

The resolving power of the instrument is illustrated by the data presented in Fig. 2. Four hundred mg of a mixture of human hemoglobins A, C and S were dissolved in 200 ml of a 1.8% (w/v) solution of Ampholine, pH range 7.2-7.6. For this experiment the machine was operated with 20 sample channels. Initially, 200 V at 22 mAmp were applied. At the completion of the 11 hour run the conditions were 1200 V at 31 mAmp. Collected fractions

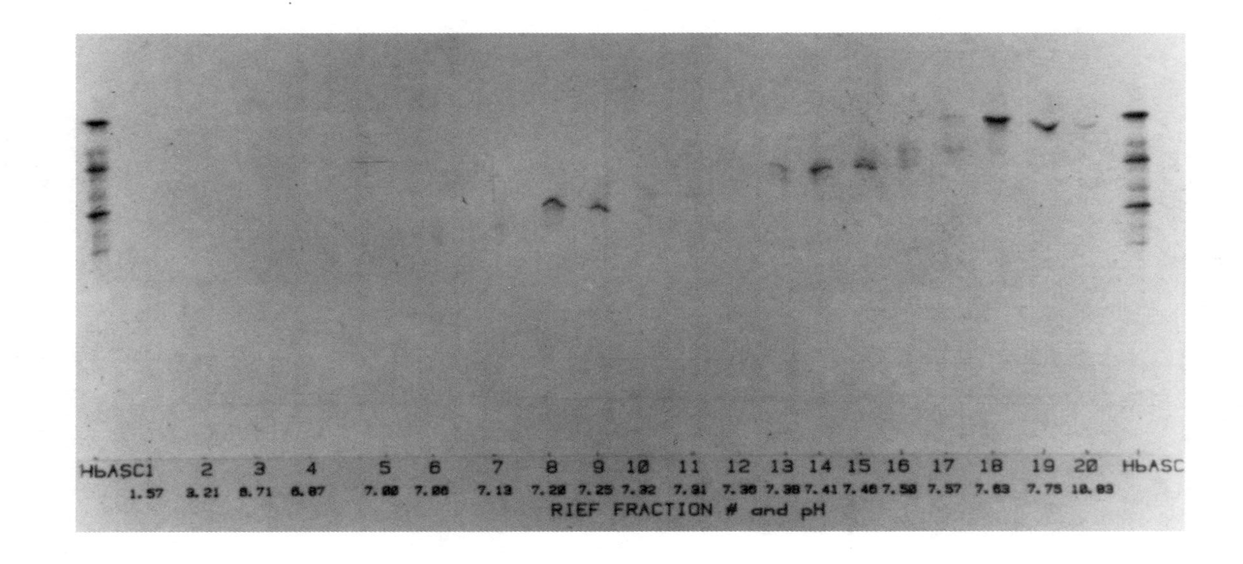

Fig. 2. IEF gel analysis of a RIEF separation of hemoglobins A, S and C by IEF. The end lanes display the starting material. The pH appears below each channel number. Minor bands are resolved in fractions 12 and 17.

were analyzed by IEF on a flat-bed polyacrylamide gel, pH range 7-9. No stains were used. The figure shows the hemoglobin A variant (pI = 7.0) focused in channels 8 and 9, the C variant (pI =7.6) focused in channels 14 and 15 and the S variant (pI = 7.8) in channels 18 and 19. The original mixture is shown at both ends of the gel. The major proteins (pI difference = 0.2 pH units) were easily resolved. In this experiment there were minor hemoglobin components which focused between the major ones, demonstrating a resolution of 0.05 pI units. This was the only experiment for which the machine was configured with 20 channels. A more efficient way to increase the resolution is to resubmit selected fractions to a second processing, using the carrier ampholyte already present to establish a narrower pH gradient. This method is described below.

The fractionation of ascites fluid containing a monoclonal antibody is shown in Fig. 3. Ten ml of ascites fluid containing 250 mg of protein, 70 mg of which were the antibody, was diluted to 100 ml using 1% Ampholine, pH 5-9 and 3 M urea. This mixture was focused in the RIEF for 3 hours and the fractions collected. Initial and final conditions were 150 V at 36 mAmp and 500 V at 29 mAmp respectively. The figure displays a Coomassie blue stained analytical IEF polyacrylamide gel. The ten RIEF fractions are bracketed by lanes containing the original sample. Ampholine with a pH range of 3.5-10 was used in the gel in conjunction with 3M urea to maintain antibody solubility. The heterogeneous monoclonal antibodies are present in fractions 5-7. Some subfractionation of the antibody has been achieved. That these bands represent the antibody is confirmed by the autoradiogram in panel B. It was produced by overlaying a gel (prepared as in panel A), containing the focused ascites fluid, with the radioactive antigen (phencyclidine). After incubation for 1 hr, the unbound antigen was removed by washing, the gel was dried and exposed to x-ray film for 24 hr.

Although increasing the resolution by resubmitting selected fractions to a second run is useful for enhancing resolution it may result in an unacceptable amount of handling of labile proteins. An alternative approach is to simply use a carrier ampholyte mixture having a narrow pH range initially. This will result in high concentrations of protein in the extreme RIEF channels, which is acceptable as long as no precipitation occurs. Narrow range carrier ampholytes can be obtained

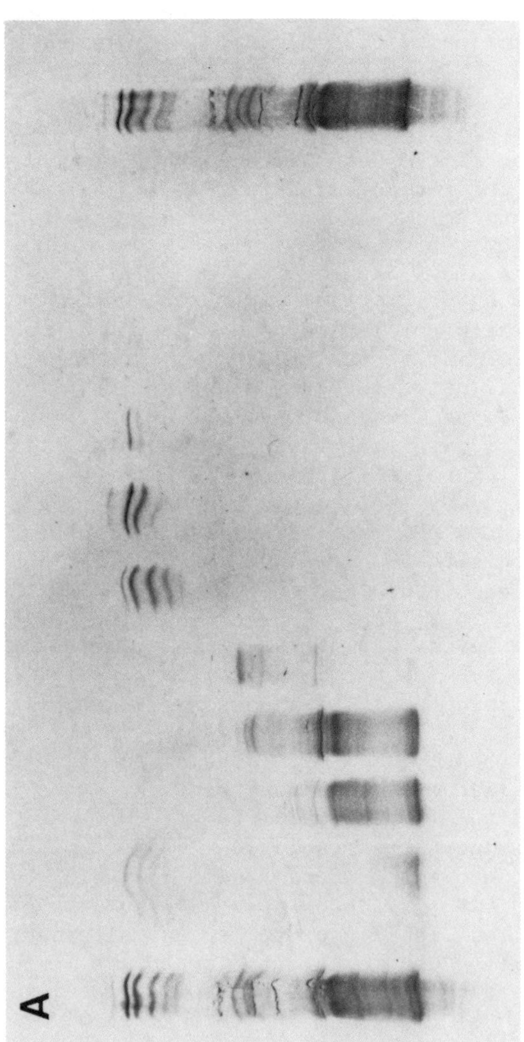

Fig. 3. IEF gel analysis of a RIEF fractionation (pH 5-9) of ascites fluid containing a monoclonal antibody. In panel A the original material, in lanes 1 and 12, brackets the 10 RIEF fractions. Panel B is an autoradiogram which was used to identify the antibody. An IEF gel pattern of the ascites proteins (pH 3.5-10) was treated with the radioactive antigen, washed and exposed to xray film.

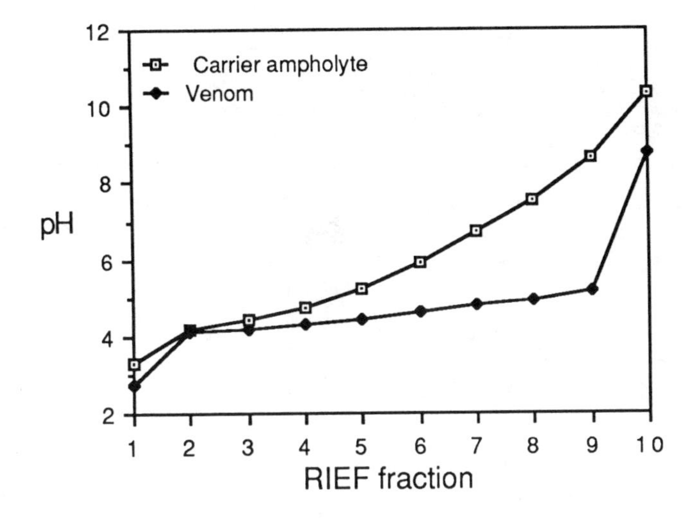

Fig. 4. The open squares present a pH gradient formed when a homemade carrier ampholyte mixture was focused in the RIEF. Fractions 2-5 of this run were pooled and used to fractionate a snake venom sample. The gradient produced is shown with the solid diamonds.

by RIEF fractionation of commercially available or homemade preparations (3,4). Fig. 4 presents 2 pH gradients. The first was obtained by focusing homemade carrier ampholytes in the RIEF. The second was generated when fractions 2-5 from this experiment were pooled and used for snake venom (Agkistrodon halys blomhoffii) fractionation. Two hundred mg of venom were fractionated in 130 ml of a 1% solution of ampholytes, pH range 4.1 to 5.1. This experiment was done at a constant power of 100 watts for 13 hrs, with initial and final voltages of 580 V and 760 V, respectively. The silver stained gel (pH 3.5-10) in Fig. 5 shows the results of the separation. Fraction 4 contains 2 proteins, and all of the phospholipase activity present in the venom. Note the large number of proteins in channel 10. These are all the venom proteins which are more basic than pH 5.1.

The flexibility of the RIEF is illustrated both by its easily scaled process volume and by the capacity to

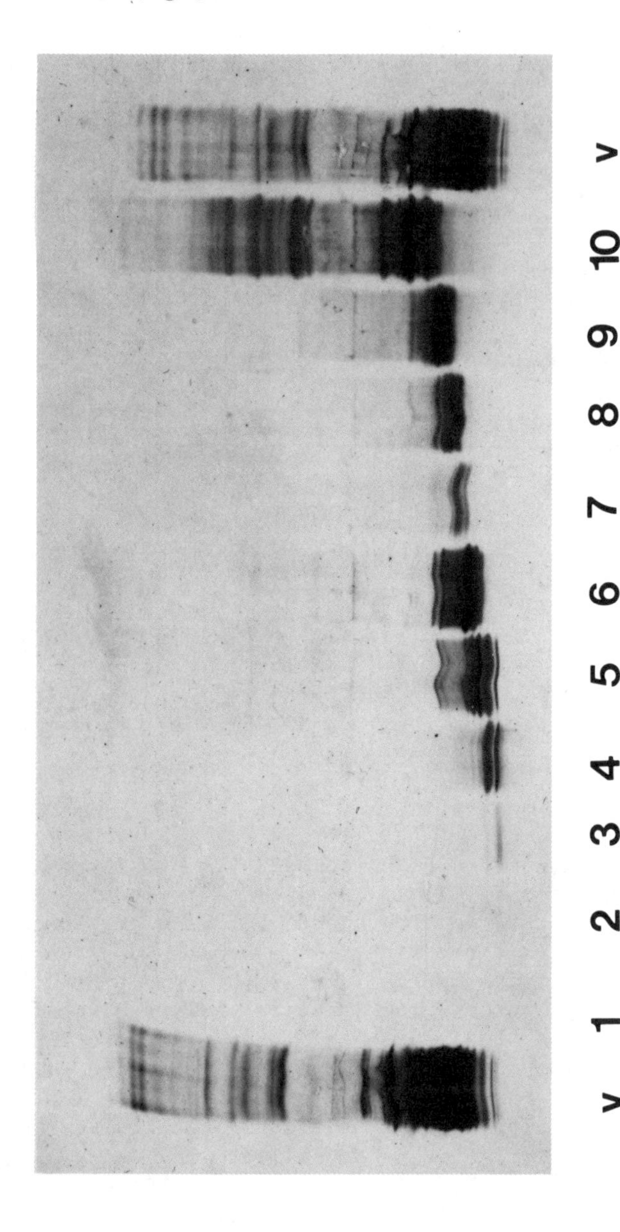

Fig. 5. An IEF gel was used to evaluate a RIEF fractionation of snake venom proteins (pH 4.1–5.1). The gradient was tailored to isolate a phospholipase (fraction 4). The 10 RIEF channels are bracketed by the original sample (V).

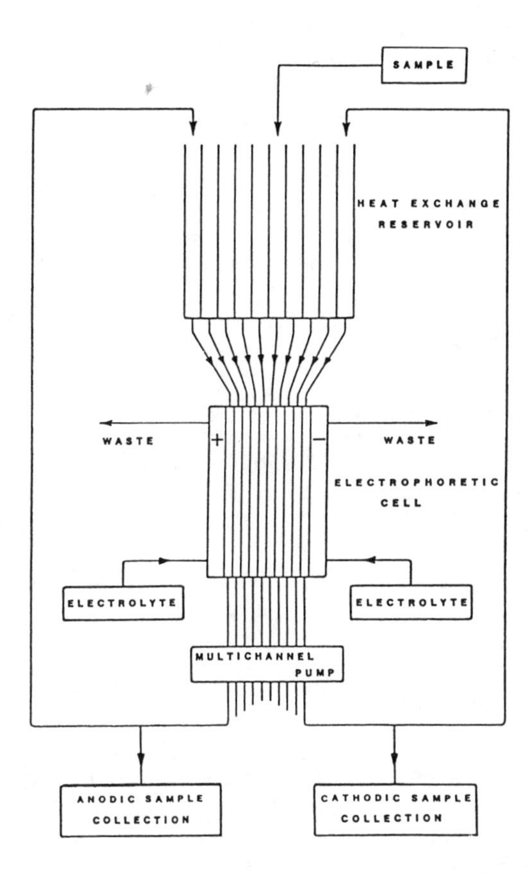

Fig. 6. Schematic of the fluid flows in the RIEF apparatus in the feed and bleed mode.

operate in a second electrophoretic mode. This high throughput method is called feed and bleed. The instrument is configured differently for this mode, as is illustrated in Fig. 6. This schematic should be compared to Fig. 1. In both modes the 10 sample channels flow from the heat exchange reservoir into the top of the focusing cell and the pump recycles the fluid back to the reservoir. Electrolyte flow is always upward through the electrolyte compartments to flush the gases generated by electrolysis.

In IEF the electrolytes are recycled whereas in feed and bleed they are discarded. No fluid is added or discarded during an IEF experiment. The feed and bleed procedure involves a continuous infusion of sample to a central reservoir of the heat exchanger and continuous withdrawal of product streams from channels 1 and 10. A uniform buffer is used in this mode and the separation is based on zone electrophoretic principles. This method produces only 2 fractions, those sample components more acidic, and those more basic, than the buffer pH. The results of a fractionation of human serum by this mode are shown in Fig. 7. The sample was dialyzed, diluted 1:5 in the background buffer (10 mM cycloserine, 3 M urea, pH 6.0) and infused at 10 ml/min. The power applied was 75 watts, at 1000 V. Fractions were withdrawn from channels 1 and 10 (5 ml/min) and analyzed by gel IEF, pH 3.5-10. The original sample has been applied in lane 1. Serum was processed at the rate of 9 g/hr in this experiment.

DISCUSSION

Recycling free fluid IEF is a high resolution method for protein fractionation, appropriate for the so called "polishing" step in purification protocols. It has been successfully applied to a wide variety of samples including mono- and polyclonal (5) antibodies and the products of recombinant DNA technology (6). An important characteristic of this instrument is its flexibility. It can operate in the high resolution batch IEF mode with an easily changed capacity, or a high throughput continuous zone electrophoretic mode. The RIEF has also been complimented with pH and UV sensors for each sample channel (7). The outputs of these sensors are fed to an analog to digital converter and captured by a computer in real time. This allows concurrent process monitoring and provides a measure of feed back control of the separation.

While there has been little demand for large scale research instruments, a smaller scale instrument called the Rotofor, with a volume of 50ml, has been developed (8). The focusing chamber is an annulus which has 20 membrane-defined subcompartments. Recycling is not used, the device rotates around a center cooling tube to insure adequate mixing and cooling. This instrument is available from BioRad.

The large scale electrophoresis equipment which is

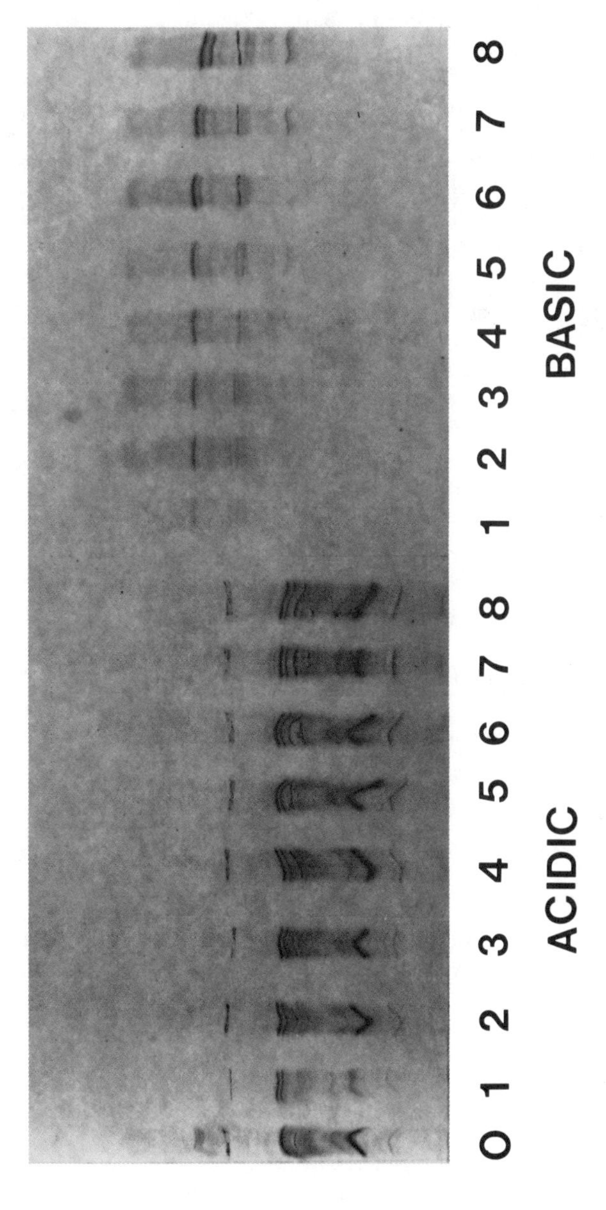

Fig. 7. An IEF gel showing the composition, at 10 min. intervals, of acidic and basic fractions obtained from a feed and bleed separation of human serum. The leftmost lane shows the original sample.

commercially available covers a wide range of throughputs, offers high resolution and in some cases very flexible operation. Included are the Biostream device marketed by CJB Developments of Portsmouth, England, the Elphor VaP manufactured by Bender and Hobein of Munich and the IsoPrep (based on the RIEF) by Ionics Corp. of Watertown, MA. The first two are present in our laboratory and in conjunction with our own preparative technology constitute the largest collection of this kind of equipment we are aware of. In addition our lab is currently developing new methods of preparative electrophoresis such as recycling isotachophoresis. This work is described elsewhere in this volume (Sloan et al.).

ACKNOWLEDGMENTS

This work was sponsored by NASA grant NAGW-693.

REFERENCES

1. Hannig, K (1961). Die tragerfreie kontinuierliche elektrophorese und ihre anwendung. Z Anal Chem 181:244.
2. Bier M, Mosher RA, Palusinski OA (1981). Computer simulation and experimental validation of isoelectric focusing in ampholine-free systems. J Chromatogr 211:313.
3. Binion SB, Rodkey, LS (1981). Simplified method for synthesizing ampholytes suitable for use in isoelectric focusing of immunoglobulins in agarose. Anal Biochem 112:362.
4. Binion SB, Rodkey LS, Egen NB, Bier M (1983). Properties of narrow-range ampholytes isolated from wide-range ampholyte preparations. Anal Biochem 128:71.
5. Binion SB, Rodkey LS, Egen NB, Bier M (1982). Rapid purification of antibodies to single band purity using recycling isoelectric focusing. Electrophoresis 3:284.
6. Nagabhushan TL, Sharma B, Trotta PP (1986). Application of recycling isoelectric focusing for purification of recombinant interferons. Electrophoresis 7:552.
7. Bier M, Egen NB, Allgyer TT, Twitty GE, Mosher RA (1979). New developments in isoelectric focusing. In Gross E, Meienhofer J (eds): "Peptides: structure and biological function" Rockford: Pierce Chemical Co., p 79.
8. Egen NB, Thormann W, Twitty GE, Bier M (1984). A new preparative isoelectric focusing apparatus. In Hirai H (ed): "Electrophoresis '83" Berlin: de Gruyter, p 547.

Protein Purification: Micro to Macro, pages 329–335

MATHEMATICAL MODELING AND ANALYTICAL ISOTACHOPHORESIS AS PREDICTORS FOR PREPARATIVE RECYCLING ISOTACHOPHORESIS[1]

Jeffrey E. Sloan, Richard A. Mosher, Wolfgang Thormann
Millicent A. Firestone and Milan Bier

Center for Separation Science, University of Arizona
Tucson, Arizona 85721

ABSTRACT: Computer simulation and analytical capillary isotachophoresis have been used to determine proper electrolytes and spacer molecules for preparative recycling isotachophoresis.

INTRODUCTION

Current preparative protein separation techniques in the biotechnology industry include precipitation, centrifugation, filtration, chromatography, and others. Electrophoresis, although a common analytical technique, remains largely unexploited for large scale bioseparations.

Isotachophoresis (ITP) is a high resolution electrophoretic technique which has primarily been utilized for the analysis of low molecular weight compounds. In contrast to the more widely used zone electrophoresis which has a uniform background buffer, ITP is characterized by a discontinuous buffer system consisting of a leading electrolyte of high net mobility and a terminating electrolyte of much lower net mobility. Sample compounds with electrophoretic mobilities between these two are injected at the interface. The unique feature of ITP is the attainment of a moving steady state consisting of adjacent sample zones (the stack) migrating at a constant relative velocity. The length of each zone is proportional to the quantity of component present.

1. This work was supported in part by NASA Grant NAGW-693 and NSF Grant CBT-8311125-01.

Amino acid and small peptide spacers have been used in ITP of complex protein samples (1). A selective spacer allows two proteins which would migrate in adjacent zones to be separated more distinctly. Two spacers may be used to bracket the protein of interest in a complex sample. Such spacers are particularly advantageous because they are simple to remove from the mixture by any technique based on molecular size (dialysis, gel filtration, etc).

The concepts of recycling isoelectric focusing and continuous flow electrophoresis are successful approaches to preparative electrophoresis (2-4) and instruments of both types are commercially available. Recycling ITP (RITP) is the process under investigation for which computer simulation and analytical ITP are shown to elucidate information important to the determination of operating parameters.

METHODS

Computer Simulation

The mathematical model used was that of Bier et. al.(5) which is capable of predicting the dynamic behavior of electrolyte systems. This model represents all four modes of electrophoresis and is revolutionary in recognizing that the essential differences between these modes resides in the initial and boundary conditions. The fundamental principles of the model are: i) any component flux is solely a result of electromigration and diffusion, ii) charge and mass are conserved, iii) chemical equilibria are considered, and iv) electroneutrality is satisfied. This model comprises the following assumptions: i) there is an absence of bulk flow, ii) the separation is performed under isothermal conditions, and iii) the separation geometry is one dimensional. Even in the light of these simplifications the set of equations generated is non-linear and must be solved numerically.

At the start of the simulation the column length, current density, initial distribution of each component, and amount of electrophoresis time are specified. The column length is divided into a grid of uniform density, typically 50 segments per centimeter. All components are defined by their electrophoretic mobility and pK value(s). The model will predict the concentration of each component and the pH and conductivity at each grid point for each requested time point.

Capillary Isotachophoresis

All of the predictive capillary separations were performed on the LKB Tachophor 2127. This device utilizes a variable length open bore teflon capillary with a 500 micron lumen and large electrode reservoirs at each end. The components migrate under the influence of the electric field through a conductivity detector and an ultraviolet absorbance detector at the end of the capillary. The wavelength to be monitored is user selectable through the use of filters which allows for detection of specific components. The entire time of analysis is approximately 20 minutes, and the amount of sample required is minimal (0.5-1 nmol of each component). The advantage of this method of predicting electrolytes is that the same solutions may be used preparatively.

Recycling Free Fluid Isotachophoresis

RITP was performed in the Recycling Free Fluid Focusing (RF3) apparatus which was originally developed for isoelectric focusing (4). The geometry of the separation chamber is that of a thin film of fluid of rectangular cross section (350mm x 55mm x 0.75mm) which is subdivided into 48 individual channels at the inlet and outlet. One wall of the chamber is cooled and maintained at a temperature near 0° C. After exiting the chamber, each channel is passed through an individual heat exchanger and reinjected in the corresponding position. The heat exchanger also serves as a pulse and bubble trap in which the majority of the process fluid volume resides. This allows for simple scale up using additional heat exchange modules.

The automated system which we have developed for RITP consists of the separation cell along with its accessory components (recirculation pump, heat exchanger, and power supply), a UV sensor for determining the position of the front interface between the leading electrolyte and the sample components, a method of sample injection, a method of applying a bulk counterflow of leading electrolyte, and a computer for data gathering/data treatment and process control. A 6502 microprocessor based data acquisition/control system is used to control the rate of counterflow thereby modifying the position of the front. Once the boundary is detected, bulk counterflow can retard or even stop the migration. This system also provides storage of process data, a real time monitoring system, and the capacity for producing hard copy of intermediate or final results.

The entire process volume is collected at the end of the run in 48 discrete fractions by the use of a multichannel valve to switch all of the recycling channels to a collection apparatus. The processing time required is between one and three hours depending upon the composition of the system and the degree of resolution required. The loading capacity is approximately one gram per run. Under the proper conditions however RITP could be made into a continuous process.

RESULTS

The model system had as leading electrolyte 10 mM HCl titrated to pH 9.5 with 2-amino 2-methyl 1,3 propanediol (ammediol), and a terminating electrolyte of 10 mM epsilon amino caproic acid (EACA) adjusted to pH 10.5 with ammediol and $Ba(OH)_2$. Albumin (Alb) and hemoglobin (Hb) were used as model proteins in this system.

The Tachophor was used to determine a proper spacer molecule for this system (Fig. 1). A series of experiments revealed that a number of neutral amino acids would space the two proteins to varying degrees. Glycine was chosen as the spacer by virtue of its low price, reasonable solubility and relative spacing characteristics. In the computer simulation the behavior of glycine was modeled in the same system and

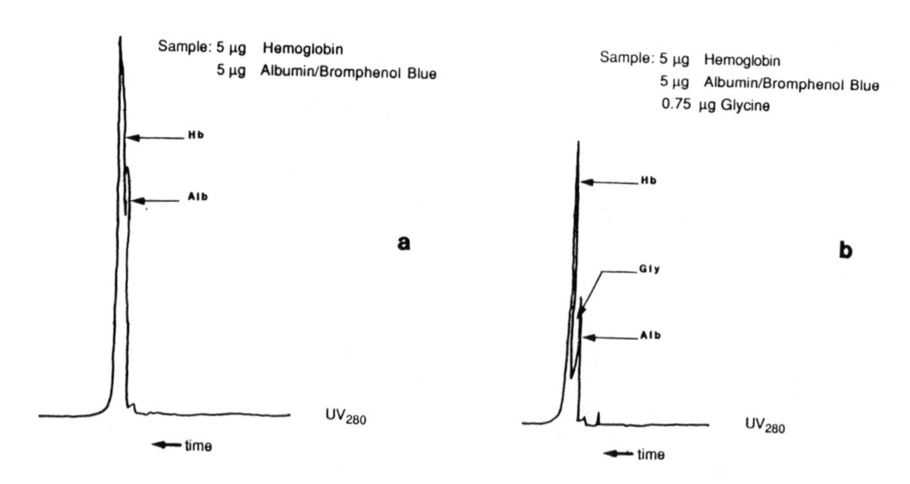

FIGURE 1. Capillary ITP data for Alb/Hb (a) and Alb/Gly/Hb (b).

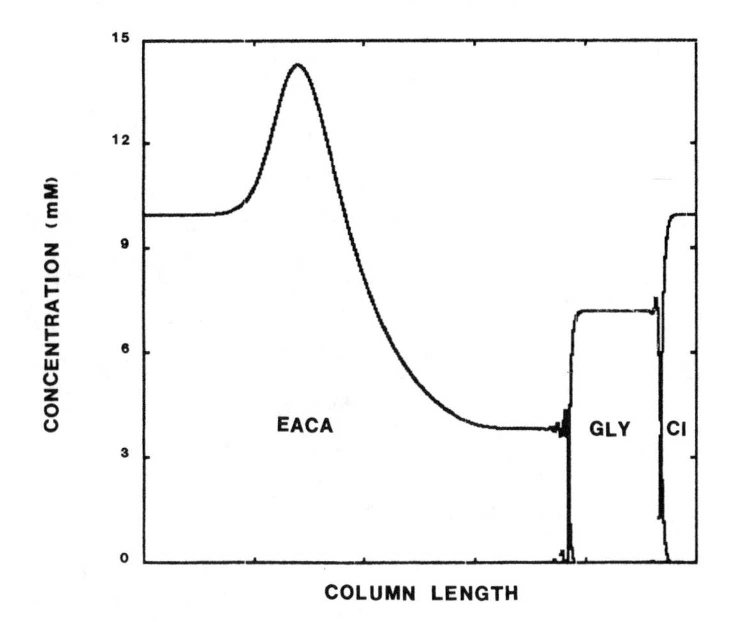

FIGURE 2. Computer simulation data depicting the steady state Gly zone between Cl and EACA.

was shown to migrate in an ITP zone between the two electrolytes (Fig. 2). The preparative ITP experiments validated the usefulness of glycine as a spacer molecule for this system (Fig. 3).

DISCUSSION

Analytical ITP can demonstrate the order of the individual components in the stack, whether a given pair of amino acids will indeed bracket the protein in the stack, and whether the choice of leading and terminating electrolytes is appropriate for the system under consideration.

Optimal use of computer simulation would require the knowledge of the electrokinetic parameters of each component in the system. Realistically these data exist for few proteins and may be best generated using isotachophoresis. Mathematical modeling is not the method of first choice to determine which electrolyte system should be employed. After the selection of a system however, computer simulation can be utilized as a powerful method for optimizing electrophoretic

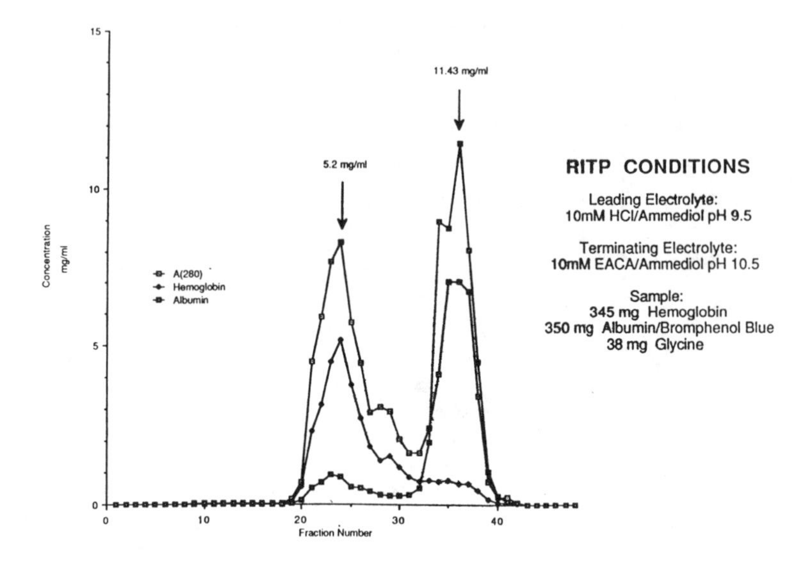

FIGURE 3. RITP separation of Alb and Hb with Gly as spacer.

buffer systems. The simulations provide quantitative as well as qualitative data regarding the interactions of the system components. The model can also accurately predict the steady state zone concentrations, this information can be used to determine if the limits of solubility will be exceeded or if it is possible to further concentrate these zones.

The separations in the RF3 and the capillary were carried out using the same ratio of sample to spacer, the same concentrations of electrolytes and the same order of magnitude of current density. The absorbance at 280 nm was continuously monitored in the capillary, and individual fractions were assayed offline for the RF3. The profiles obtained demonstrated quantitatively that the separation using RITP was well modeled in the capillary. The scale up is linear in respect to the ratio of sample to spacer molecules required. This is quite remarkable considering the fact that the increase in scale is over six orders of magnitude.

Computer controlled RITP represents a new preparative protein separation methodology which can serve as a unit operation for the emerging biotechnology industry. This automation is the initial step in the development of a continuous process which can be monitored and controlled in real time.

REFERENCES

1. Kopwillem A, Merriman WG, Cuddeback RM, Smolka AJK, Bier M (1976). Serum protein fractionation by isotachophoresis using amino acid spacers. J Chromatogr. 118:35.
2 Bier M, Egen NB, Allgyer TT, Twitty GE, Mosher RA (1979). New developments in isoelectric focusing. In Gross E, Meienhofer J (eds): "Peptides Structure and Biological Function," Rockford: Pierce Chemical Co, p 79.
3. Wagner H, Mang V (1985). Voruntersuchungen und Empfehlungen zum Einsatz und zur wirtschaftlichen Nutzung der Elektrophorese under Weltraumbedingungen, Universitaet des Saarlandes, Saarbruecken.
4. Bier M, Egen NB, Twitty GE, Mosher RA, Thormann W (1986). Preparative electrophoresis comes of age. In King CJ, Navratil JD (eds): "Chemical Separations, Vol 1: Principles," Denver: Litarvan Literature, p 133.
5. Bier M, Palusinski OA, Mosher RA, Saville DA (1983). Electrophoresis: Mathematical Modelling and Computer Simulation. Science 219:1281.
6. Sloan JE, Thormann W, Mosher R, Twitty G, Bier M (1986). Recycling isotachophoresis: A novel approach to preparative protein fractionation. In Dunn MJ (ed): "Electrophoresis '86," Deerfield Beach: VCH Publishers, p 696.

Protein Purification: Micro to Macro, pages 337–354

PURIFICATION AND CHARACTERIZATION OF RECOMBINANT HUMAN MALARIA VACCINE CANDIDATES FROM E. COLI

G. Folena Wasserman, R. Inacker, C. Cohen Silverman and M. Rosenberg

Biopharmaceutical R&D, Smith Kline and French Labs, Swedeland, PA 19479

The pAS1 vector system was used for cloning and expression of the immunodominant repeat region of the human malaria circumsporozoite protein of *P. falciparum*. Efficient expression in *E. coli* was coupled to a high yield purification scheme for 3 CS protein constructs. The constructs were composed of 32 tetrapeptide repeat sequences from the CS protein fused to a 2, 32, and 81 amino acid C-terminal peptide. Using temperature induction, expression levels of 4-11% of total cellular protein were achieved. Each protein was isolated using several precipitation steps and then purified by cation exchange chromatography and reversed-phase HPLC. The purified products were characterized by SDS-PAGE; amino acid composition; N-terminal sequence analysis; and analytical size exclusion and reversed-phase HPLC. This approach enabled the production of gram quantities of highly purified CS antigens to study as human malaria vaccine candidates.

INTRODUCTION

The circumsporozoite (CS) protein is the predominant coat protein of the sporozoite, or infectious stage of the malaria parasite. Immunization of humans and animals with irradiated sporozoites produces a protective response that is correlated with antibody against the CS protein [1-8]. The primary structure of the CS protein includes a large central repeating polypeptide. Although there is no sequence homology between species, this repeat domain contains the immunodominant epitope(s) of the CS protein [9,10], and has been identified as a possible vaccine target [9,11,12]. Attempts to develop a human anti-sporozoite vaccine have failed because sporozoites can not be cultured _in vitro_, and because adequate amounts of the CS protein can not be isolated from the available sporozoites. Since the complete CS gene of the most common human malaria parasite, _Plasmodium falciparum_, has been cloned and sequenced [9], the recombinant approach is an attractive alternative for vaccine production.

This paper reviews our progress on the development of a recombinant human sporozoite vaccine. The actual utility of this approach depends on, first, the ability to express a stable CS product in high yield; and, second, the ability to isolate large amounts of highly purified antigen. The pAS1 expression vector system has been used to efficiently express many heterologous gene products in _E. coli_ [13], including a series of _P. falciparum_ CS protein constructs containing the immunodominant tetrapeptide repeat Asn-Ala-Asn-Pro [14]. Evidence from this study that the number of linear epitopes is important for immunological response, led us to further develop 3 _P. falciparum_ CS proteins--R32tet$_{32}$, R32Leu-Arg, and R32NS1$_{81}$. Each construct is a hybrid protein containing 30 Asn-Ala-Asn-Pro repeats and 2 Asn-Val-Asp-Pro repeats from the CS protein fused to a C-terminal peptide. Tet$_{32}$ represents 32 amino acids derived from the tetracycline resistance gene on pAS1 [14], and NS1$_{81}$ represents 81 amino acids derived from the influenza NS1 gene [15].

Expression yields of 4-11% of total cellular protein were achieved with pAS1 in an _E. coli_ lysogenic host using thermal induction. A flexible purification scheme incorporating several precipitation steps (heat, polyethyleneimine, ammonium sulfate, and pH 2 treatments), cation exchange chromatography and reversed-phase HPLC (RP-HPLC) was adapted to each gene product. Purification yields ranged from 11-65%. The purification schemes were scaled up to produce 1-2 g of highly purified proteins to test as potential vaccine candidates.

METHODS

Figure 1 illustrates the pAS1 plasmid that was transformed into _E. coli_ for expression of R32tet$_{32}$.

Details of the construction method are described elsewhere [14]. Similar pAS1 derivatives were made for expression of R32Leu-Arg and R32NS1$_{81}$. (M. Gross and J. Young, unpublished results). For R32Leu-Arg, the C-terminal dipeptide was derived from a synthetic linker. For R32NS1$_{81}$, the NS1$_{81}$ fragment was derived from the influenza virus NS1 gene [15].

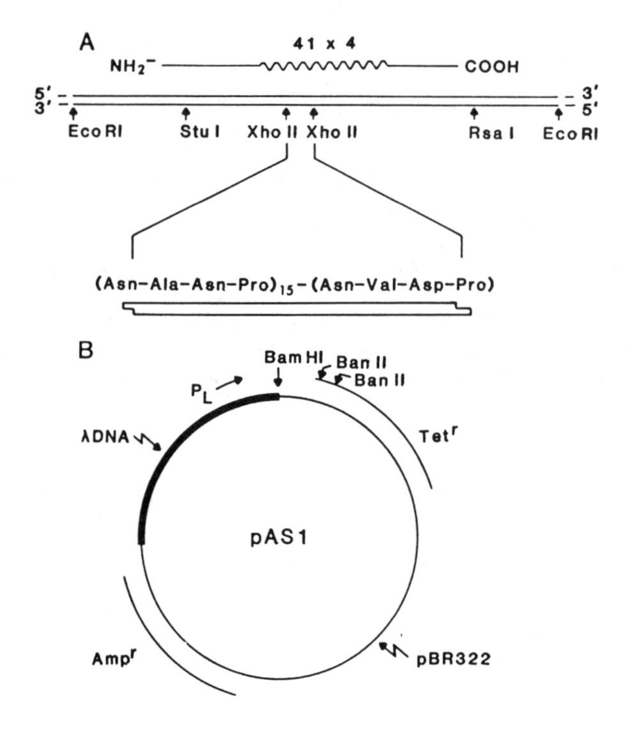

FIGURE 1. (A) Illustration of the CS gene showing the 41 tetrapeptide repeats in the central domain. A 192 base pair Xho II fragment coding for 16 repeats was excised from the CS gene and 3 fragments were ligated into the Bam HI site of the pAS1 expression plasmid (B). (Taken from Young _et al_., 1985, [14]).

E. coli cells were lysed in 50mM Tris, 10 mM EDTA, 5% glycerol, 0.1% deoxycholate (DOC), pH 8 using a Manton-Gaulin homogenizer and the lysate centrifuged at 12,000 x _g_ for 1 hr. The supernate was either heated to 80°C or treated with 0.5% polyethyleneimine (PEI) and then centrifuged for 30 min as described. Crude antigen was precipitated from the supernate with ammonium sulfate; resolubilized in 50mM Tris, pH 8; and then acidified to pH 2 with trifluoroacetic acid (TFA). The solution was clarified by centrifugation and the product purified by chromatography on either CM-Trisacryl M or Sulfopropyl Sepharose at pH 5 followed by RP-HPLC on a C4 Vydac support using a TFA/acetonitrile mobile phase.

RESULTS

Figure 2 shows the primary sequence of the 3 CS protein constructs expressed in E. coli using the various pAS1 vector derivatives. Each recombinant protein has 30 Asn-Ala-Asn-Pro repeats and 2 Asn-Val-Asp-Pro repeats fused to a C-terminal peptide of 32 ($R32tet_{32}$), 2 (R32Leu-Arg), or 81 ($R32NS1_{81}$) amino acids. Following temperature induction, immunoblots of cell lysates show that these constructs are stable and, in each case, are expressed as a single major immunoreactive species.

MET-ASP-PRO-[(ASN-ALA-ASN-PRO)$_{15}$(ASN-VAL-ASP-PRO)$_{1}$]$_{2}$-

+

A. LEU-ARG-ARG-THR-HIS-ARG-GLY-ARG-HIS-HIS-ARG-ARG-HIS-ARG-CYC-GLY-CYC-TRP-ARG-LEU-TYR-ARG-ARG-HIS-HIS-ARG-TRP-GLY-ARG-SER-GLY-SER

B. LEU-ARG

C. ASN-THR-VAL-SER-SER-PHE-GLN-VAL-ASP-CYS-PHE-LEU-TRP-HIS-VAL-ARG-LYS-ARG-VAL-ALA-ASP-GLN-GLU-LEU-GLY-ASP-ALA-PRO-PHE-LEU-ASP-ARG-LEU-ARG-ARG-ASP-GLN-LYS-SER-LEU-ARG-GLY-ARG-GLY-SER-THR-LEU-GLY-LEU-ASP-ILE-GLU-THR-ALA-THR-ARG-ALA-GLY-LYS-GLN-ILE-VAL-GLU-ARG-ILE-LEU-LYS-GLU-GLU-SER-ASP-GLU-ALA-LEU-LYS-MET-THR-MET-LEU-VAL-ASN

FIGURE 2. Primary sequence of recombinant P. falciparum CS proteins. A) $R32tet_{32}$, B) R32Leu-Arg, and C) $R32NS1_{81}$.

Figure 3 outlines the general purification scheme. This scheme is based on the high solubility and the acid pH and temperature stability of the R32 region of each molecule. The cell lysis supernate is carried through several precipitation steps to remove the majority of host cell proteins and nucleic acids. The crude antigen is then purified by cation exchange chromatography and/or RP-HPLC. Results of the R32tet$_{32}$ purification will be discussed in detail. Results for R32Leu-Arg and R32NS1$_{81}$ are given to emphasize various aspects of the scheme, and to indicate where changes were made in the process.

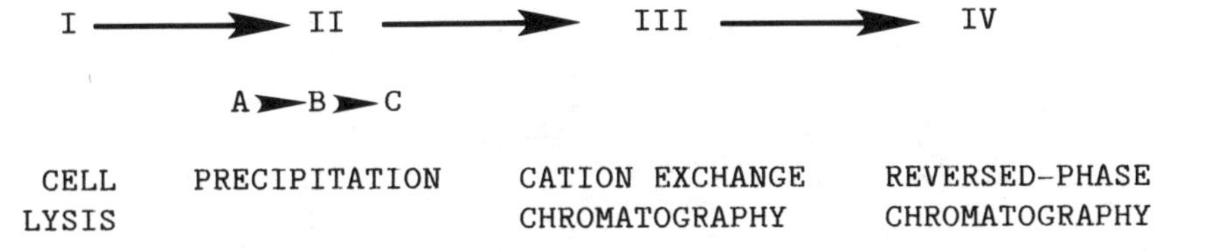

FIGURE 3. General purification scheme for recombinant malaria proteins.

Purification of R32tet_{32}

Figure 4 illustrates the specific purification scheme for R32tet_{32}. In the initial isolation steps the majority of host cell proteins are precipitated by heat treatment at 80°C. Most E. coli proteins precipitate between 40-80°C, whereas, R32tet_{32} is relatively heat stable and remains soluble. R32tet_{32} is concentrated, and the major contaminating proteins removed, by ammonium sulfate precipitation.

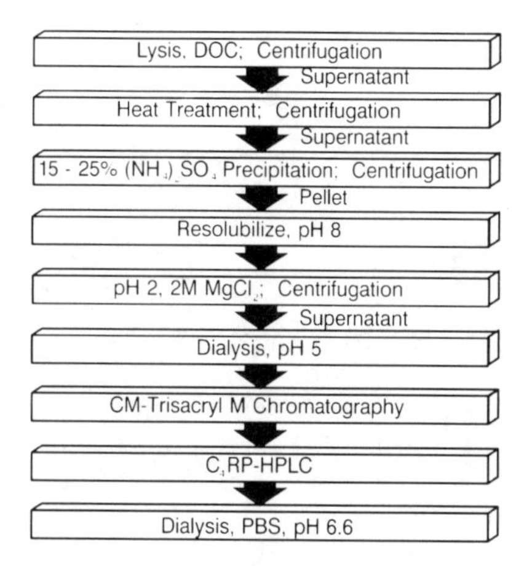

FIGURE 4. R32tet_{32} purification scheme.

Figure 5 shows the corresponding SDS-PAGE analysis of each isolation step. Although the final product (Lanes 11-14) appears homogeneous, further purification was required to remove small amounts of E. coli proteins carried through in larger scale preps, as well as, endotoxins and nucleic acids.

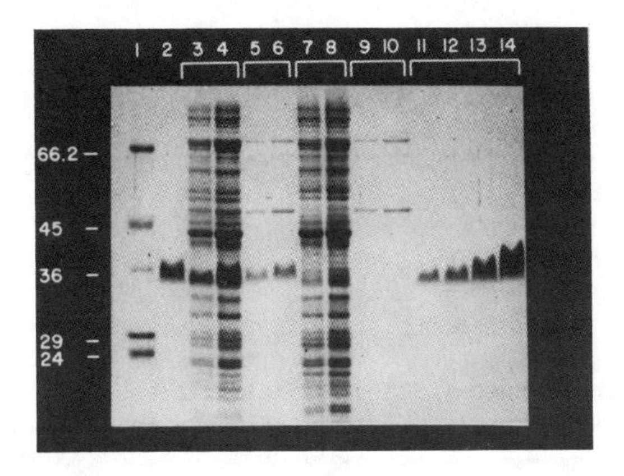

FIGURE 5. SDS-PAGE analysis of the $R32tet_{32}$ purification from E. coli. Proteins were separated on a 10% acrylamide gel [16] and stained with Coomassie Brilliant Blue R250. Lane 1, molecular weight markers; Lane 2, $R32tet_{32}$ standard; Lanes 3-4, lysis supernate; Lanes 5-6, heated supernate; Lanes 7-8, heated pellet; Lanes 9-10, ammonium sulfate supernate; and Lanes 11-14, resolubilized ammonium sulfate pellet representing 2.5, 5, 10, and 20 ug of protein, respectively.

Initial purification attempts were designed to take advantage of the high isoelectric point of $R32tet_{32}$ (calculated pI=12.8) by using cation exchange chromatography; and the acid stability and low molecular weight of the protein by using RP-HPLC. Unfortunately, the number and distribution of arginine residues in the tet_{32} tail results in the formation of a tightly associated protein-nucleic acid complex that is dominated by the physical properties of the nucleic acids. The complex copurifies by RP-HPLC; binds to anion exchange, but not cation exchange, supports at pH values from 6-8; and migrates at a molecular weight greater than 300 kd during size exclusion chromatography. Although disruption methods such as urea and guanidine-HCl treatments and nuclease digestions were tried, nucleic acids were more easily removed by acid pH treatment. Low pH alone precipitates R32tet$_{32}$ as the complex, but increasing the ionic strength of the solution with $MgCl_2$ results in specific precipitation of only the nucleic acids. Subsequent chromatography steps then proceed as expected.

Although the heat treatment is highly efficient at removing almost all E. coli proteins from the cell lysate (Figure 5), it also produces a variety of R32-containing degradation products. The extent of cleavage depends on the scale of the preparation and the severity of the heat treatment. Amino acid analysis of several truncated products indicates that all of the cleavage points are in the C-terminal tet$_{32}$ region of the molecule. Since most of these products are far less basic than the full length protein they are easily removed by cation exchange chromatography. Chromatography on CM-Trisacryl M at pH 5 removes the less basic by-products from the temperature treatment in the 0.1 and 0.2M ammonium chloride washes, and R32tet$_{32}$ elutes with 0.6M salt (Figure 6). Final purification is by RP-HPLC using a C4 Vydac solid support and a TFA/acetonitrile mobile phase.

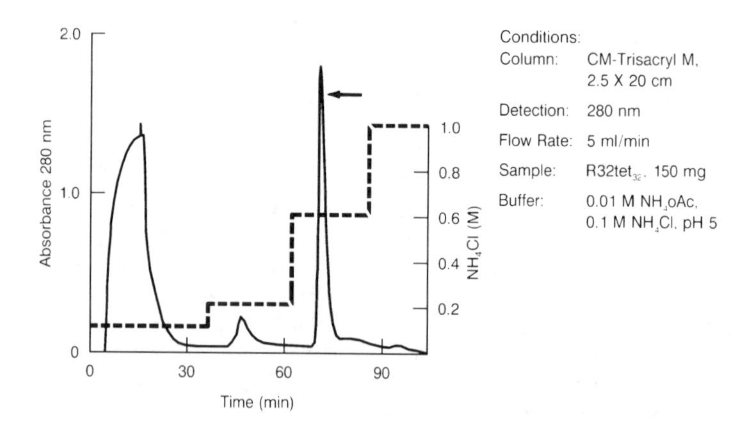

FIGURE 6. CM-Trisacryl M chromatography of R32tet$_{32}$. The arrow indicates elution of R32tet$_{32}$.

Table 1 lists the recovery data for $R32tet_{32}$ from a 20L fermentation or, approximately 300 g of wet cell paste. A monoclonal based immunoassay [14] was used for measuring product recovery in the crude cell lysate. Because it measures the concentration of all peptides that contain the R32 repeat (including temperature degradation products), rather than full length $R32tet_{32}$; this assay is less useful following the heat treatment step. Based on the acid stability of $R32tet_{32}$, a RP-HPLC assay was used to monitor product recovery for each purification step listed in Table 1. By HPLC analysis, the overall process recovery is 11%. The most product loss is during the heat precipitation step (66%).

TABLE 1
RECOVERY OF $R32TET_{32}$ [a]

Step	Total Protein[b] (mg)	$R32tet_{32}$ [c] (mg)	Step Recovery (%)
Lysis	28,000	3,100[d]	-
Heat	11,600	1,050	34
$(NH_4)_2SO_4$	1,100	540	51
pH 2/1M $MgCl_2$/ dialysis	600	500	93
CM-Trisacryl	320	420	84
RP-HPLC/dialysis	270	340	85
		Overall recovery = 11%	

a From 300 g cell pellet
b Lowry protein assay
c HPLC assay
d Immunoassay

Purification of R32 Leu-Arg

The expression and purification protocols used for R32tet$_{32}$ were modified to prepare R32Leu-Arg, a truncated version of the antigen that is mainly composed of tetrapeptide repeats (Figure 2). Removal of the tet$_{32}$ tail increases the temperature stability of the molecule and, also, eliminates the nucleic acid binding properties of the recombinant protein. The purification scheme is similar to that for R32tet$_{32}$ except that $MgCl_2$ is not required in the acid treatment step and the ion exchange chromatography is eliminated. The product is purified using 3 precipitation steps (heat treatment, ammonium sulfate precipitation, and acid pH treatment, respectively); and RP-HPLC as the single chromatography step.

The final HPLC step separates closely related antigen species that are inseparable by ion exchange or size exclusion chromatography techniques. Figure 7A illustrates the elution profile of a semi-preparative RP-HPLC run. Components 1 and 2 both react with monoclonal antibody to the tetrapeptide repeat and have been further characterized by amino acid analysis. The relative ratio of component 1 to component 2 varies for each preparation, and is always <5%. The amino acid composition of component 2 agrees with the predicted composition for R32Leu-Arg. The absence of 1 Pro, 1 Leu, and 1 Arg residue in the analysis of component 1 suggests that this peptide is the C-terminal cleavage product missing the -Pro-Leu-Arg sequence (Figure 2). Figure 7B illustrates the analytical RP-HPLC analysis of the crude cell lysate vs. purified R32Leu-Arg. Since this protein does not stain in SDS-gels with colorimetric or silver stains, the HPLC assay was particularly helpful during purification for quantitating full length product recovery, and for detecting R32Leu-Arg . Using the RP-HPLC assay, the average step recovery is 90% (including the heat treatment), and the overall process recovery is about 65%.

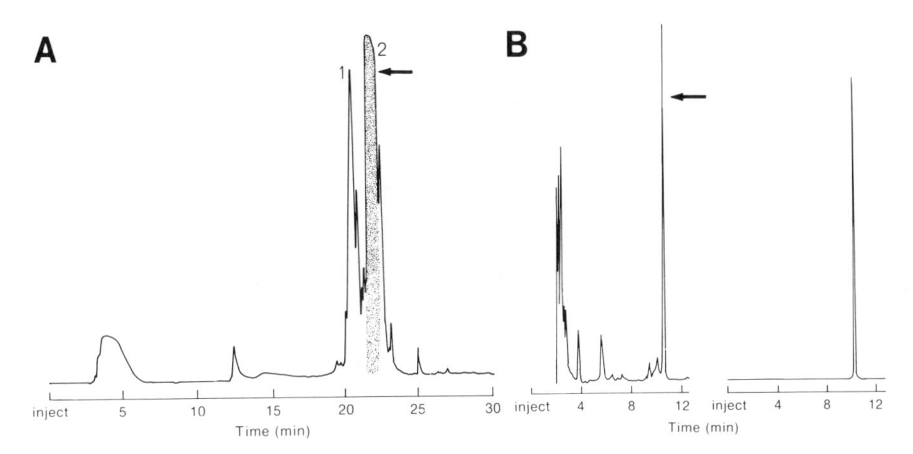

FIGURE 7. (A) Semi-preparative RP-HPLC of R32Leu-Arg on a 1 X 25 cm column of Vydac C4; flow rate 5 ml/min; UV 229 nm; 0–40% acetonitrile/0.05% TFA over 30 min. The shaded region indicates the pooled product. (B) Analytical HPLC analysis of the crude cell lysate (left) and the final product (right) using a 4.6 X 250 mm Vydac C4 column; flow rate 1 ml/min; UV 214 nm; 0–40% gradient over 12 min. The arrow indicates elution of R32Leu-Arg.

Purification of $R32NS1_{81}$

The R32 repeat was also expressed as a fusion product with the first 81 amino acids of the influenza NS1 gene [15] (Figure 2). Although $R32NS1_{81}$ retains the temperature stability characteristics of the R32 repeat segment, the heat treatment step was removed and replaced with a non-destructive polyethyleneimine precipitation of the crude cell lysate. This step removes most of the *E. coli* nucleic acids and some of the soluble protein. The majority of other host proteins are eliminated by ammonium sulfate precipitation, pH 2 treatment and chromatography on Sulfopropyl Sepharose at pH 5. Figure 8 shows the SDS-PAGE analysis of samples from each purification step. This scheme yields homogeneous product with a 32% overall yield.

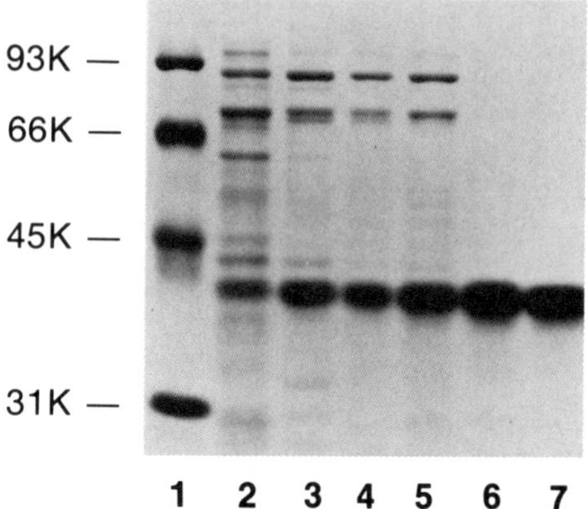

FIGURE 8. SDS-Page of R32NS1_{81} purification from E. coli. Gel conditions are indicated in Figure 5; 10 ug protein loaded/lane. Lane 1, molecular weight markers; Lane 2, lysis supernate; Lane 3, PEI supernate; Lane 4, ammonium sulfate pellet; Lane 5, pH 2 supernate; Lane 6, S-Sepharose product; and Lane 7, C4 RP-HPLC product.

Table 2 summarizes the expression and purification yields for the individual recombinant proteins. Each process was scaled to accommodate 1 Kg of wet cell paste (~100 L fermentation) and produces 1-2 g of CS protein. Both the tet_{32} and NS1_{81} C-terminal additions enhance the expression level of the repeat segment by about 2-fold. All 3 products are single peaks by analytical RP-HPLC. In each case, the amino acid composition and the N-terminal sequence is consistent with the predicted sequence. The tet_{32} and NS1_{81} fusion proteins show single bands on SDS gels stained with Coomassie Blue R250 or silver stain.

TABLE 2
COMPARISON OF CS PROTEIN YIELDS

Construct	Expression		Purification	
	% Yield[a]	Total Yield[b]	% Yield[c]	Total Yield[d]
R32tet$_{32}$	11	1100	11	120
R32Leu-Arg	4	300	65	200
R32NS1$_{81}$	9	700	32	230

a % of total cell protein (Lowry protein assay).
b Mg product/100 g wet cell pellet (HPLC assay).
c Overall purification process yield (HPLC assay).
d Mg final product/100 g wet cell pellet (HPLC assay).

Aberrant migration behavior was observed for all 3 proteins in both electrophoretic and gel filtration studies (Table 3). The apparent molecular weight of each protein, calculated by SDS-PAGE (reducing conditions), is greater than the calculated molecular weight; and increases as the percentage of proline in each molecule increases. The apparent molecular weights of R32Leu-Arg and R32NS1$_{81}$ are also greater than the calculated molecular weights when measured using size exclusion chromatography on a TSK 3000SW column. R32tet$_{32}$ does not elute from this support at pH 6.5.

TABLE 3
SIZE CHARACTERISTICS OF CS PROTEINS

Construct	% Pro	Calculated MW	Apparent MW SDS-PAGE[a]	Apparent MW SEC[b]
R32tet$_{32}$	20	17,300	36,000	NM[c]
R32Leu-Arg	25	13,400	90,000	58,000
R32NS1$_{81}$	16	22,300	35,500	168,000

a Immunoblot of SDS-gels using P. falciparum CS antibody [14].
b Size exclusion chromatography on a 7.5 X 600 mm TSK 3000SW column in 0.2M sodium phosphate, pH 6.5.
c Not measurable; does not elute from support at pH 6.5.

DISCUSSION

The results of this study demonstrate that highly purified potential vaccines can be produced from E. coli expression systems. In spite of the unusual amino acid composition of the P. falciparum CS constructs, each product accumulates to 4-11% of total cellular protein and is stable during and after heat induction. The flexibility of the purification scheme results from the dominating properties of the R32 peptide sequence that is found in all 3 recombinant proteins. These properties (temperature and acid pH stability) result in substantial purification by using only precipitation methods. Although the tet$_{32}$ peptide is somewhat unstable to 80°C treatment, the purification factor was so high that protein loss was tolerable for initial production of this material. Had expression levels of R32tet$_{32}$ been lower, the use of temperature precipitation would have been far less attractive.

RP-HPLC was important both analytically, for antigen detection and accurate quantitation of full length product recovery, as well as preparatively, for resolution of closely related antigens in the product mixture. A good example of component resolution was described for the protein R32Leu-Arg. In this example, even though the major species in the mixture was in excess of 95%, separation of 2 proteins differing by 3 amino acid residues is achieved. We observed similar separations during preparation of the other CS proteins where truncated or oxidized C-terminal derivatives were removed by RP-HPLC.

Electrophoretic and gel filtration mobility artifacts are commonly observed for proteins that are enriched in proline [17-20]. This behavior is apparently due to extended or elongated regions in these molecules that are not adequately described by globular protein standards during size exclusion chromatography or by SDS-denatured globular proteins during gel electrophoresis. CD and NMR experiments are in progress in order to study the solution conformation of R32Leu-Arg, and to develop a structural correlation between this subunit and the full length CS coat protein of _P. falciparum_.

A previous study by Young _et al_. [14] provided the first evidence that (for a subunit sporozoite vaccine) the number of tetrapeptide repeats may play a critical role in determining immunogenicity. In this study the proteins R32- and R48tet$_{32}$ were more immunogenic than the R16-fusion. This data suggests that our approach may have an advantage over a classical synthetic path because the recombinant molecules have a greater number of linear epitopes than the molecules currently available through synthesis [21,22]. Furthermore, R32Leu-Arg can be modified by the same methods used for synthetic peptides to study the effects of epitope presentation on the production of an effective vaccine.

Malaria vaccine development is partly limited by the species and stage specificity of the various coat proteins. One solution to this problem is development of a multivalent vaccine composed of protective epitopes from several proteins. Application of our approach to the _P. vivax_ CS protein and, also, a variety of merozoite stage antigens is presently underway. The ability to express and purify specific regions within each surface antigen gives us the potential ability to develop essentially any region of any cloned malaria protein for vaccine studies. In addition, hybrid molecules, previously produced by chemical coupling methods, have the potential of being produced by recombinant methods, provided that the gene sequences are known or the genes are available.

ACKNOWLEDGEMENTS

The authors wish to thank Dr. J. Young and Mr. M. Gross for the cloning and expression of the R32 constructs; Dr. D. Zabriskie and Mr. D. Wareheim for the fermentations; Dr. R. Sitrin and Mr. J. Dingerdissen for help with the large scale purification processes; Dr. J. Strickler, Ms. Lyn Gorniak and Ms. Lynette Miles for amino acid composition and N-terminal analyses; and Ms. J. Rosenbloom for her help with the purification.

REFERENCES

1. Nussenzweig RS, Vanderberg J, Most H (1969). Protective immunity produced by the injection of x-irradiated sporozoites of Plasmodium berghei. IV. Dose response, specificity and humoral immunity. Milit Med 134:1176.
2. Clyde DF, McCarthy V, Miller RM, Woodward WE (1975). Immunization of man against falciparum and vivax malaria by use of attenuated sporozoites. Am J Trop Med Hyg 24:397.
3. Rieckmann KH, Beaudoin RL, Cassells JS, Sell KW (1979). Use of attenuated sporozoites in the immunization of human volunteers against falciparum malaria. Bull WHO 57:261.
4. Vanderberg J, Nussenzweig RS, Most H (1969). Protective immunity produced by the injection of x-irradiated sporozoites of Plasmodium berghei. V. In vitro effects of immune serum on sporozoites. Milit Med 134:1183.
5. Cochrane AH, Nussenzweig RS, Nardin EH (1980). Immunization against sporozoites. In Kreier JP (ed): "Malaria in Man and Experimental Animals," New York: Academic Press, p 163.
6. Potocnjak P, Yoshida N, Nussenzweig RS, Nussenzweig V (1980). Monovalent fragments (Fab) of monoclonal antibodies to a sporozoite surface antigen (Pb44) protect mice against malarial infection. J Exp Med 151:1504.
7. Yoshida N, Nussenzweig RS, Potocnjak P, Nussenzweig V, Pikawa M (1980). Hybridoma produces protective antibodies directed against the sporozoite stage of malaria parasite. Science 207:71.

8. Nardin EH, Nussenzweig V, Nussenzweig RS, Collins WE, Harinasuta KT, Tapchaisri P, Chomcharn Y (1982). Circumsporozoite proteins of human malaria parasites Plasmodium falciparum and Plasmodium vivax. J Exp Med 156:20.
9. Dame JB, Williams, JL, McCutchan TF, Weber JC, Wirtz RA, Hockmeyer WT, Sanders GS, Reddy EP, Maloy WL, Haynes JD, Schneider I, Roberts D, Diggs CL, Miller LH (1984). Structure of the gene encoding the immunodominant surface antigen on the sporozoite of the human malaria parasite Plasmodium falciparum. Science 225:593.
10. Zavala F, Cochrane AH, Nardin E, Nussenzweig RS, Nussenzweig V (1983). Circumsporozoite proteins of malaria parasites contain a single immunodominant region with two or more identical epitopes. J Exp Med 157:1947.
11. Hockmeyer WT, Dame JB (1984). Recent efforts in the development of a sporozoite vaccine against human malaria. In Zouhair Atassi M, Bachrach HL (eds): "Immunobiology of Proteins and Peptides-III Viral and Bacterial Antigens," New York: Plenum, p 233.
12. Weber JL, Hockmeyer WT (1985). Structure of the circumsporozoite protein gene in 18 strains of plasmodium-falciparum. Molecular Biochem Parasitol 15:305.
13. Rosenberg M, Ho YS, Shatzman A (1983). The use of pKC30 and its derivatives for controlled expression of genes. Methods Enzymol 101:123.
14. Young JF, Hockmeyer WT, Gross M, Ballou WR, Wirtz RA, Trosper JH, Beaudoin RL, Hollingdale MR, Miller LH, Diggs CL, Rosenberg M (1985). Expession of P. falciparum circumsporozoite protein derivatives in E. coli for development of a human malaria vaccine. Science 228:958.
15. Young JF, Desselberger V, Palese P, Ferguson B, Shatzman A, Rosenberg M (1983). Efficient expression of influenza virus NS1 nonstructural proteins in E. coli. Proc Nat'l Acad Sci 80:6105.
16. Laemmli UK (1970). Cleavage of structural proteins during the assembly of the head of bacteriophage T4. Nature 227:680.

17. Bhandari DG, Levine BA, Trayer IP, Yeadon ME (1986). ^{1}H-NMR study of mobility and conformational constraints within the proline-rich N-terminal of the LC1 alkali light chain of skeletal myosin. Eur J Biochem 160:349.
18. Evans JS, Levine BA, Trayer IP, Dorman CJ, Higgins CF (1986). Sequence-imposed structural constraints in the TonB protein of E. coli. Febs Lett 208:211.
19. Oppenheim FG, Offner GD, Troxler RF (1982). Phosphoproteins in the parotid saliva from the subhuman primate Macaca fascicularis. J Biol Chem 257:9271.
20. Furthmayr H, Timpl R (1971). Characterization of collagen peptides by sodium dodecylsulfate-polyacrylamide electrophoresis. Anal Biochem 41:510.
21. Ballou WR, Rothbard J, Wirtz RA, Gordon DM, Williams JS, Gore RW, Schneider I, Hollingdale MR, Beaudoin RL, Maloy WL, Hockmeyer WT (1985). Immunogenicity of synthetic peptides from circumsporozoite protein of Plasmodium falciparum. Science 228:996.
22. Zavala F, Tam JP, Hollingdale MR, Cochrane AH, Quakyi I, Nussenzweig RS, Nussenzweig V (1985). Rationale for development of a synthetic vaccine against Plasmodium falciparum malaria. Science 228:1436.

Protein Purification: Micro to Macro, pages 355–365

GENETICALLY ENGINEERED SECRETION OF FOREIGN PROTEINS FROM ASPERGILLUS SPECIES

David I. Gwynne, Frank P. Buxton, M. A. Gleeson and R. Wayne Davies

Biochemicals Division
Allelix Inc.
Mississauga, Ontario
Canada
L4V 1P1

ABSTRACT We have achieved the secretion of significant amounts of biologically active human interferon α2 and a prokaryotic endoglucanase from *Aspergillus nidulans* and *Aspergillus niger*. The sequences encoding these proteins were linked to NH_2-terminal fungal secretion signal sequences and expression was controlled by the efficient, controllable promoters of the *A. nidulans* alcohol dehydrogenase I gene (*alcA*) or the *A. niger* glucoamylase gene. Both the human and the bacterial proteins are produced under the appropriate induction condition. ADHI driven expression of interferon α2 (before growth optimization) produces secreted levels of product (1 mg/litre) within range of the best available optimized microbial system. A growth and induction optimization program for ADHI controlled expression has demonstrated the secretion of 500 mg/l of *A. niger* glucoamylase from *A. nidulans*.

INTRODUCTION

Filamentous fungi of the genus *Aspergillus* are major sources of industrial enzymes[1], primarily because they are able to secrete large quantities of certain proteins. Filamentous fungi are eukaryotes, several species have FDA approval (GRAS status) and large-scale fermentation technology is well established so that if recombinant DNA technology can be used to induce them to secrete heterologous proteins, they represent ideal host systems for the production of a wide range of proteins. Secretion of many pharmaceutically useful proteins at even a fraction of the level achieved by strains specialized for industrial enzyme production is an attractive proposition. As a first step toward achieving this goal, we have developed a series of vector molecules which allow controlled expression and secretion of heterologous proteins after transformation of appropriate strains of *Aspergillus nidulans*.

Here we present results obtained using the glucoamylase gene promoter of *A. niger* in *A. niger* and *A. nidulans* and the promoter of the *alcA* gene (coding for alcohol dehydrogenase I) of *A. nidulans* in *A. nidulans*. In general, the excellent genetics and developing molecular biology of *A. nidulans* make this species the organism of choice for secretion system development, during which technology is constantly transferred into the important industrial species *A. niger*. In the work reported below, we have constructed vector molecules in which these controllable promoters are combined with secretion signal sequences linked in frame with heterologous coding regions, and report the successful, controlled secretion of a bacterial endoglucanase and human interferon α2 from *Aspergillus niger* and *nidulans*.

Two series of secretion vector constructions were used one based on the use of the cis-acting control region and promoter of the glucoamylase gene of *A. niger*, the other based on the use of the analogous region of the *alcA* gene of *A. nidulans*.

GLUCOAMYLASE VECTORS

The construction of the glucoamylase promoter vectors was done as follows. The glucoamylase gene was isolated from a *Sau3A* partial gene library of *A. niger* strain ATCC 22343 in pUC12 by hybridization with oligonucleotide probes designed on the basis of 'reverse translation' of the published amino acid sequence of *A. niger* glucoamylase[2]. Synthetic linker sequences containing a unique *EcoRV* site were inserted downstream of the secretion signal peptide sequence so that the signal peptidase cleavage site was recreated yielding the vector pGL2.

ALCOHOL DEHYDROGENASE I VECTORS

The *alcA* gene of *A. nidulans* has been cloned[3] and completely sequenced together with its flanking region[4], allowing the location of introns, of the messenger start sites and of the putative TATA box to be determined. A plasmid containing all of the *alcA* promoter region and messenger RNA leader up to 50bp 5' to the natural *alcA* ATG codon was obtained by resection of the *alcA* gene. The TATA box and mRNA start sites remain. The resulting vector (pALCA1) has *SalI* and *XbaI* sites immediately downstream of the *alcA* promoter; a synthetic DNA fragment comprising a region coding for a translation start followed by a consensus fungal secretion signal peptide was inserted between the *SalI* and *XbaI* sites, yielding pALCA1S, which is the vector used for expression and secretion of interferon $\alpha 2$ and the other proteins.

EXPRESSION AND SECRETION OF ENDOGLUCANASE AND INTERFERON

Two non-fungal genes were selected for expression and secretion studies. The endoglucanase of *C. fimi* is a bacterial secreted protein, and the C-terminal coding fragment of this gene used in this series of experiments was known to be efficiently secreted into the periplasm of *E. coli*[5] and the extracellular medium by *Saccharomyces cerevisiae*[6]; this protein is thus relatively well secreted by heterologous systems. Human interferon $\alpha 2$ in the other hand, is inefficiently secreted by heterologous systems, including yeast (Skipper, N., unpublished results) and *B. subtilis*[7]. The endoglucanase thus represented a good model for system development while interferon $\alpha 2$ served as a prototype for solving the difficulties that might be associated with producing certain mammalian proteins in this system.

Fig. 1 shows the interferon $\alpha 2$ gene incorporated into the vectors pGL2B and pALCAIS. In pGL2BIFN the fragment containing the interferon $\alpha 2$ gene is linked to the region of pGL2B coding for the glucoamylase signal sequence. The interferon gene was taken out of pGL2BIFN as a *BglII* - *EcoRI* fragment and cloned into *BamHI*, *EcoRI* cut pALCAIS to produce pALCAISIFN in which the interferon $\alpha 2$ gene is fused in frame to the synthetic signal peptide coding sequence.

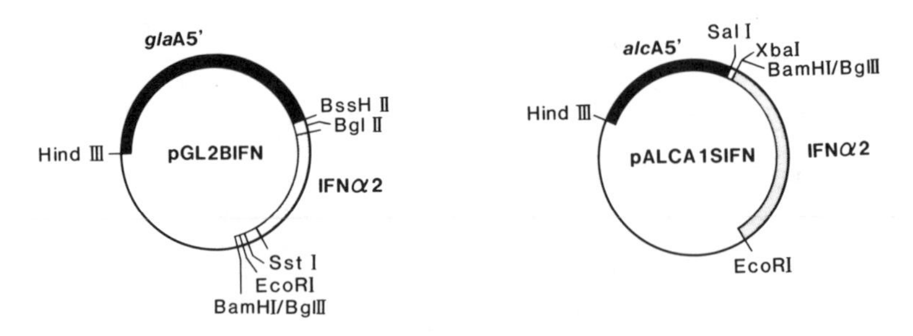

FIGURE 1. Restriction maps of plasmids pGL2BIFN and pALCA1SIFN

Transformants were obtained in A. niger and A. nidulans after cotransformation of argB⁻ strains with either pGL2BIFN and a plasmid (pDG2[8]) carrying the argB gene, or pALCAISIFN and (pDG2[8]). ArgB⁺ transformants were selected, and those (typically 80-90%) containing singly or multiple copies of the human interferon α2 gene as identified by Southern blot analysis (Fig. 6). Most of the transformants which contained the human interferon α2 gene showed some level of secreted interferon activity (data not shown).

The plasmid pGL2C was used to express and secrete C. fimi endoglucanase, the gene for which was introduced into the BglII site pGL2C as a BamHI fragment (Fig. 2). pALCAISENDO was made by taking an EcoRI fragment containing the C. fimi endoglucanase gene and cloning it into the EcoRI site of pALCAIS (Fig. 2); Transformants were obtained as described for the interferon α2 vectors, except that they were recognized by cloning of growth medium applied to CMC plates before staining with Congo red.

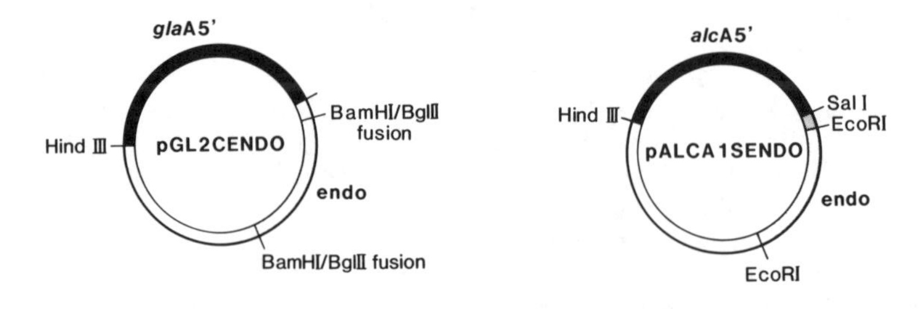

FIGURE 2. Restriction maps of plasmids pGL2CENDO and pALCA1SENDO

Typical results of Southern transfers probed with human interferon DNA are shown in Fig. 3 (transformed with pGL2BIFN). Seventeen out of twenty cotransformants contain sequences which are homologous to human interferon DNA. These sequences are present at a range of one (379) to approximately twenty (378) copies per genome. Most of the multiple integrated copies are found in tandem (data not shown) with no detectable rearrangement within the 1.8kb BamHI-SstI fragment (containing the coding region derived from human interferon α2 cDNA fused to 1.0kb of A. niger glucoamylase 5' flanking DNA). On longer exposure (Fig. 3) other integrated fragments of variable size are visible indicating some rearrangement and integration at several other chromosomal sites.

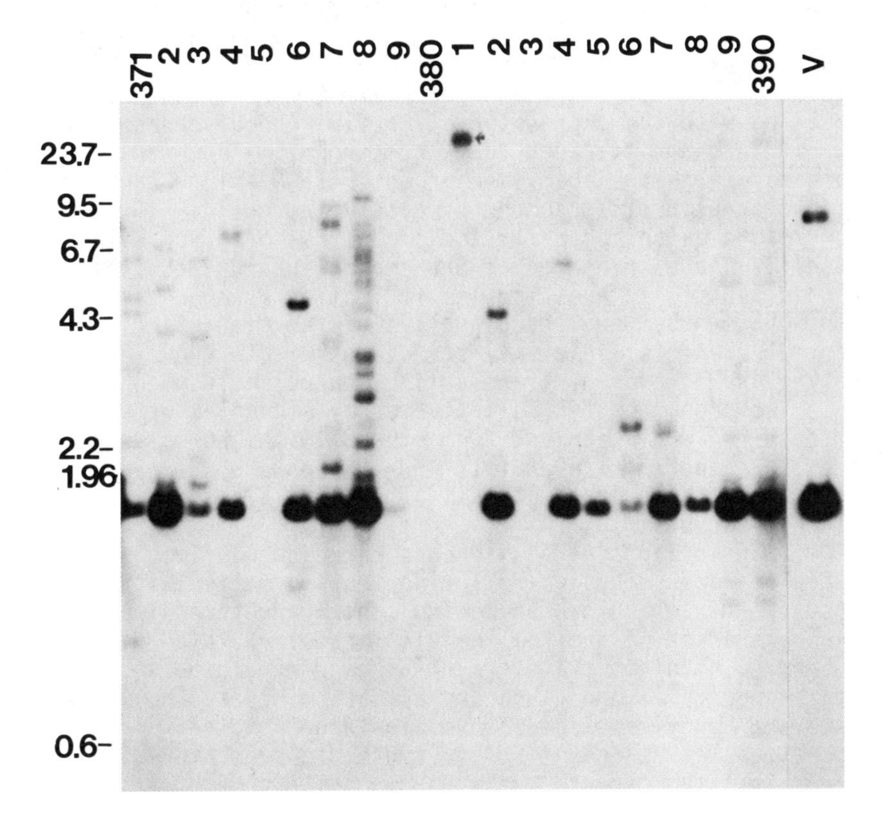

FIGURE 3. Southern transfer of DNA from Aspergillus nidulans interferon transformants. Equal quantities of DNA were cut at sites flanking the interferon insert (BamHI and SstI), electrophoresed and transferred to Genescreen. The blot was probed with the same DNA. Twenty

transformants are shown (371 to 390). Vector pGL2BIFN(V) was cut with the above enzymes and used as reference.

SECRETION OF INTERFERON α2 AND BACTERIAL ENDOGLUCANASE FROM ASPERGILLUS NIGER AND NIDULANS UNDER THE CONTROL OF THE GLUCOAMYLASE PROMOTER AND THE alcA PROMOTER

The glucoamylase promoter, when in A. niger, is induced in the presence of starch, and is subject to glucose repression; the molecular bases of these controls are not understood. Fig. 4 shows however that secretion of interferon α2 by transformants A. nidulans was obtained, and that the relative levels of secretion of these proteins under non-inducing, inducing and glucose repression conditions were as expected from the behaviour of the glucoamylase promoter when driving expression of glucoamylase itself in A. niger. Similar results were obtained with the expression of C. fimi endoglucanase in A. nidulans and A. niger (data not shown). Thus we are able to use the glucoamylase promoter in A. niger to make heterologous proteins under precisely the optimal conditions established for the production of glucoamylase itself. The maintenance of the control phenotypes in A. nidulans is interesting, as it implies strong conservation between these species of the cis-acting control elements and of key structures within RNA polymerase II and any trans-acting control molecules.

The alcA promoter is activated by a complex of a small effector molecule derived from ethanol with the product of the alcR gene[9,10] i.e. it is under a known positive control system. The synthesis of ADHI is also glucose-repressible.

Transformants of A. nidulans carrying C. fimi endoglucanase sequences following transformation with pALCA1SENDO were grown on medium containing threonine to induce transcription from the alcA promoter. Nine out of twenty A. nidulans transformants showed regulated secretion of the bacterial endoglucanase as determined by the plate clearing assay described above, (data not shown). Secretion was dependent on the synthetic secretion signal, since in transformants containing plasmid pDG6 (alcA-endoglucanase fusion) which does not contain the synthetic secretion signal, endoglucanase activity is detectable internally and is not detectable in the medium (data not shown). An antibody raised against C. fimi endoglucanase was used in an ELISA assay. Those transformants showing regulated synthesis of endoglucanase activity also showed positive results with the ELISA assay. No cross reacting activity to endoglucanase antibodies was detectable in an untransformed strain grown under identical conditions (data not shown).

Transformants of *A. nidulans* carrying human interferon $\alpha2$ gene sequences after *transformation* with pALCAISIFN were grown on medium containing threonine to induce transcription from the *alcA* promoter. Eleven of twenty *A. nidulans* *transformants* showed regulated secretion of *interferon* $\alpha2$, both inducer-dependent secretion and glucose repression being observed as expected from the normal control of ADHI expression in *A. nidulans*. Quantitative data from one transformant is *shown in Fig*. 4. No A. niger transformants have been found that express the interferon $\alpha2$ gene from this promoter, but this is not surprising in the absence of a known supply of the *alcR* gene product. Secreted interferon was shown to be *biologically* active (plaque reduction assay).

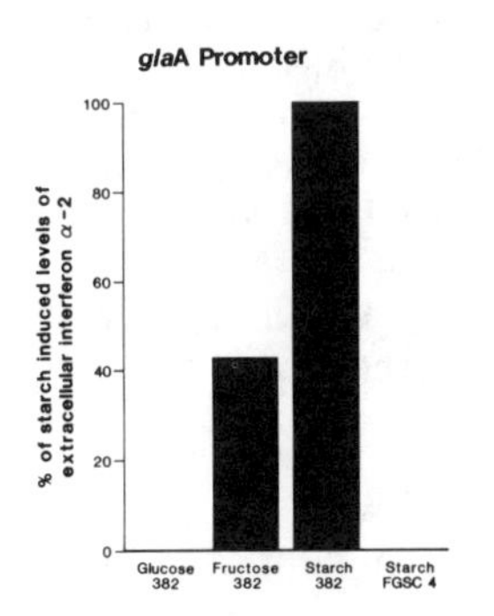

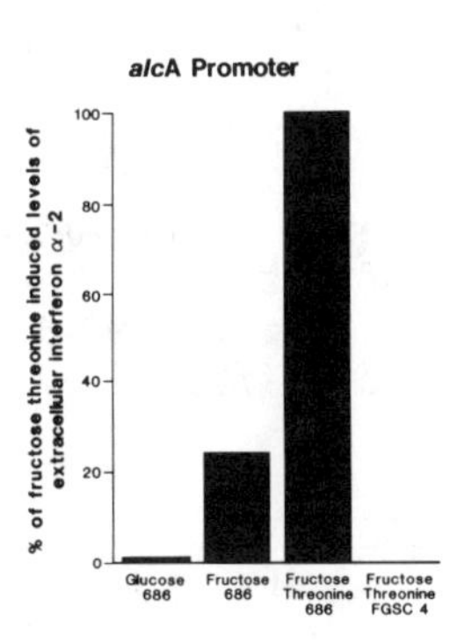

FIGURE 4. Regulation of interferon $\alpha2$ secretion in transformants carrying the interferon gene under glucoamylase and *alcA* promoter control.

Transformants were grown at 37°C on 1% glucose, 1% fructose, or 0.1% fructose, 100 mm threonine (to induce *alcA* or 1% soluble starch (to induce glucoamylase). Samples of growth medium were removed at 48 hours and tested for interferon activity with an immunoassay kit. Data is expressed as percent of maximally induced specific activity on fructose threonine medium (for *alcA*) and starch medium (for glucoamylase).

OPTIMIZATION OF TRANSLATION INITIATION SEQUENCE

It has been established[11,12] in eukaryotic organisms other than fungi, that the sequences flanking the AUG initiator codon modulate translation. The initial construction of pALCAIS was such that the SalI site used to insert the synthetic signal peptide downstream of the alcA promoter provided the sequence immediately preceding the initiation codon, and 52 bp of the original alcA untranslated mRNA leader were deleted. It was therefore important to determine the effects of these alterations on the efficiency of the system. A synthetic DNA fragment was inserted into the SalI site, restoring the DNA sequence coding for the alcA untranslated mRNA leader sequence except for the 8 bases immediately before the ATG.

These 8bp were then replaced by oligonucleotide directed in vitro replacement mutagenesis. The effects of these two steps on the yield of secreted interferon α2 from a transformant containing integrated copies of pALCA1SIFN as follows: restoring most of the mRNA leader made no difference to the yield of interferon, but restoring the sequence immediately before the ATG to that found in the alcA gene caused an average five-fold increase in yield of interferon protein. This was calculated from a repeated analysis of twenty transformants. The alcA sequence contains the A at position -3 that is known to be characteristic of efficiently translated genes[11,12].

DIRECT FUSION OF THE SYNTHETIC SIGNAL TO THE MATURE INTERFERON PROTEIN SEQUENCE

The original fusion of the synthetic signal to the interferon coding sequence in plasmid pALCA1SIFN provides 16 amino acids between the presumed last amino acid of the synthetic signal and the first amino acid of interferon mature protein sequence. This insert contains a portion of the interferon signal peptide sequence as well as sequence which accumulated during various vector constructions (encoding the first seven amino acids of the 16 amino acid region described above). Removal of the region encoding this 16 amino acid insert using oligonucleotide-directed in vitro mutagenesis results in a 15 fold increase in secreted interferon protein as detected immunologically. The deleted area contains a 48 base pair region which consists of sequence accumulated during successive cloning steps as a 27 base pair region which encodes the final nine amino acids of the human interferon secretion signal. The extra

16 amino acids may confer a novel structure at the N terminus of the interferon protein which interferes with its translocation or a later event in the secretion process.

EXPRESSION IN STRAINS CONTAINING MULTIPLE COPIES OF THE REGULATORY GENE (alcR)

When strains containing multiple copies of the alcA promoter linked to human IFα2 were examined for expression of alcA (from the endogenous single copy gene) certain characteristics (low ADHI levels, low alcA mRNA levels) of these strains indicated that the levels of the trans acting positive regulatory protein encoded by the alcR gene were limiting. When the copy number of alcR was increased in the strains described above the level of expression of the endogenous alcA gene was increased significantly. The level of expression of interferon α2 (under the control of the alcA promoter) was also increased significantly in those strains containing multiple copies (Fig. 5).

It is believed that the strains containing multiple copies of the alcR gene produce the alcR product at levels which are no longer limiting inside the cell.

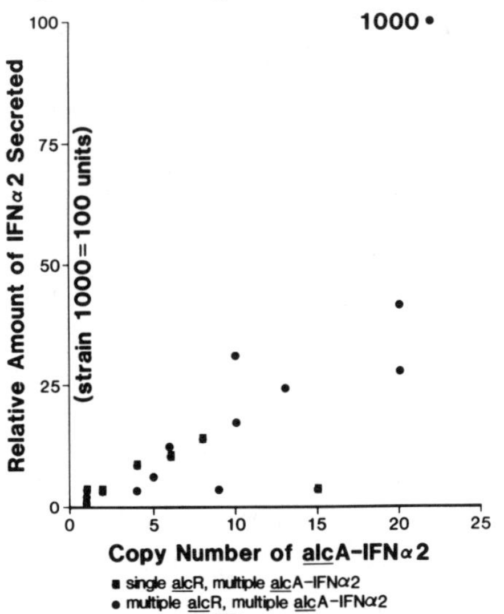

FIGURE 5. Relative amounts of interferon secreted from A. nidulans transformants containing variable copy numbers of alcA-interferon fusions in a single copy alcR background and a multiple alcR background and a multiple copy alcR background.

DISCUSSION

The yields obtained from the optimized alcA vectors in A. nidulans are approximately 1 mg/litre for interferon α2. The C. fimi endoglucanase protein is secreted at levels which were approximately 30 fold higher. These levels are obtained from relatively poorly growing cultures on minimal medium. Interferon α2 itself is a poorly secreted protein. Recent work in Bacillus subtilis[7] showed that interferon α2 is produced at as level (approximately 10 mg/litre) which is 100 to 400 fold below tht of B. subtilis α-amylase from the identical strain. Clearly, each protein will have a different characteristic behaviour in a secretion pathway and hence a different efficiency of secretion. Preliminary experiments aimed at optimization of growth and induction of the alcA promoter in A. nidulans have demonstrated the secretion of more than 500 mg/litre of an efficiently secreted heterlogous protein. In this particular case the glucoamylase coding region from A. niger was expressed in the alcA vector in A. nidulans. The accumulation of improvements in the vector, the host strains and the growth and induction conditions should allow the rational design of industrial fungi which are specialized for the secretion of a variety of proteins.

REFERENCES

1. Barbesgaard, P. (1977). Industrial enzymes produced by members of the genus Aspergillus. In: Genetics and Physiology of Aspergillus. Smith, J.E. and Pateman, J.A. (eds) Academic Press, London.
2. Svensson, B., Pederson, T., Svendsen, I., Sakai, T., and Ottesen, M. (1982). Characterisation of two forms of glucoamylase from Aspergillus niger. Carlsberg Res. Commun. 47:55-69.
3. Lockington, R.A., Sealy-Lewis, H.M., Scazzocchio, C. and Davies, R.W. (1986): Cloning and characterization of the ethanol utilization regulon in Aspergillus nidulans. Gene 33:137-149.
4. Gwynne, D.I., Buxton, F.P., Sibley, S., Davies, R.W., Lockington, R.A., Scazzocchio, C. and Sealy-Lewis, H.M. (1987). Comparison of the cis-acting control regions of two coordinately controlled genes involved in ethanol utilization in Aspergillus nidulans. Gene, in press.

5. Wong, W.K.R., Gerhardt, B., Guo, Z.M., Kilburn, D.G., Warren, R.A.J., and Miller, R.C. Jr. (1986). Characterization and structure of an endoglucanase gene of Cellulomonas fimi. Gene 44:315-344.
6. Skipper, N., Sutherland, M., Davies, R.W., Kilburn, D., Miller, R.C. Jr., Warren, A. and Wong, R. (1985). Secretion of a bacterial cellulase by yeast. Science 230:958-960.
7. Schein, C.H., Kashiwagi, K., Fujisawa, A. and Weissmann, C. (1986). Secretion of mature IFN-$\alpha 2$ and accumulation of uncleaved precursor by Bacillus subtilis transformed with a hybrid α-amylase signal sequence IFN-$\alpha 2$ gene. Bio/Technology 4:719-725.
8. Buxton, F.P., Gwynne, D.I. and Davies, R.W. (1985). Transformation of Aspergillus niger using the argB gene of Aspergillus nidulans. Gene 37:207-214.
9. Pateman, J.H., Doy, C.H., Olson, J.E., Norris, U., Creaser, E.M. and Hynes, M. (1983). Regulation of alcohol dehydrogenase (ADH) and aldehyde dehydrogenase (AldDH) in Aspergillus nidulans. Proc. Roy. Soc. Lond. B217:243-264.
10. Sealy-Lewis, H.M. and Lockington, R.A. (1984). Regulation of two alcohol dehydrogenases in Aspergillus nidulans. Curr. Genet. 8:253-259.
11. Kozak, M. (1984). Compilation and analysis of sequences upstream from the translational start site in eukaryotic mRNAs. Nucleic Acids Res. 12:857-872.
12. Kozak, M. (1986). Point mutations define a sequence flanking the AUG initiator codon that modulates translation by eukaryotic ribosomes. Cell 44:283-292.

Protein Purification: Micro to Macro, pages 367–373

PURIFICATION OF RECOMBINANT EPIDERMAL GROWTH FACTOR FROM A FUSION PROTEIN EXPRESSED IN ESCHERICHIA COLI

Geoffrey Allen and Cora A Henwood

Department of Protein Chemistry, Wellcome Biotech, Langley Court, Beckenham, Kent BR3 3BS, UK

ABSTRACT Epidermal growth factor (murine sequence) has been purified in gram quantities following chemical synthesis of the gene and expression in *E coli* as a fusion protein with part of the Trp E gene product. The fusion protein was mainly present as a insoluble aggregate within the *E coli* cells, and was pelleted by centrifugation after cell lysis. Following solubilization in 8M urea under reducing conditions, the EGF polypeptide was liberated by specific proteolysis and allowed to reform disulphide bonds by controlled oxidation. EGF was purified by sequential chromatography on columns of reverse-phase (C18) silica gel, DEAE-cellulose and BioGel P-10. Chemical and biological properties were indistinguishable from those of the major component of EGF extracted from submaxillary glands.

INTRODUCTION

Epidermal growth factor is a small (53-residue) protein first isolated from mouse submaxillary glands over 25 years ago (1) and of established primary structure (2). The protein is a potent mitogen for a number of cell lines in culture, and *in vivo* several activities have been described, including inhibition of gastric acid secretion, inhibition of hair growth, but acceleration of tooth eruption and eyelid opening in neonatal mice, and a weakening of growing wool fibres in sheep. The latter observation has led to an interest in the potential development of EGF as a wool harvesting agent (3). In order to produce sufficient EGF for more extensive studies in sheep, a recombinant DNA approach was used (4), a synthetic gene for mouse EGF was inserted into plasmids, one of which directed the synthesis of a fusion protein consisting of part of the *E coli* TrpE gene product, a lysine codon and EGF. This fusion protein could be digested with a lysine-specific protease and EGF isolated following refolding of the denatured polypeptide chain and reoxidation of the six cysteine residues to form the correct three disulphide bonds.

METHODS

Oligodeoxynucleotides up to 27-mers for the construction of the EGF gene were synthesized by a manual solid-phase phosphate triester technique. Protected mono- and di-nucleotide phosphate diester intermediates were synthesized from nucleosides. The oligodeoxynucleotides were purified by ion-exchange hplc at 55°C in 30% (v/v) formamide in gradients of 10-700mM KH_2PO_4 on a Partisil SAX (Whatman) column, and were shown to be pure by polyacrylamide gel electrophoresis of kinase-labelled samples.

Sixteen partially complementary oligodeoxynucleotides were ligated both in groups, with gel electrophoretic purification of intermediates, and in a single reaction, to form a 174 bp gene for EGF with flanking regions coding for a BamHI restriction site sticky end, a lysine codon immediately 5' to the EGF coding region, a stop codon 3' to the EGF gene, and an EcoRI sticky end. This gene was ligated into the BamHI-EcoRI large fragment of pAT153, and ampicillin-resistant colonies were selected for tetracycline sensitivity. 11 out of 15 colonies tested by restriction mapping had the correct insert size and the genes in two isolates were completely sequenced.

For expression of EGF as a fusion protein, the gene was inserted into the 5' BglII site of the Trp E gene in a plasmid containing the *trp* promoter, attenuator, Trp E and part of the Trp D gene, yielding plasmid pWRL 500. For higher level expression, this gene construct was transferred into a plasmid, pMM1, with temperature-dependent runaway-copy-number phenotype, yielding pWRL505. The gene was also placed under the control of the *tac* promoter (A Makoff, Wellcome Biotech).

Alternative fusion protein gene constructs were made by resynthesising the linking region between the BglII site in the Trp E gene and the EcoRI site in codons 1-2 of the EGF gene. The EGF gene was also placed under the control of the *tac* promoter and an initiator methionine codon. In some of these experiments, oligonucleotides synthesized on a Bio-search SAM-1 automated instrument were used.

On the small scale (3 litres), E coli HB101 bearing pWRL500 was grown at 37°C in M9 salts with casamino acids to an OD_{600}1. Indole acrylic acid was added to 0.1 mg/ml, and fermentation continued to OD_{600}8. Harvested cells were disrupted with lysozyme/EDTA and freeze-thawing, or by X-Press treatment. After DNAse treatment to reduce the viscosity the suspension was centrifuged at 40,000 x g for 1 h at 20°C. The pellet was homogenized in 8M urea, 1mM EDTA, 1mM 2-mercaptoethanol, 50mM NH_4HCO_3, pH8.3, and the mixture diluted 5-fold into 50mM NH_4HCO_3. The opalescent mixture was centrifuged, and the supernatant treated with endoproteinase Lys-C (Boehringer-Mannheim) (0.75 U) for 24 h at 37°C, followed by the addition of a further 0.75 U and continuing digestion for 3 days. The digest was dialysed against 1mM NH_4HCO_3, 0.2mM EDTA, pH 8 at 4°C for 2 h, followed by 50mM HCl, 0.1M NaCl for 2 h. The mixture was centrifuged at 40,000 x g for 10 min at 4°C. The supernatant was concentrated (Amicon UM2 membrane) and the EGF was purified by BioGel P-10 chromatography followed by DEAE-cellulose chromatography as described by Savage and Cohen (5). EGF was identified by radioimmunoassay during the purification, using EGF and rabbit anti-EGF antiserum from BRL (Bethesda, MD).

On the larger scale (200 l fermentations), E coli (bearing pWRL505) cell paste was processed in 1Kg aliquots. The cells were disrupted with lysozyme/EDTA, and nonidet P40 detergent in the presence of phenylmethanesulphonylfluoride and mercaptoethanol. After DNAse treatment the pellet fraction was collected by centrifugation at 27,000 x g for 45 min, and was washed by resuspension in 50mM NH_4HCO_3, 1mM EDTA, 1mM 2-mercaptoethanol, pH8.3, and re-centrifugation. Urea (0.9 g/g pellet) was added, and the pellet homogenized at 37°C. The clear viscous mixture was diluted 2-fold and centrifuged as above. The supernatant was diluted 2.5-fold with 50mM NH_4HCO_3. Hydroxyethyl disulphide (1-5mM) and endoproteinase Lys-C were added. After 3 days at 37°C acetonitrile was added to 15% (v/v) and the pH was reduced to 3.75 with HCl. After filtration or centrifugation the supernatant was applied to a column (150ml total) of reverse-phase silica gel (Waters Prep RP18). EGF was eluted as the first major A_{280} peak in a gradient from 15-50% acetronitrile in 0.1% (v/v) trifluoroacetic acid. The crude EGF was diluted two-fold, neutralised, and chromatographed on a column (30 x 200mm) of DEAE-cellulose (Whatman DE-52) in a gradient of ammonium acetate at pH 5.6, totalling 1.5 litres. The EGF peak was acidified and purified by chromatography on a column of BioGel P-10 as above, but 50mm x 250mm in size. Some samples were further chromatographed on a column of SP-Sephadex C-25 in a pH gradient with citrate-phosphate buffers at pH 2.5-4.5.

The EGF was characterised by amino-acid analysis, N-terminal amino acid analysis and partial sequencing, reverse-phase hplc, thermolysin peptide mapping on thin-layers and by biological activity on cell monolayers, in mice (eyelid opening) and in sheep.

Small-scale experiments were also performed in attempting to isolate EGF or EGF-related proteins after expression of other fusion proteins with differing linker regions, including Factor X_a-specific, collagenase-specific, partial acid hydrolysis-specific (Asp-Pro dipeptide), and submaxillary arginine protease-specific (Arg), without success. An EGF derivative, EGF(Leu^{21}) was purified from a fusion protein, Trp E'-Lys-EGF(Leu^{21}), the gene for which was constructed by site-specific mutagenesis of the EGF gene to replace the single methionine residue in the EGF sequence by a leucine residue. The linking region was also replaced by a sequence coding for a methionine residue 5' to the EGF gene, to allow production of EGF(Leu^{21}) by CNBr cleavage of the fusion protein TrpE'-Met-EGF(Leu^{21}).

Further details of methods are given in ref (6).

RESULTS AND DISCUSSION

The manual synthesis of oligodeoxnucleotides yielded pure products, and once a library of intermediate dimers was prepared was relatively rapid, two oligomers being synthesized per day. Some problems were experienced during ligations, primarily related to the design of the gene, which included sticky ends that, although not phosphorylated initially, were to some degree ligated, leading to polynucleotides of higher molecular weight than expected. This appeared to be related to residual activity in the polynucleotide kinase not completely destroyed by heating before oligonucleotides were mixed for ligation.

The gene sequence of one isolated clone had a single base change (C to T) at the third base of the codon for Cys-42, but the other had the correct sequence (6). There have been several reports in the literature of point mutations in genes cloned from synthetic DNA; this was the only example seen in the work from our laboratory and is ascribed to a chance event, presumably related to some chemical side reaction during synthesis.

Attempts to express EGF directly from the tac promoter and a methionine initiator codon were unsuccessful; this was not surprising, since small foreign polypeptides are generally rapidly proteolysed in E coli. The fusion protein TrpE'-Lys-EGF was well expressed, around 10% of total cell protein in E coli harboring pWRL 500, up to 28% in cells bearing pWRL505, and as high as 38% in E coli JM103 with a pAT153-based plasmid containing the fusion protein under the control of the tac promoter, as measured by densitometry after Coomassie-blue stained SDS-PAGE. Most of the fusion protein was present in insoluble 'inclusion bodies' as detected by immunoelectron microscopy and SDS-PAGE analysis of pellet and supernatant fractions of cell lysates. This formation of 'inclusion bodies' when such fusion proteins are expressed in E coli is commonly seen (7). The fusion protein dissolved in 8M urea under reducing conditions; if autoxidation was allowed to occur at this stage, however, an insoluble gum was formed, but this redissolved following further reduction.

A large proportion of the EGF released by lysine-specific proteolysis appeared able to refold under the conditions of digestion to give the correct three-dimensional structure as assessed by hplc elution and biological activity. The overall yield from the content of fusion protein in the cells (assessed by Coomassie blue stained SDS-PAGE) to the final purified EGF was in some cases as high as 30%, although 15% was more typical. Occasionally lower yields were experienced, and these appeared to be associated with lower levels of expression, although variation in folding conditions cannot be ruled out.

The EGF was estimated to be about 90% pure by reverse-phase hplc analysis, and only a single band was seen on SDS-PAGE of reduced protein at 10μg loading with Coomassie blue staining. Much of the additional material seen on hplc was EGF-related, and some had EGF activity. EGF purified from mouse submaxillary glands by the method of Savage and Cohen (5) consists of a number of components separable by hplc, some of which are artefacts produced during the isolation (8). The EGF produced in E coli comigrated with the major component of submaxillary gland EGF.

In EGF are two sequences, Asn^{1}-Ser and Asn^{16}-Gly that may be particularly susceptible to deamidation, with the resultant formation of imide structures or β-aspartyl peptide bonds. Such deamidation may occur under rather mild conditions, and indeed in some preparations of crude EGF from E coli a significant EGF-like component of higher affinity for DEAE-cellulose was observed. Although this was not a homogeneous material, peptide mapping on thin layers clearly showed that the expected N-terminal thermolytic peptide was essentially absent, and was replaced by a more acidic peptide of the same amino acid composition after acid hydrolysis, but with N-terminal aspartic acid rather than asparagine.

The EGF could be further purified by reverse-phase hplc, but for the required purpose in studying sheep defleecing its purity was adequate. Although no particular steps were taken to remove endotoxin during the purification, levels in the purified EGF were insufficient to cause concern.

Protein chemical studies, including amino acid analysis and analysis of all thermolysin-derived peptides from the performic acid-oxidised protein, were fully consistent with the expected structure, and the biological activity was indistinguishable from that of the major component (EGF-α) of submaxillary gland EGF.

Some batches of endoproteinase Lys-C obtained from Boehringer (Mannheim) were not effective at specific digestion, and investigation showed that the relative specific activity of different batches of the enzyme on different substrates varied significantly, and that components in the protease preparations differed according to SDS-PAGE analysis. A number of alternative cleavage possibilities for releasing EGF from a fusion protein was therefore investigated. These included the use of the Ile-Glu-Gly-Arg sequence, specific for blood-clotting factor X_a, the Gly-Pro-Leu-Gly-Pro sequence, intended to be a site for cleavage by Clostridial collagenase, and the Asp-Pro dipeptide sequence, for relatively specific acid hydrolysis. None of these approaches gave any significant yield of EGF-like polypeptide material. In addition, the use of submaxillary gland arginine-specific protease for digesting the factor X_a-specific site was unsuccessful; in this case little digestion was seen, possibly due to inhibition of the protease by EGF, with which it, or a related protease, is firmly bound _in vivo_.

The conversion of the EGF gene to a gene for EGF(Leu^{21}) using standard site-specific mismatch repair mutagenesis in M13 proceeded smoothly. EGF(Leu^{21}) was purified using the same methods as for EGF from TrpE'-Lys-EGF(Leu^{21}) fusion protein, in similar yield, but with a small but detectably different elution position in reverse-phase hplc. When this gene was linked through a methionine codon to the TrpE' carrier gene, the resulting fusion protein, TrpE'-Met-

EGF(Leu^{21}) was expressed in high yield. The protein was cleaved with CNBr, and the EGF derivative isolated following various methods designed to allow refolding of the EGF(Leu^{21}) polypeptide. However, only low yields of incompletely purified product were obtained in preliminary experiments. It was found possible to isolate sufficient of a lysine-specific protease preparation from cultures of *Lysobacter enzymogenes* to allow preparation of gram quantities of EGF by the methods described above.

ACKNOWLEDGEMENTS

We thank Hugh Spence and Dave Brown for assistance and Andrew Makoff, Mike Winther and Philip Moore for collaborating in this work.

REFERENCES

1. Cohen S (1962). Isolation of a mouse submaxillary gland protein accelerating incisor eruption and eyelid opening in the new-born animal. J Biol Chem 237:1555.
2. Savage CR, Inagami T, Cohen S (1972). The primary structure of epidermal growth factor. J Biol Chem 247:7612.
3. Moore GOM, Panaretto BA, Scott TW (1983). Biological Wool Harvesting. In "Outlook for Australia's Natural Fibres" Symp Aust Acad Technol Sci, Parkville, Victoria, p85.
4. Allen G, Paynter CA, Winther MD (1985). Production of epidermal growth factor in Escherichia coli from a synthetic gene. J Cell Sci Suppl 3:29.
5. Savage CR, Cohen S (1972). Epidermal growth factor and a new derivative. Rapid isolation procedures and biological and chemical characterization. J Biol Chem 247:7609.
6. Allen G, Winther MD, Henwood CA, Beesley J, Sharry LF, O'Keefe J, Bennett JW, Chapman RE, Hollis DE, Panaretto BA, Van Dooren P, Edols RW, Inglis AS, Wynn PC, Moore GPM (1987). Synthesis and cloning of a gene coding for a fusion protein containing mouse epidermal growth factor. Isolation from transformed *E coli* and some physical, chemical and biological characteristics of the growth factor. J Biotechnol, In Press.
7. Marston FAO (1986). The purification of eukaryotic polypeptides synthesized in *E coli*. Biochem J 240:1.
8. Smith JA, Ham J, Winslow DP, O'Hare MJ, Rudland PS (1984) The use of high-performance liquid chromatography in the isolation and characterization of mouse and rat epidermal growth factors and examination of apparent heterogeneity. J Chromatog Biomed Appl 305:295.

Protein Purification: Micro to Macro, pages 375–381

PRODUCTION OF PEPTIDE HORMONES IN *E.COLI* VIA MULTIPLE JOINED GENES

Stephen Cockle, Michael Lennick and Shi-Hsiang Shen

Connaught Research Institute, Willowdale, Ontario, Canada M2R 3T4

ABSTRACT Over-production of human proinsulin and cardionatrin in *E.coli* was achieved by generating them as fused polypeptides consisting of a minimal bacterial leader and multiple copies of the target sequence, separated by a short removable linker. Such precursors formed stable, insoluble, easily-isolated aggregates within the host cell. Chemical and enzymic methods were employed to fragment the precursors, trim the monomers to the correct size, and introduce the appropriate disulfide bonds. Biologically active insulin and cardionatrin were obtained in good yield.

INTRODUCTION

Efficient biosynthesis of small eukaryotic proteins in bacteria is often limited by the instability of the product in the hostile environment of the host cell (1,2). This problem is commonly addressed by fusing the foreign DNA coding sequence to a large segment of an endogenous structural gene, resulting in expression of a hybrid polypeptide (3,4). However, a major disadvantage of this approach for large-scale production is that the desired product constitutes only a small portion of the precursor. Thus much of the cell's synthetic potential is wasted in making unwanted bacterial leader, which itself complicates purification of the eukaryotic protein.

A novel strategy has been developed in this laboratory to overcome these drawbacks, applied originally to the case of human proinsulin expressed in *E.coli* (5-7). Multiple joined proinsulin genes under the control of an inducible *lac* promoter were constructed to direct synthesis of fused polypeptides comprising several proinsulin moieties linked in tandem, behind a minimal prokaryotic leader. Such precursors, containing about 90% proinsulin, were produced in high yield and accumulated as insoluble aggregates, or inclusion bodies, protected from the action of intracellular proteases. To demonstrate the wider applicability of this concept, we have now shown that fused polypeptides comprising multiple copies of the smaller peptide hormone cardionatrin are also produced in stable form in *E.coli*. In this paper we outline the procedures developed to fragment the respective multi-unit precursors and obtain mature insulin and cardionatrin. Full experimental details will be published elsewhere.

HUMAN INSULIN

Since insulin contains two disulfide-linked polypeptide chains, it is more convenient to produce it via single-chain proinsulin, which has 86 amino acids and three disulfide bridges. Human proinsulin was generated in *E.coli* as precursors with the following primary structure:

T M I T D S L A M (Proinsulin R R N S M)$_n$ G S

The nine-amino-acid leader corresponds to the first eight residues of *E.coli* β-galactosidase plus a methioinine linker, while individual proinsulins are separated by a five-amino-acid linker also ending in methionine. As proinsulin contains no internal methionines, cyanogen bromide can be used to fragment the molecule into monomers of proinsulin "analog" (PIA). The double arginine site allows the C-terminal linker to be removed later by digestion with trypsin and carboxypeptidase B, which concurrently convert proinsulin to insulin.

Expression of precursors comprising from two to seven copies of the proinsulin sequence was assessed by SDS-gel electrophoresis (Figure 1). Optimum levels (around 25% of total cellular protein) were reached at four to five copies; thus the construct containing four copies was chosen for full process development. Cells were homogenized in aqueous buffer at pH8, and the insoluble precursor was isolated by centrifugation and washed with 3M guanidinium chloride (GdmCl), by which time it constituted at least 75% of the stainable protein. Although the precursor could be purified to near homogeneity by gel filtration on Sephacryl S-200 in 7M GdmCl, this was not routinely performed. Instead, the crude material was treated directly with cyanogen bromide in 75% formic acid. Almost quantitative monomer yields were

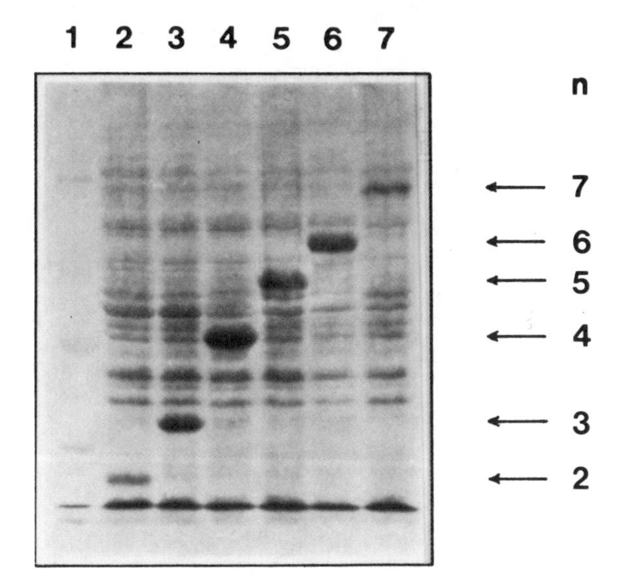

FIGURE 1. Expression of proinsulin precursors containing 2-7 copies of the proinsulin sequence, as judged by SDS-gel electrophoresis.

observed by electrophoresis and immunoblotting.

To ensure chemical homogeneity of cysteine residues and to prevent random formation of disulfide bridges, the polypeptide mixture was subjected to oxidative sulfitolysis with sodium sulfite and sodium tetrathionate in 6M GdmCl. The resulting proinsulin analog S-sulfonate was readily assayed in situ by reverse phase HPLC (RPLC)(Figure 2); yields were typically 35-40 mg per g dry cell mass. The derivatized protein was purified by anion exchange chromatography in 25% v/v formamide at pH8.5. About 2000 mg of PIA could be processed on a 5x15 cm column of DEAE-Sepharose FF or QMA-Accell, eluted with a sodium chloride gradient.

Disulfide bonds were introduced by a thiol-mediated exchange (8). The S-sulfonate, preferably at concentrations below 0.5 mg/ml, was incubated at pH9.0 with a three-fold molar excess of glutathione with respect to $-SSO_3$ groups for 20 hours at $4^{o}C$. At 0.2 mg/ml protein, 65% yields as determined by RPLC were reached with both PIA and human proinsulin itself, indicating that the C-terminal linker had no adverse effect on chain folding. Higher yields were possible at

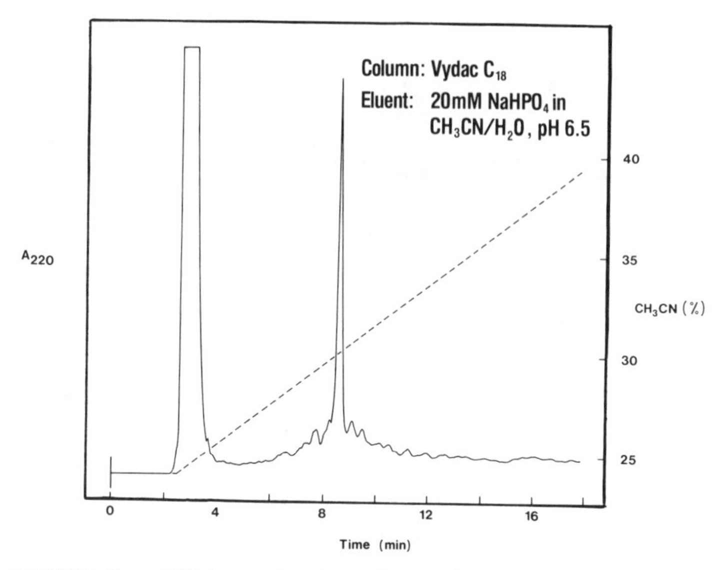

FIGURE 2. RPLC analysis of crude sulfitolysed proinsulin analog measured *in situ*. The large injection peak contains GdmCl, sulfitolysis reagents and non-retained cellular material.

lower protein concentrations, but working volumes then became inconveniently great. PIA was purified by hydrophobic interaction chromatography on Octyl Sepharose equilibrated with 0.4M potassium acetate at pH7.3 (up to 1000 mg of protein on a 5x15 cm gel bed). Nearly homogeneous folded PIA was obtained by elution with a shallow isopropanol gradient (0-20% v/v). The column was then stripped with 60% isopropanol to recover disulfide-bonded oligomers, which could be recycled by sulfitolysis. RPLC peptide mapping after digestion with *S.aureus* V8 protease established that PIA indeed contained the correct disulfide bonds.

Proinsulin can be converted to insulin by brief exposure to trypsin and carboxypeptidase B, in a process resembling the natural mode of activation (9). Rapid tryptic cleavage occurs at dibasic ArgArg and LysArg sites bounding the C peptide region; carboxypeptidase B then eliminates residual basic amino acids from each of the two newly created C-termini to give mature insulin and C peptide (Figure 3). The C-terminal linker of PIA was designed to be removed by the same enzymes. In practice this step was slower than anticipated, and losses were suffered due to further tryptic

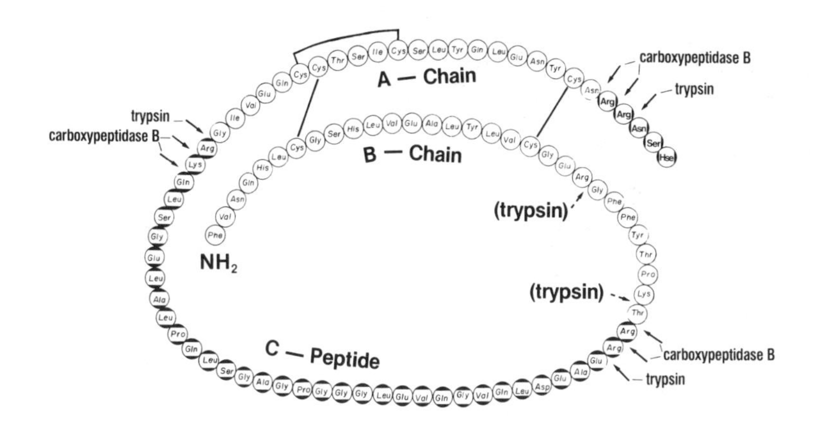

FIGURE 3. Primary structure and enzyme cleavage sites in human proinsulin analog.

degradation of the insulin B chain. Under optimum conditions at pH7.3 and 25°C the yield of insulin reached 60-65% as judged by RPLC, compared with 82% from human proinsulin: thus the relative efficiency was around 75%. We have purified the final product by preparative RPLC on a C_{18} column, and characterized it by isolation of insulin A and B chains, by amino acid sequence and by biological potency.

HUMAN CARDIONATRIN

Cardionatrin is a 28-amino-acid peptide containing one internal disulfide bridge. It was synthesized in E.coli by a similar expression system to that used for proinsulin, giving multi-unit precursors of structure:

T M I T D S L D F R S $(\text{K Cardionatrin K G S})_n$

As before the leader comprises the first few amino acids of β-galactosidase followed by a synthetic linker. Cardionatrin sequences are flanked at each end by lysine, an amino acid which the hormone lacks. Fragmentation can thus be achieved by digestion with endoproteinase Lys-C which cleaves uniquely at the carboxyl side of lysine (10). The resulting cardionatrin "analog" is extended by a single lysine residue, easily removable by the action of carboxypeptidase B.

Expression of precursors containing fewer than six cardionatrin sequences could not be detected by gel electrophoresis. However, the eight-copy precursor was observed as a prominent band accounting for up to 10% of the stainable protein. Since further elongation did not lead to increased protein production, the eight-copy construct was used for all further work. As previously, the cardionatrin precursor remained in the insoluble phase after cell disruption. Purification to near homogeneity was achieved by extraction into 10M urea followed by size exclusion chromatography on Fractogel TSK HW-65F in 8M urea. The precursor was then digested at 37oC with endoproteinase Lys-C in 4M urea at pH8.5. Almost quantitative monomer yields were observed within two hours by either gel electrophoresis or RPLC. After desalting, the cardionatrin analog was trimmed back to the correct C-terminus by brief treatment with carboxypeptidase B at pH8.5 and 37oC. The product was fully reduced with dithiothreitol and purified by preparative RPLC. Gentle oxidation with potassium ferricyanide at pH6.5 (11) served to introduce the single disulfide bridge in near-quantitative yield. The folded cardionatrin was purified to homogeneity by RPLC, and characterized by HPLC behavior in comparison with the chemically synthesized hormone, by amino acid sequence and by biological activity.

DISCUSSION

These studies illustrate the advantages of the multiple gene concept for producing small recombinant eukaryotic proteins in E.coli. A primary product is generated in high yield which is at the same time stable in vivo, readily isolated by virtue of its insolubility, yet consists almost entirely of the desired amino acid sequence. Proinsulin and cardionatrin are particularly suited to this approach because their lack of certain amino acids (methioinine and lysine respectively) permit comparatively simple processing schemes to be developed for their multi-copy precursors. More generally, it will be necessary to exploit a much wider range of sequence-specific cleavage methods to obtain the correct amino acid at both ends of the target protein.

ACKNOWLEDGEMENTS

We are grateful to our many colleagues at Connaught Research Institute for their assistance with these projects. Part of this work was supported by the National Research Council of Canada.

REFERENCES

1. Itakura K, Hirose T, Crea R, Riggs AD, Heyneker HL, Bolivar F, Boyer HW (1977). Expression in Escherichia coli of a chemically synthesized gene for the hormone somatostatin. Science 198:1056.
2. Davis AR, Nayak DP, Ueda M, Hiti AL, Dowbenko D, Kleid DG (1981). Expression of antigenic determinants of the hemaglutinin gene of a human influenza virus in Escherichia coli. Proc Natl Acad Sci USA 78:5376.
3. Goeddel DV, Kleid DG, Bolivar F, Heyneker HL, Yansura DG, Crea R, Hirose T, Kraszewski A, Itakura K, Riggs AD (1979). Expression in Escherichia coli of chemically synthesized genes for human insulin. Proc Natl Acad Sci USA 76:106.
4. Shine J, Fettes I, Lan NCY, Roberts JL, Baxter JD (1980) Expression of cloned β-endorphin gene sequences by Escherichia coli. Nature 285:456.
5. Shen S-H (1984). Multiple joined genes prevent product degradation in Escherichia coli. Proc Natl Acad Sci USA 81:4627.
6. Cockle SA, Loosmore SM, Wosnick MA, James E, Shen S-H (1984). The production of human insulin from Escherichia coli. Proc Biotech 84 USA p.393.
7. Shen S-H, Cockle SA, Loosmore SM (1985). Production of human insulin from E.coli by multiple joined proinsulin genes. Proc Bio-Expo 85 p.169.
8. Frank BH, Pettee JM, Zimmerman RE, Burck PJ (1981). The Production of human proinsulin and its transformation to human insulin and C-peptide. Proc VII American Peptide Symposium p.729.
9. Kemmler W, Peterson JD, Steiner DF (1971). Studies on the conversion of proinsulin to insulin. J Biol Chem 246:6786.
10. Jekel PA, Weijer WJ, Beintema JJ (1983). Use of endoproteinase Lys-C from Lysobacter enzymogenes in protein sequence analysis. Anal Biochem 134:347.
11. Felix AM, Jimenez MH, Wang C-T, Meienhofer J (1980). Synthesis of somatostatin and (D-Trp_8)somatostatin. Int J Peptide Protein Research 15:342.

Protein Purification: Micro to Macro, pages 383-391

ISOLATION OF *E. COLI* DERIVED MURINE INTERLEUKIN-2 FROM INTRACELLULAR AND SECRETORY EXPRESSION SYSTEMS

Hung V. Le, Rosalinda Syto, Carol Mays, Paul Reichert, Satwant Narula, Keith Gewain, Robert Greenberg, Robert Kastelein, Anita Van Kimmenade, Tattanahalli L. Nagabhushan and Paul P. Trotta

Schering-Plough Corp., Bloomfield, New Jersey 07003 and DNAX, Palo Alto, California 94304

ABSTRACT The mature form of recombinant murine interleukin-2 (muIL-2) expressed in *E. coli* with a secretory vector was extracted by osmotic shock and purified to homogeneity by sequential DEAE-Sepharose and Sephadex G-100 chromatographies. The purified protein exhibited a specific activity of 3.5×10^7 units/mg in an HT-2 cell proliferation assay. The apparent molecular weight estimated by sodium dodecyl sulfate-polyacrylamide gel electrophoresis and gel filtration chromatography was 19 kd and 30 kd, respectively, suggesting that the secreted form of muIL-2 is dimeric. In parallel studies muIL-2 was also isolated by extraction of *E. coli* engineered with an intracellular expression system. MuIL-2 extracted either in the presence or absence of deoxycholate was extensively aggregated. The aggregated muIL-2 was stabilized by disulfide bonds and demonstrated significantly lower biological activity than the dimeric muIL-2. These results suggest an advantage for the secretory expression system for the isolation of native muIL-2.

INTRODUCTION

Murine interleukin-2 (muIL-2) is a lymphokine produced by mouse T cell stimulated with antigens or mitogens. Originally described as a factor capable of maintaining long-term growth of T cells in culture, this glycoprotein has been shown to mediate a large number of immunological effects (1,2,3). Characterization of the gene by cDNA cloning and sequencing has recently been described (4,5). The mature protein sequence of muIL-2 is predicted from the cDNA to consist of 149 amino acids, including 3 cysteine residues. A high degree of homology (63%) exists between the predicted human and murine amino acid sequences although the latter contains a unique stretch of 12 consecutive glutamines near the N-terminus. We report here a comparative study of the isolation of muIL-2 from two strains of E. coli that have been genetically engineered with intracellular and secretory vectors, respectively. Our data indicate that muIL-2 expressed intracellularly under the control of a TAC promoter cannot be readily isolated in its native form principally due to aggregation involving intermolecular disulfide bond formation. In distinction, muIL-2 that was expressed at a similar level in the same host using a secretion vector was released from the periplasm in a more native state and was purified to apparent homogeneity by conventional column chromatography.

MATERIALS AND METHODS

MuIL-2 was expressed in E. coli 294 using both an intracellular expression plasmid (exTAC-mIL-2) and a secretory plasmid (pOmpA-mIL-2). The construction of these plasmids is described elsewhere (6,7). E. coli 294/exTAC-mIL-2 was grown overnight at 37°C followed by inactivation with 1% toluene. The resuspended cells were disrupted by Manton-Gaulin homogenization. E. coli 294/pOmpA-mIL-2 grown under similar conditions, was collected by centrifugation and the muIL-2 was released from the periplasm by osmotic lysis (8). MuIL-2 activity was assayed employing the murine T-cell line HT-2 (9) and the MTT colorimetric assay (10). One unit is defined as the amount of factor in 0.1 ml that produces 30% of the maximal stimulation attainable under the same conditions. Quantitation of sulfhydryl groups and disulfide bonds was performed employing 4-4′-dithiodipyridine (11) and 2-nitro-5-thiosulfobenzoic acid (12). Protein concentration was determined as described by Bradford (13) with bovine serum albumin as standard.

All chromatographic resins were purchased from Pharmacia (Piscataway, NJ). Polyethyleneimine was obtained from Sigma Chemical Co. (St. Louis, MO).

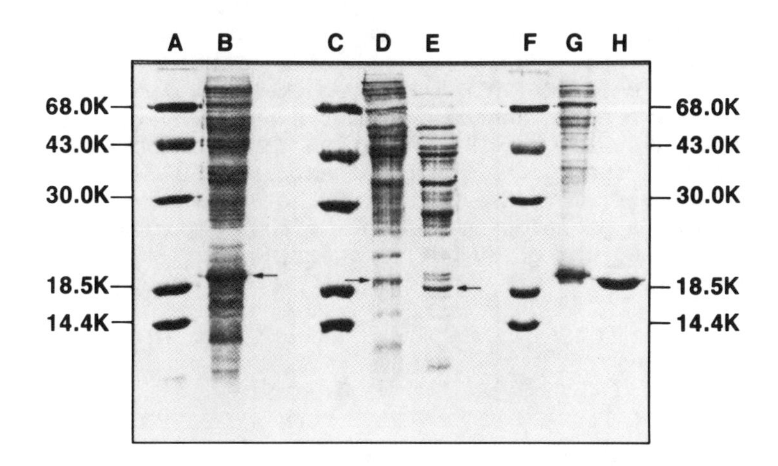

FIGURE 1. SDS-PAGE of E. coli extracts and purified recombinant muIL-2. All samples were boiled with 1% 2-mercaptoethanol for 2 min prior to electrophoresis. A,C,F: molecular weight standards (bovine serum albumin, ovalbumin, carbonic anhydrase, β-lactoglobulin and lysozyme); B: 1% deoxycholate extract of E. coli 294/exTAC-mIL-2; D: phosphate-buffered saline extract of E. coli 294/exTAC-mIL-2; E: osmotic lysate of E. coli 294/pOmpA-mIL-2; G: muIL-2 partially purified from a deoxycholate extract of E. coli 294/exTAC-mIL-2; H: purified muIL-2 derived from E. coli 294/pOmpA-mIL-2. The gel was stained with Coomassie Brilliant Blue R-250. MuIL-2 is identified with an arrow.

RESULTS

The supernatant obtained after centrifugation of the osmotic lysate of the *E. coli* secretory strain 294/pOmpA-mIL-2 constituted the starting material for purification of muIL-2. Sodium dodecyl sulfate-polyacrylamide gel electrophoresis (SDS-PAGE) indicated a prominent band with an apparent molecular weight of 18,000 (Fig. 1, lane E). This species was identified as muIL-2 by comparison to the extract obtained from the parental *E. coli* strain that did not contain the muIL-2 gene (data not shown). Purification to a high degree of homogeneity was achieved by DEAE-Sepharose and Sephadex G-100 chromatographies (Table 1). The purified protein exhibited a specific activity of 3.5×10^7 units/mg. SDS-PAGE indicated that the purity of the final preparation was greater than 90% and migrated with a molecular weight similar to that observed for the muIL-2 species in the crude extract (Fig. 1, lane H). However, muIL-2 eluted from the Sephadex G-100 column at a position corresponding to an apparent molecular weight of 30,000, suggesting that the quaternary structure is a dimer.

TABLE 1

PURIFICATION OF RECOMBINANT MUIL-2 DERIVED FROM *E. COLI* STRAIN 294/pOmpA-mIL-2[a]

Step	Protein mg	Units x 10^6 Per mg	Units x 10^6 Total	Recovery %
Osmotic lysate	665	4.2	280	(100)
DEAE-Sepharose	31	50	160	57
Sephadex G-100	7.5	35	26	9.7

[a]. All steps were performed at 4°C.

In a parallel study muIL-2 was extracted from E. coli engineered with the intracellular expression system by homogenization in both the presence and absence of 1% deoxycholate. SDS-PAGE of these extracts indicated a band with an apparent molecular weight of 19,200 (Fig. 1, lanes B and D) that did not appear in an extract of the parental E. coli strain not containing the muIL-2 gene (data not shown). The deoxycholate extract exhibited a specific activity of 5 x 10^5 units/mg compared to 1 x 10^5 units/mg for the phosphate-buffered saline extract. Thus, a significant portion of the expressed muIL-2 required treatment with detergent for solubilization.

Purification of both extracts was attempted by treatment with 0.1% polyethyleneimine followed by DEAE-Sepharose and Sephacryl S-200 chromatographies performed in the absence of detergent. However, only a 1.5-2.0 fold increase in specific activity could be achieved (data not shown). Similarly, SDS-PAGE of the Sephacryl S-200 eluate (Fig. 1, lane G) showed significant contamination of the partially purified muIL-2 with E. coli proteins. Interestingly a comparison of the migration on SDS-PAGE of purified muIL-2 from the secretion and intracellular strains indicated that the former had a distinctly faster mobility (Fig. 1, lanes G and H). Edman degradation of the N-terminus of muIL-2 derived from the 294/ex-TAC-mIL-2 strain demonstrated the presence of an N-terminal methionine, whereas muIL-2 derived from the pOmpA-mIL-2 strain was found to lack this amino acid (14). This observation may account for the somewhat different apparent migrations observed on SDS-PAGE. Of special importance was the fact that muIL-2 eluted in the excluded volume of the Sephacryl S-200 column, supporting an apparent molecular weight in excess of 250,000. Thus, the failure to remove contaminating proteins by chromatography is likely related to a strong interaction between muIL-2 and E. coli proteins.

Since the cDNA sequence of muIL-2 predicted the presence of three cysteine residues (4,5), we performed experiments to determine whether intermolecular disulfide bonds might be associated with the high molecular weight muIL-2 aggregates. SDS-PAGE was performed on the phosphate-buffered saline extract of

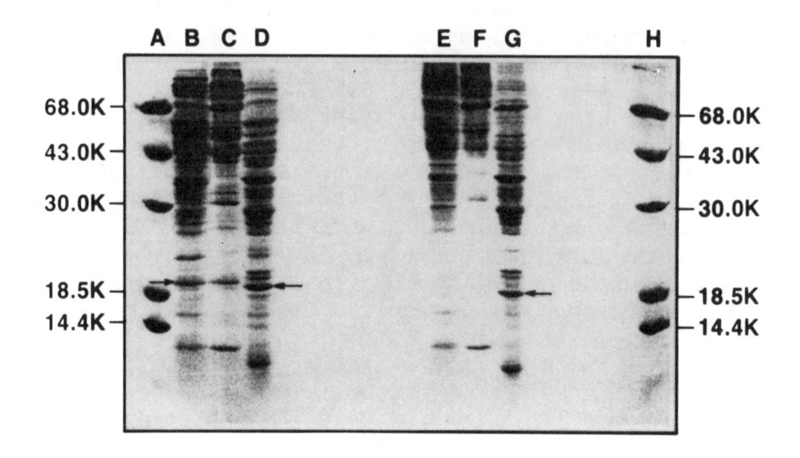

FIGURE 2. SDS-PAGE of E. coli extracts of muIL-2 with and without prior reduction with 2-mercaptoethanol. Samples in lanes A, B, C, D and H were boiled for 2 min. with 1% 2-mercaptoethanol compared to samples in lanes E, F and G, which were not treated with 2-mercaptoethanol. A, H.: molecular weight standards; B,E: phosphate-buffered saline extract of E. coli 294/exTAC-mIL-2; C,F: phosphate-buffered saline extract of E. coli 294/exTAC-mIL-2 partially purified; G,D: osmotic lysate of E. coli 294/pOmpA-mIL-2. The gel was stained with Coomassie Brilliant Blue R-250. MuIL-2 is identified with an arrow.

strain 294/exTAC-mIL-2 both with and without prior reduction with 2-mercaptoethanol. As shown in Fig. 2 (lanes B and E) the species corresponding to muIL-2 was absent in the non-reduced sample in comparison to the presence of a prominent band with a 19,200 molecular weight in the reduced extract. A similar experiment performed on a preparation of muIL-2 partially purified from this extract yielded the same results (Fig. 2, lanes C and F). In both cases the loss of the muIL-2 band in the non-reduced samples was associated with the appearance of high molecular weight aggregates at the top of the 15% polyacrylamide gel. Similar results were observed for the deoxycholate extract (data not shown). In contrast, as illustrated in Fig. 2 (lanes D and G), SDS-PAGE of a reduced and non-reduced osmotic lysate of strain 294/pOmpA-mIL-2 were virtually identical.

TABLE 2

SULFHYDRYL AND DISULFIDE CONTENTS OF RECOMBINANT MUIL-2 DERIVED FROM E. COLI STRAIN pOmpA-mIL-2

Sulfhydryls (moles/mole)	Disulfides (moles/mole)
1.8 ± 0.1[a]	0.7 ± 0.1[a]

[a] ± standard deviation. Data represent the average of 3-5 determinations.

We have also characterized the state of oxidation of the cysteine residues in the purified preparation of muIL-2 from strain 294/pOmpA-mIL-2. As shown in Table 2, titration with 4,4′-dithiodipyridine indicated the presence of 1.8 moles of titratable sulfhydryl groups per mole of protein. Titration with 2-nitro-5-thiosulfobenzoic acid in the presence of sodium sulfite indicated 0.7 moles of disulfide bond present per mole of protein. The total cysteine content derived from the summation of these values (i.e., 3.2 moles) is in good agreement with the cDNA prediction of 3.0. (4,5).

DISCUSSION

The capacity to produce a biologically active recombinant protein with the proper primary structure and three-dimensional folding is highly dependent on the choice of expression vectors and host organisms (15). We report here characteristics of the physico-chemical properties of E. coli-derived recombinant muIL-2 that support the conclusion that a secretory vector that results in the export of mature muIL-2 into the periplasmic compartment provides distinct advantages over an intracellular expression system for the preparation of native muIL-2. The most striking difference between the muIL-2 preparations isolated from the two expression systems is related to the quaternary structure; i.e., muIL-2 from strain 294/pOmpA-mIL-2 is an apparent dimer whereas muIL-2 from strain 294/exTAC-mIL-2 is present in high molecular weight aggregates. The fact that the muIL-2 aggregates cannot be readily separated from *E. coli* proteins by conventional chromatography suggests an interaction with host proteins, which we have shown to be stablized at least in part by disulfide bonds. These bonds are capable of being formed within the bacterial cytoplasm (16). We cannot exclude, however, the possibility that a portion of muIL-2 also self-associates to high molecular weight aggregates. Purification of biologically active recombinant protein from inclusion bodies has required strong chaotropic agents followed by renaturation (17, 18). This observation is consistent with our inability to purify muIL-2 from a detergent or saline extract.

MuIL-2 isolated from the secretory system appears to be a dimer that is stabilized by non-covalent interactions. It is notable that naturally occurring muIL-2 has also been reported to be a dimer (19). Similarly, the biological activity of the purified recombinant muIL-2 (i.e., 3.5×10^7 units/mg) is comparable to that reported for muIL-2 isolated from EL4 cells (20). We were able to identify the presence of an intramolecular disulfide bond, which has been shown to be essential for human IL-2 activity by site-directed mutagenesis (21). However, the fact that only 0.7 moles of disulfide were titratable per mole of muIL-2 supports the conclusion that a subpopulation exists in which all the cysteine residues are reduced. It has not yet been established whether the reduced muIL-2 is biologically active.

The data presented here support a significant difference in protein processing in the intracellular and periplasmic compartments. Further studies with other recombinant proteins will be required to establish whether the periplasmic expression system is consistently advantageous for purification of native proteins.

REFERENCES

1. Morgan DA, Ruscetti FW and Gallo R (1976) Science 193: 1007.
2. Gillis S, Ferm MW, On W and Smith KA (1978) J Immunol 120: 2027.
3. Smith KA (1984) Ann Rev Immunol 2: 319.
4. Yokota T, Arai N, Lee F, Rennick D, Mosman T and Arai K (1985) Proc Natl Acad Sci USA 82: 68.
5. Kashima N, Nishi-Takaota C, Fujita T, Taki S, Yamada G, Hamuro J and Taniguchi J (1985) Nature 313: 402.
6. Zurawski SM, Mosmann TR, Benedick M and Zurawski G (1986) J Immunol 137: 3354.
7. Kastelein R (1987) manuscript submitted.
8. Neu HC and Heppel LA (1965) J Biol Chem 240: 3685.
9. Watson J (1979) J Exp Med 150, 1510.
10. Mosmann TR (1983) J Immunol Methods 65: 55.
11. Grassetti DR, Murray JF Jr (1967) Arch Biochem Biophys 119: 41.
12. Thannhauser TW, Konishi Y, Scheraga H (1984) Anal Biochem 138: 181.
13. Bradford M (1976) Anal Biochem 72: 248.
14. Bond M, personal communication.
15. Nicaud J-M, Mackman N, Holland IB (1986) J Biotechnology 3: 255.
16. Shoemaker JM, Brasnett AH, Marston FAO (1985) EMBO J 4: 775.
17. Winkler ME, Blaber M, Bennett GL, Holmes W, Vehar GA (1985) Bio/Technology 3: 990.
18. Marston FAO, Lowe PA, Doll MT, Schoemaker JM, White S, Angal S (1984) Bio/Technology 2: 800.
19. Caplan B, Clifford G, Paetkau V (1981) J Immunol 126: 1351.
20. Riendau D, Harnish DG, Bleackley RC, Paetkau V (1983) Nature 258: 12, 114.
21. Wang A, Lu S, Mark D (1984) Science 224: 1431.

Protein Purification: Micro to Macro, pages 393–399

ENGINEERING PROTEIN EXPORT IN *ESCHERICHIA COLI* : EXPRESSION AND EXCRETION OF MUTANT CLOACIN MOLECULES[1]

Arnold J. van Putten and Bauke Oudega

Department of Molecular Microbiology, Vrije Universiteit,Amsterdam,The Netherlands

ABSTRACT The bacteriocin cloacin DF13 is excreted into the culture medium of the gram-negative bacterium *Escherichia coli*. In this study we have investigated the expression and excretion of mutant cloacin molecules. We demonstrate that large deletions or additions at the carboxyl-terminus of the cloacin affect both expression and release of the polypeptide.

INTRODUCTION

Protein export by bacteria has been extensively studied during the past years. Only a few proteins such as bacteriocins (1) and hemolysins (2) are excreted efficiently into the culture medium of the gram-negative bacterium *Escherichia coli*. One of these bacteriocins, denominated cloacin DF13, is encoded by the small non-conjugative plasmid CloDF13 (3). The bacteriocin is released from the host cells as an equimolar complex of two polypeptides, namely cloacin (M_r 59,293) and its immunity protein (M_r 9,974).

[1]This work was supported by the Netherlands Organisation for the Advancement of Pure Research (Z.W.O.).

The immunity protein protects bacteriocin producing cells against the endoribonucleolytic activity of the bacteriocin. The export process requires the presence of a third protein, the bacteriocin release protein (BRP, M_r 2871;4).

We are studying the structure/function relationship of the cloacin protein with respect to high level expression and efficient excretion. For that purpose we have analyzed several CloDF13::Tn*901* mutants encoding different truncated cloacin polypeptides. We observed that the expression of the smallest truncates was drastically reduced as compared to the cloacin present in the complex. Furthermore, we found that only small alterations in the carboxyl-terminus of the cloacin are tolerated in order to allow excretion into the culture medium. Our data suggest the presence of topogenic sequences within the primary structure of cloacin which are somehow required for high level expression and efficient excretion.

METHODS

Experiments were carried out with *Escherichia coli* strain N3406 as host bacterium (5). The construction of the plasmids used in this study was carried out by using standard recombinant-DNA techniques as described elsewhere (6). Immunological analysis of the proteins (ELISA, Western-blotting) has been described previously (7).

RESULTS

We have analyzed different CloDF13::Tn*901* mutants (8). These mutants, except pJN67, contain a transposon Tn*901* insertion within the gene encoding the cloacin. Consequently, no expression of the downstream located immunity and BRP gene was observed in these cases. The plasmids used are listed in Table 1.

TABLE 1
EXPRESSION OF CLODF13::Tn*901* MUTANTS

Plasmid	Polypeptide[a] (Md)	Expression[b] (μgr/ml)	Excretion[c]
pJN67	59	788	+
pJN81	58	260	+
pJN82	55	260	-
pJN76	52	422	-
pJN78	47	2	n.d.
pJN80	39	16	n.d.

[a]The size of the truncated cloacin polypeptides was determined by SDS-PAGE.

[b]Expression of the polypeptides was determined by an ELISA using monoclonal antibodies raised against the amino-terminal part of cloacin.

[c]Excretion of the polypeptides was studied in the presence of a complementing plasmid expressing the BRP gene.

n.d. not detectable

The data listed in Table 1 show that plasmid pJN67, encoding the complete cloacin and the immunity protein, gives rise to a high level of expression of cloacin DF13 (the equimolar complex of cloacin and the immunity protein). The expression of the polypeptides encoded by the plasmids pJN81, pJN82 and pJN76 is only 2-4 fold reduced as compared to pJN67 (the cloacin complex). Apparently, the removal of less than 15% of the cloacin at the COOH-terminal end hardly affects the expression of the protein. However, in the case of pJN78 and pJN80,the expression of the cloacin truncate was greatly reduced (50-400 fold) as compared to the cloacin present in the bacteriocin complex. With respect to excretion we found that only the polypeptides encoded by pJN67 and pJN81 were efficiently excreted. This indicated that the removal of more than 10 amino acid residues from the carboxyl-terminus of cloacin blocks excretion.

To further investigate the expression and excretion of mutant cloacin molecules we have constructed a plasmid which encodes a polypeptide consisting of the "core" protein of plasmid pJN81 (appr. 552 aa) fused to a stretch of heterologous amino acid residues. The construction of this plasmid, designated pAP6, is given in Figure 1.

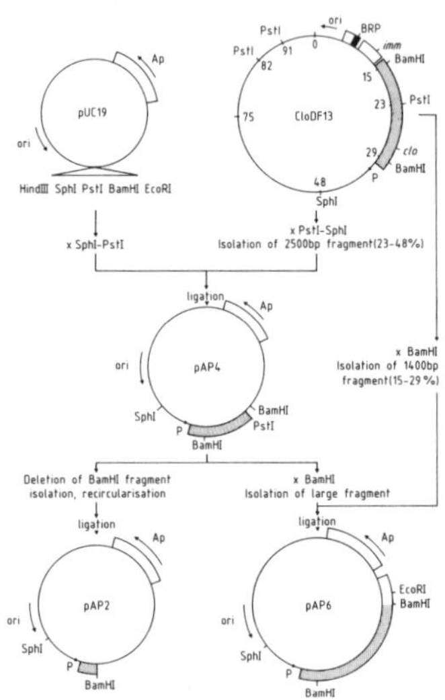

FIGURE 1. Construction of plasmid pAP6. Indicated are cloacin (▦) and other (□) protein encoding sequences. Relevant restriction sites are shown. *clo*, cloacine gene; *imm*, immunity gene; BRP, bacteriocin release protein; P, cloacin promoter; Ap, ampicillin resistance gen.

Plasmid pUC19 was used as a cloning vehicle in the construction of plasmid pAP6. We analyzed the expression and release of the truncated cloacin molecules encoded by pAP6 and that encoded by the intermediate plasmids pAP2 and pAP4. No or little expression was observed for the cloacin truncates encoded by these latter plasmids (Figure 2, lane 1 and 2, respectively), which is in good agreement with the data obtained with pJN78 and pJN80. The truncated cloacin molecule encoded by plasmid pAP6 was strongly expressed. This protein is slightly larger than wildtype cloacin (see Figure 2, lanes 3 and 5), which can be explained by the additional 82 amino acid residues resulting from read-through translation. Analysis of the excretion of this truncated protein in the presence of a BRP complementing plasmid demonstrated that less then 10% of the protein was excreted into the culture medium.

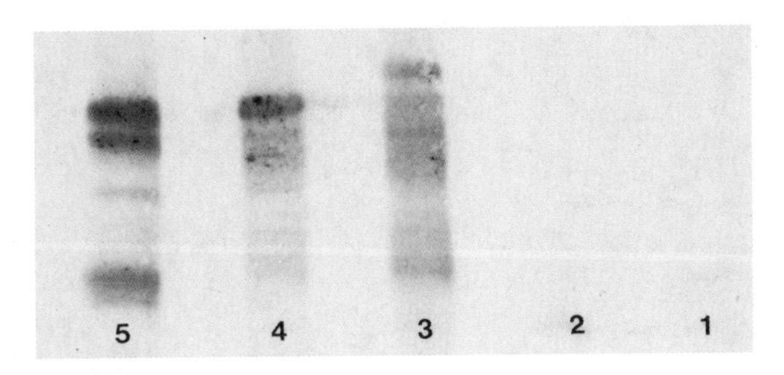

FIGURE 2. Immunological analysis of mutant cloacin polypeptides using Western-blotting. Lane 1: pAP2; lane 2: pAP4; lane 3: pAP6; lane 4:pAP7; lane 5: cloacin DF13 (marker).

Since we assumed that the 82 additional carboxyl-terminal amino acid residues in this truncated protein hamper the excretion, we constructed plasmid pAP7. This plasmid was made by the introducion of a frame-shift mutation by filling in the unique EcoRI-site of plasmid pAP6 (Figure 1). The mutant cloacin protein encoded by this plasmid consists of 560 amino acids instead of 643 of pAP6 and possesses no endoribonucleolytic activity (see discussion).As shown in Figure 2 (lane 4) the molecular weight of this protein is comparable to that of wildtype cloacin (561 aa). We observed no difference between pAP6 and pAP7 with respect to expression.

DISCUSSION

In this paper we described the expression and excretion of different mutant cloacin polypeptides. We analyzed cloacin fragments obtained by transposon mutagenesis and by specific cloning experiments. Our data clearly demonstrated that the removal of a large part of the carboxyl-terminus of cloacin strongly affects the expression of the protein. We infer that the low level expression of these cloacin truncates results from a rapid proteolytic degradation. Earlier, De Graaf et al. (9) showed that the removal of the immunity protein, which binds to the carboxyl-terminal part of the cloacin molecule diminishes the stability of cloacin and results in an increased proteolytic degradation.

From our experiments we concluded that the presence of a cloacin "core" fragment is essential for stable expression. Fusions to the carboxyl-terminal end of this fragment hardly influence protein expression. However,with respect to excretion we found that only small alterations at the carboxyl-terminus of cloacin are tolerated, since only the mutant polypeptides encoded by pJN81 and pAP7 could be detected in the culture medium. This indicated an involvement of the carboxyl-terminal part of cloacin in the excretion process. This region might contain so-called topogenic sequences which are essential either for the recognition of the excretion sites by the cloacin or for correct folding of the cloacin prior to excretion. A comparable suggestion was done by Baty et al. (10) for the excretion of colicin A.

An interesting feature of the polypeptide encoded by pAP7 is the absence of any endoribonucleolytic activity. Normally, cloacin producing cells lacking the immunity protein are not viable. Since this polypeptide shows the same properties as its wildtype equivalent with respect to size, expression and excretion, this "redesigned" protein may form the basis for the study of the cloacin excretion system in the absence of the immunity protein.

REFERENCES

1. De Graaf FK, Oudega B (1986). Production and release of cloacin DF13 and related colicins. Curr Topics in Microbiol and Immunol 125 : 183.
2. Cavalieri SJ, Bohach GA, Snyder IS (1984). *Escherichia coli* α-hemolysin: characteristics and probable role in pathogenicity. Microbiol Rev 48: 326.
3. Tieze GA, Stouthamer AH, Jansz HS, Zandberg J, van Bruggen EFJ., (1969). A bacteriocinogenic factor of *Enterobacter cloacae*. Mol Gen Genet 106: 48.
4. Oudega B, Stegehuis F, Van Tiel-Menkveld GJ, De Graaf FK. (1982). Protein H encoded by plasmid CloDF13 is involved in excretion of cloacin DF13. J Bacteriol 150: 1115.
5. Van Tiel-Menkveld GJ, Rezee A, De Graaf FK (1979). Production and excretion of cloacin DF13 by *Escherichia coli* harboring plasmid CloDF13. J Bacteriol 140:415.
6. Van Putten AJ (1986). "The bacteriocinogenic plasmid CloDF13: plasmid maintenance and transfer". Thesis, Vrije Universiteit, Amsterdam, The Netherlands.
7. Krone WJA, De Vries P, Koningstein G, De Jonge AJR, De Graaf FK, Oudega B (1986). Uptake of cloacin DF13 by susceptible cells: removal of immunity protein and fragmentation of cloacin molecules. J Bacteriol 166: 260.
8. Andreoli PM, Overbeeke N, Veltkamp E, Van Embden JDA, Nijkamp HJJ (1978). Genetic map of the bacteriocinogenic plasmid CloDF13 by insertion of transposon Tn*901*. Mol Gen Genet 160:1.
9. De Graaf FK, Stukart MJ, Boogerd FC, Metselaar K (1978).Limited proteolysis of cloacin DF13 and characterisation of the cleavage products. Biochemistry 17:1137.
10. Baty D, Knibiehler M, Verheij H, Pattus F, Shire D, Bernadac A, Lazdunski C (1987). Site-directed mutagenesis of the COOH-terminal region of colicin A: Effect on secretion and voltage-dependent channel activity. Proc Natl Acad Sci USA 84: 1152.

Protein Purification: Micro to Macro, pages 401–407

SIMPLE THREE STEP BATCH PURIFICATION OF *ESCHERICHIA COLI* TRYPTOPHAN APOREPRESSOR[1].

Dennis N. Arvidson, Andrew A. Kumamoto, and Robert P. Gunsalus

Department of Microbiology
and Molecular Biology Institute
University of California, Los Angeles, CA 90024

ABSTRACT We have developed a simple three step procedure for large scale purification of the *Escherichia coli* tryptophan (Trp) aporepressor, a small dimeric (12.5-kilodalton/subunit) DNA-binding regulatory protein. Aporepressor was overproduced to 0.75% of the total cellular protein via an *E. coli* strain containing an expression plasmid. Clarified lysate was heated to 85°C for 10 min in 0.5% streptomycin sulfate. Precipitate was removed by centrifugation and the supernatant was batchwise salt eluted from a heparin-agarose column. Pooled aporepressor fractions were then batchwise salt eluted from a Cibacron-Blue-agarose column. This three step batch procedure routinely gave 150 mg of aporepressor from 300 g of cell paste (approximately 55% yield). Aporepressor prepared by this method was greater than 99% pure. Activity for the binding of tryptophan and operator was nearly 100%. Parts of this procedure may be generally applicable to purification of other DNA-binding proteins.

[1]This work was supported in part by Public Health Service grant GM 29456 (R.P.G.) from the National Institutes of Health. D.N.A. was supported in part by Cellular and Molecular Biology Training Grant GM 07185 from the National Institutes of Health. A.A.K. was supported in part by Genetics Training Grant GM 07104 from the National Institutes of Health.

INTRODUCTION

Trp aporepressor negatively regulates transcription of the trpEDCBA and aroH biosynthetic operons as well as the trpR operon which encodes the aporepressor (1, 2). The aporepressor dimer binds the corepressor, L-tryptophan, at two sites to form the Trp repressor complex (3). The repressor binds the operator DNA of the three operons preventing initiation of transcription by RNA polymerase (1, 2). Trp aporepressor thus responds to the level of intracellular L-tryptophan and adjusts transcription of these operons to maintain the concentration of L-tryptophan in the cell within levels needed for efficient protein synthesis.

We have developed a purification procedure for the Trp aporepressor which exploits the high stability of the protein and its affinity for DNA by employing a heat step and two affinity columns for DNA-binding proteins. This procedure was developed from a more time consuming procedure which required gradient elution of columns and lengthy assay of column fractions by RsaI restriction enzyme protection assay for trp operator DNA binding activity (1, 4), and by SDS-PAGE for visualization of aporepressor protein. Only one protein peak (which contained aporepressor activity) was observed upon gradient elution of heparin-agarose (contaminating protein did not adsorb) or Cibacron-Blue-agarose (contaminating protein remained bound) columns. This led to development of batch procedures that made activity assays and protein gels of column fractions unnecessary. The procedure reported here was rapid and efficient, a 55% yield of greater than 99% pure Trp aporepressor was routinely obtained. The pure protein retained nearly 100% activity for L-tryptophan binding and greater than 90% activity for operator DNA binding. This method will facilitate the purification of some mutant forms of Trp aporepressor protein.

EXPERIMENTAL PROCEDURES

Materials.

Buffer components were purchased from Sigma Chemical Co. (St. Louis, Mo.). The restriction enzyme RsaI was prepared as described (5, 6). Heparin-agarose was prepared as described (7). Cibacron-Blue-agarose (Affi-Gel Blue) was obtained from Bio-Rad Inc. (Emeryville, Cal.). Bacterial growth media was from Difco Laboratories (Detroit, Mich.).

Methods.

Analysis of Aporepressor Purity. Aporepressor purity was analyzed by denaturing polyacrylamide gel electrophoresis (19% acrylamide, 0.75% bis, 6 M Urea, 0.1 % SDS). Purity of the native aporepressor was analyzed by high performance liquid chromatography gel filtration as described (3).

Storage of purified aporepressor. Purified aporepressor was stored at -70°C in buffer II (see below) plus 0.2 M KCl at concentration greater than 2 mg/ml (8). For use, a sample of the frozen material was thawed rapidly by gently mixing the solution under tepid running water. Prolonged storage in buffers of low ionic strength (less than 0.1 M KCl) or low pH (less than 7) was avoided due to low solubility of the protein under these conditions.

Protein concentration. Concentration of pure aporepressor was determined from the absorbance at 280 nm using the reported extinction coefficient of 1.2 cm^{-1} mg^{-1} ml (4). Protein concentration was determined by the method of Bradford (9) using purified Trp aporepressor as a standard.

RESULTS AND DISCUSSION

Trp Aporepressor Production.

Trp aporepressor was produced in *E. coli* strain LE392 containing the multiple copy *trpR*$^+$ plasmid pRPG47 (3). This plasmid, a pBR322 derivative, contains the *trpR* structural gene downstream from the *E. coli* *lac*UV5 promoter and ribosome binding site. Cell cultures were grown to late log phase (O.D. at 600 nm of 4.0) in the following medium (0.75% Bacto-Tryptone, 0.5% yeast extract, 0.5% lactose, 0.5% NaCl) in a New Brunswick 200 liter fermentor, and harvested by Sharples continuous centrifugation. Thin slabs of cell paste were wrapped in plastic wrap, frozen and stored at -20°C until used. Trp aporepressor was stable in the frozen cell paste for more than a year.

Trp Aporepressor Purification.

Figure 1 shows an SDS-PAGE gel of repressor fractions at each step of purification. All steps were carried out at 4°C unless otherwise indicated. Frozen cell paste (50 g) was thawed by stirring at room temperature in 250 ml of 0.1 M Tris HCl pH 7.8, 1 mM EDTA. The suspended cells were broken by passage through a French Pressure cell at 12,000 to 16,000 psi and the lysate was clarified by centrifugation at 14,500 rpm for 30 min in a Sorvall SA-600 rotor. Streptomycin sulfate (10% [wt/vol]) was added slowly to the supernatant with stirring to a final concentration of 0.5 %. The mixture was stirred for an additional 30 min, rapidly heated to 85°C by shaking in a flask in a boiling water bath, held at 85°C for ten min, then cooled rapidly to 4°C in ice water. The precipitate was removed by centrifugation at 8,500 rpm for 20 min in a Sorvall SA-600 rotor. The supernatant was then applied to a heparin-agarose column (2.5 cm x 20 cm, 100 ml bed vol) equilibrated with buffer II (10 mM potassium phosphate, pH 7.8; 0.1 mM EDTA; 5% [vol/vol] glycerol). The column was washed with buffer II plus 0.1 M KCl until no protein was detected in the flow-through (Bradford protein assay; ~five column volumes). The column was batchwise salt eluted with buffer II plus 0.6 M KCl. Fractions (4.5 ml) which gave blue color when 100 μl aliquots were tested in the Bradford protein assay were pooled. This material was diluted with buffer II until conductivity was below 0.05 M KCl, and then applied to a Cibacron-blue-agarose column (2.5 cm x 10 cm, 50 ml bed vol). The column was washed with two volumes of buffer II plus 0.05 M KCl and then batchwise salt eluted with buffer II plus 0.6 M KCl. Fractions (4.5 ml) which gave blue color when 100 μl aliquots were tested in the Bradford protein assay were pooled. This material was dialyzed against buffer II plus 0.2 M KCl, filter sterilized (Millex-HA 0.45 μm filter unit, Millipore), dispensed into microcentrifuge tubes (0.25 ml/tube) and stored at -70°C. Table 1 summarizes the purification protocol results. We obtained 25 mg of Trp aporepressor from 50 g of cell paste (55% yield). This method was successfully scaled up to 300 g cell paste. We estimate the protein to be greater than 99% pure from analysis by SDS-PAGE and by HPLC gel sizing techniques (3). The protein elutes from a TSK-250 gel sizing column as a single dimeric species with an apparent native molecular weight of 30,000 (3).

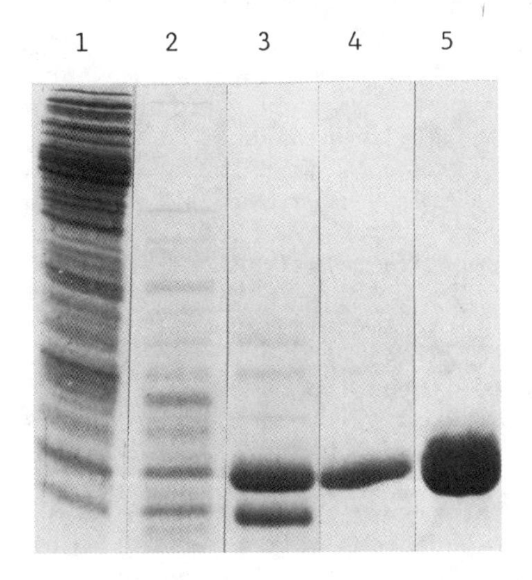

Figure 1. Purity of Trp aporepressor at each stage of purification. SDS-PAGE was performed as described in methods. Lane 1, 38 μg crude lysate. Lane 2, 9.5 μg heat step sample. Lane 3, 10.6 μg pooled heparin-agarose eluate. Lane 4, 5.1 μg pooled Cibacron-Blue-agarose eluate. Lane 5, 34 μg pooled Cibacron-Blue-agarose eluate. Protein was stained with Coomassie Brilliant Blue dye.

TABLE 1
PURIFICATION OF THE TRP APOREPRESSOR

Sample	Volume ml	Total Protein mg	Yield[a] %	Purification[a] fold
Crude	280	5300	100	1
Heated, streptomycin	255	480	75	8
Heparin-agarose	31	52	65	66
Cibacron-Blue-agarose	42	25	55	120

[a]estimated from protein gels and restriction enzyme protection assay.

Corepressor Binding Activity of Purified Trp Aporepressor.

The ability of purified Trp aporepressor to bind the corepressor, L-tryptophan, was determined by equilibrium dialysis using L-[^{14}C]tryptophan (3). Scatchard plots of the binding data indicate that the aporepressor preparations are nearly 100% active for L-tryptophan binding and show that the two binding sites on the aporepressor dimer are identical and independent (3).

Operator DNA Binding Activity of Purified Trp Aporepressor.

The activity of the Trp aporepressor for operator DNA binding was evaluated by the ability of Trp repressor to protect a *Rsa*I restriction site located within the *trp* operon operator from cleavage by the restriction enzyme *Rsa*I (1, 4). Using this assay we have shown that our aporepressor preparations were greater than 90% active for operator DNA binding and that L-tryptophan increased the relative affinity of the aporepressor for both operator DNA and nonspecific DNA (6). Our aporepressor preparations were also shown to be greater than 90% active for operator binding in a DNAse I titration assay (10).

ACKNOWLEDGMENTS

We thank Cindy Grove for expert technical assistance.

REFERENCES

1. Gunsalus RP, Yanofsky C (1980). Nucleotide sequence and expression of *Escherichia coli trpR*, the structural gene for the trp aporepressor. Proc Natl Acad Sci USA 77:7117.
2. Zurawski G, Gunsalus RP, Brown KD, Yanofsky C (1981). Structure and regulation of *aroH*, the structural gene for the tryptophan-repressible 3-deoxy-D-arabino-heptulosonic acid-7-phosphate synthetase of *Escherichia coli*. J Mol Biol 145:47.

3. Arvidson DN, Bruce C, Gunsalus RP (1986). Interaction of the Escherichia coli Trp aporepressor with its ligand, L-tryptophan. J Biol Chem 261:238.
4. Joachimiak A, Kelley RL, Gunsalus RP, Yanofsky C, Sigler PB (1983). Purification and characterization of trp aporepressor. Proc Natl Acad Sci USA 80:668.
5. Lynn SP, Cohen LK, Kaplan S, Gardner JF (1980). RsaI: a new sequence-specific endonuclease activity from Rhodopseudomonas sphaeroides. J Bacteriol 142:380.
6. Arvidson DN, Gunsalus RP (1987). Interaction of Escherichia coli Trp aporepressor with L-tryptophan and analogs. Effects on operator and nonspecific DNA binding. (in preparation).
7. Bickle TA, Pirrota V, Imber R (1977). A simple, general procedure for purifying restriction endonucleases. Nucleic Acids Res 4:2561.
8. Scopes R (1982). "Protein Purification" New York: Springer-Verlag Inc.
9. Bradford M (1976). A rapid and sensitve method for the quantitation of microgram quantities of protein using the principle of protein-dye binding. Anal Biochem 72:248.
10. Kumamoto AK, Miller WM, Gunsalus RP (1987). The Escherichia coli tryptophan repressor binds multiple, tandem sites. Genes and Development (submitted).

Protein Purification: Micro to Macro, pages 409–419

PURIFICATION OF NATIVE AND RECOMBINANT TUMOR NECROSIS FACTOR

Leo S. Lin and Ralph Yamamoto

Department of Protein Chemistry, Cetus Corporation
Emeryville, California 94608

ABSTRACT Tumor Necrosis Factor (TNF) is a cytolytic protein found in serum of lipopolysaccharide-stimulated mice and rabbits. A similar factor is produced by tumor cell lines when treated with carcinogens. This protein was purified from tissue culture supernatants and its amino acid sequence determined. Using recombinant DNA technology, the TNF gene was cloned and the recombinant protein expressed in E. coli. TNF is a 17,000 dalton nonglycosylated protein, contains a single disulfide bond, and exists soluble in aqueous buffers as a homodimer. The E. coli produced molecule is similar to the native protein in many aspects. Most importantly, it is bioactive with a similar specific activity as the native counterpart. Additionally, it contains a single disulfide bond, and exists soluble in aqueous buffers as a homodimer. The similarity between the native and recombinant molecules, allows us to apply a similar purification scheme consisting of ion exchange and hydrophobic interaction chromatography to obtain highly purified TNF protein from either source.

PRODUCTION OF NATIVE TNF

Native TNF is produced from a promyelocytic leukemia cell line HL60. Cells were grown to a density of 2 x 10^6 cells per ml in suspension culture. Growth medium used for culturing is RPMI 1640 supplemented with 20% fetal calf serum. The cells were collected by centrifugation, washed with serum free medium, and resuspended at 10^7 cells per ml in serum free medium containing 100 ng/ml phorbol myristic acid. After incubation for 30 min at 37°C, the cells were collected by centrifugation and resuspended in serum free medium containing 10 μg/ml of bacterial endotoxin and 10 μg/ml of calcium ionophore A23187. After incubation for 4 hours at 37°C with constant agitation, the cells were removed by centrifugation and the culture supernatant harvested for TNF purification.

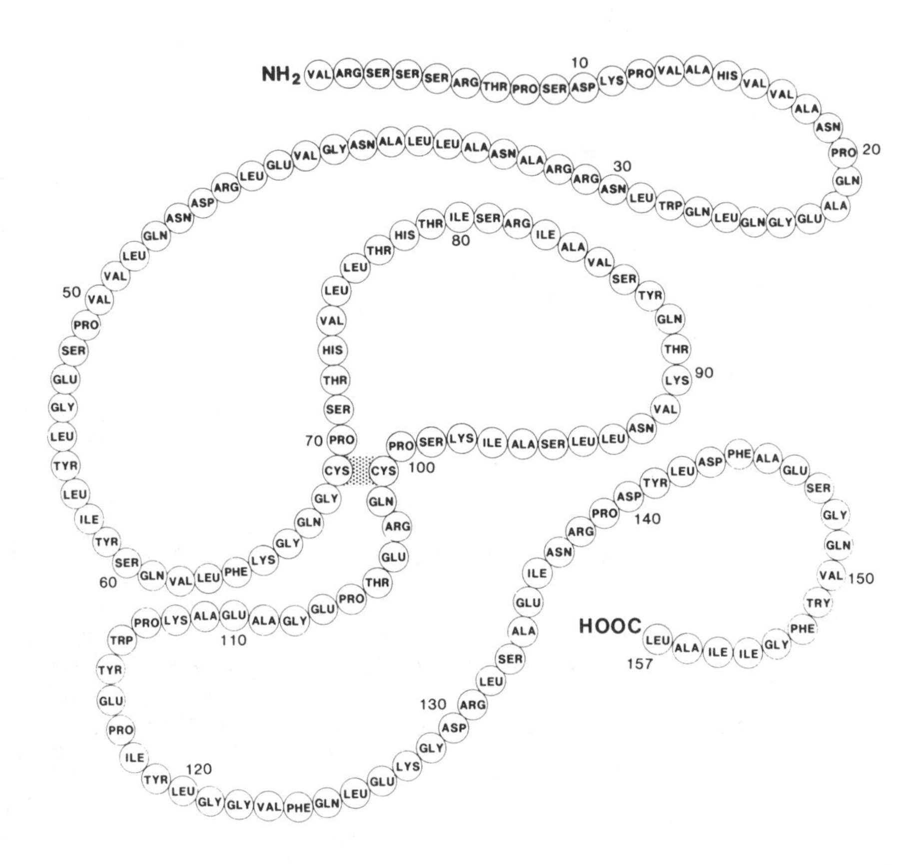

FIGURE 1. Amino acid sequence of human tumor necrosis factor. A single intramolecular disulfide bond is found between cysteine 69 and cysteine 101.

MICRO PURIFICATION OF NATIVE TNF

Concentration

Tissue culture production of native TNF is at the submicrogram level and comprises less than 0.01% of the total protein in the starting material. Components of fetal calf serum and secreted cellular proteins are the major contaminants that need to be removed in the purification of native TNF. Volume of the dilute starting material is reduced by ultrafiltration, and the salt concentration of the growth medium (approx. 0.15 M NaCl) adjusted to the desired concentration by diafiltration against the appropriate buffer using the same ultrafiltration apparatus. Monitoring the conductance of the ultrafiltrate ensures a controlled performance of the subsequent ion-exchange chromatography step.

Ion-Exchange Chromatography

The first ion-exchange chromatography step was designed to bind the majority of serum protein contaminants while allowing TNF to flow through. This is accomplished by maintaining a conductance in the sample of greater than 4.6 mS. A shortened diafiltration/concentration process time is another feature of this higher salt concentration. The flow through fractions of the first ion-exchange column (DEAE Sepharose) contains most of the TNF. The fractions are concentrated and the salt concentration adjusted to a conductance of 0.56 mS by ultrafiltration/diafiltration. At pH 8.2 and this salt concentration, TNF binds to the second ion exchange matrix (DEAE- Sepharose). Elution is accomplished by a stepwise increase of salt concentration. The bulk of the contaminating protein will not bind to the column and will be separated from TNF. The resulting TNF from the sequential ion-exchange chromatography steps is about 10-15% pure by SDS-PAGE analysis.

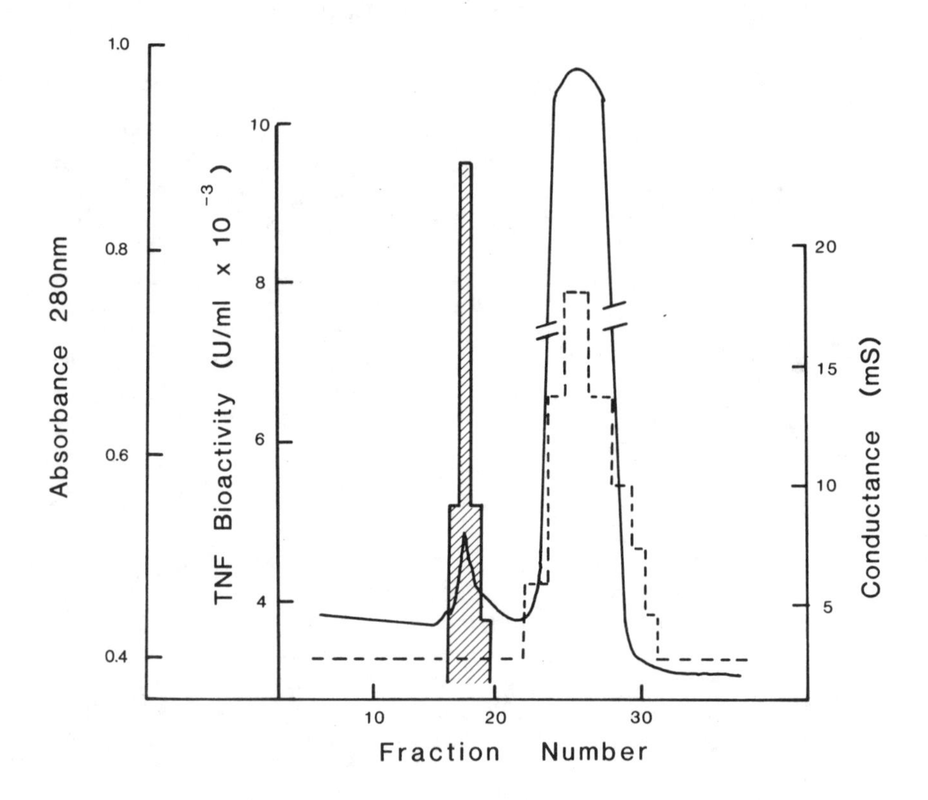

FIGURE 2. Anion Exchange Chromatography-DEAE Sepharose. Column equilibrated with 10 mM Tris-HCl/1 mM NaCl pH 8.2. Concentrated culture medium at 4.6 mS conductance was applied to the column. TNF elutes in the flow through fractions (cross- hatched area).

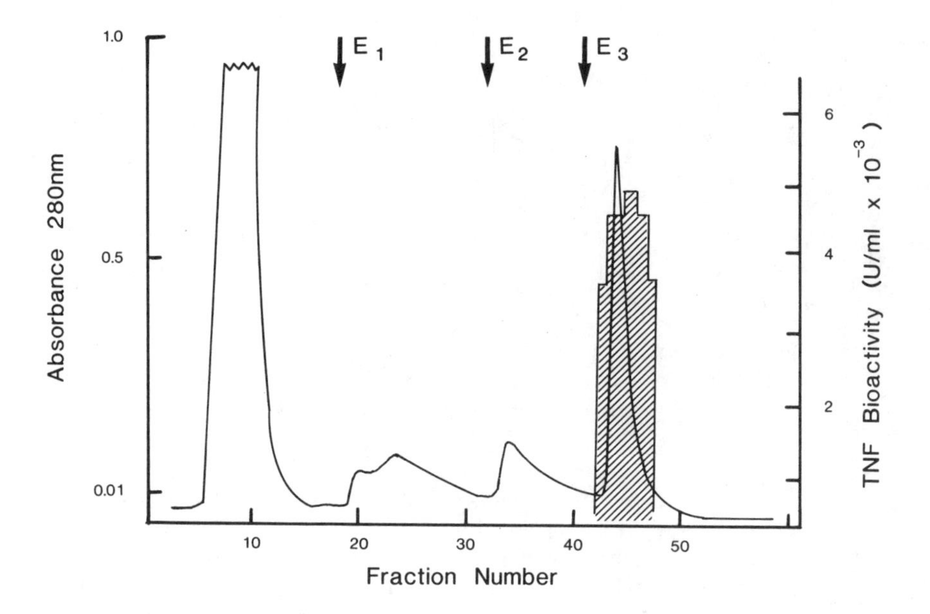

FIGURE 3: Anion Exchange Chromatography-DEAE Sepharose. Column equilibrated with 10 mM Tris-HCl/1 mM NaCl pH 8.2. TNF flow through fractions at 0.56 mS conductance was applied to the column. Elution buffers are made up of 10 mM Tris HCl pH 8.2 containing different concentrations of NaCl. Buffer E_1 contains 20 mM NaCl, buffer E_2 contains 40 mM NaCl and buffer E_3 contains 0.1 M NaCl. TNF is eluted from the column by buffer E_3.

Hydrophobic Interaction Chromatography

Fractions containing the peak TNF activity are pooled and solid ammonium sulfate is dissolved into the TNF solution to a final concentration of 1.8 M. This material is applied to a HPLC hydrophobic interaction column (TSK Phenyl 5PW), which was previously equilibrated with 0.1 M phosphate/1.8 M ammonium sulfate pH 7.0. TNF is eluted by a decreasing linear gradient of ammonium sulfate concentration. The resulting TNF is about 90% pure by SDS-PAGE analysis.

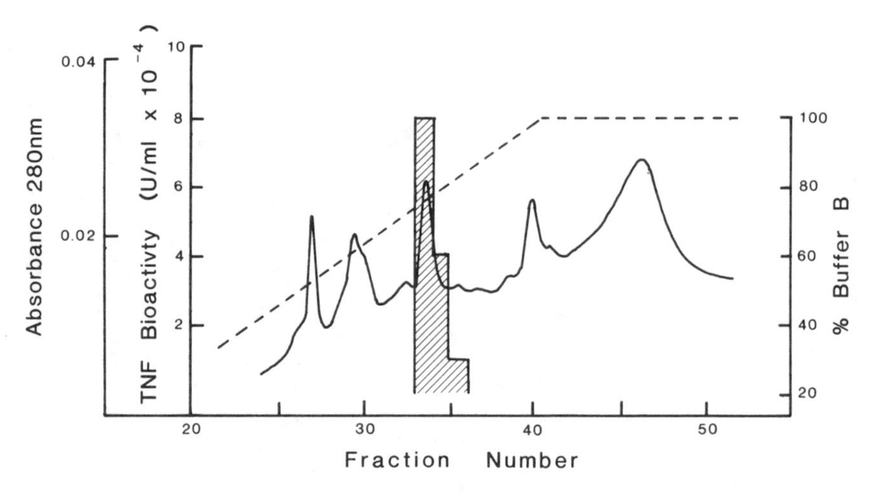

FIGURE 4. Hydrophobic Interaction Chromatography. TSK/Phenyl 5PW Column equilibrated with 0.1 M phosphate/1.8 M ammonium sulfate buffer pH 7.0. TNF solution containing 1.8 M ammonium sulfate is applied to the column. Elution is by a linear gradient of decreasing concentration of ammonium sulfate.

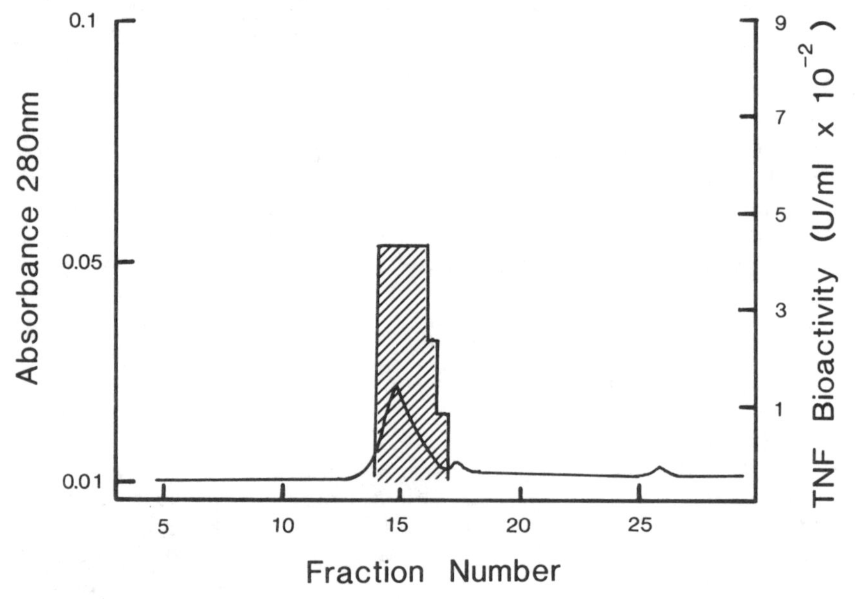

FIGURE 5. Gel Permeation Chromatography-TSK/SW2000. Column equilibrated with 30 mM ammonium bicarbonate buffer pH 7.3. Column is developed with the same buffer.

Gel Permeation Chromatography

The gel permeation chromatography step (TSK/SW2000) serves the purpose of desalting. The phosphate buffer containing ammonium sulfate is exchanged for buffers compatible for further chemical or biological characterization of the TNF molecule.

PRODUCTION OF RECOMBINANT TNF

Recombinant TNF is produced in E. coli under the control of the bacteriophage lambda pL promoter. Bacterial culture harboring the TNF expression plasmid was grown at 30°C in a fermentor to the desired cell density. Production of rTNF was initiated by increasing the culturing temperature to 42°C. After growth for several hours at the elevated temperature, bacterial cells were harvested by centrifugation. The cells were resuspended in 10 mM Tris-HCl/1 mM NaCl buffer pH 8.2 and disrupted by sonication.

MACRO PURIFICATION OF rTNF

Disruption

Production of rTNF in E. coli is approximately 100 μg per ml of bacterial culture and comprises 10% of total soluble cellular protein. Disruption of the bacterial cell by any number of mechanical means will release the rTNF from the bacteria. Removal of cell debris can be accomplished by centrifugation (30,000g x 10 min).

Ion-Exchange Chromatography

The clarified supernatant is applied to an anion-exchange column (Mono-Q/Pharmacia) previously equilibrated with 10 mM Tris-HCl pH 7.0. Some of the E. coli proteins will not bind to the ion-exchange matrix under these conditions. Elution of TNF is achieved by applying an increasing linear gradient of NaCl. The fractions containing TNF activities are pooled and adjusted to 1.8 M by the addition of solid ammonium sulfate.

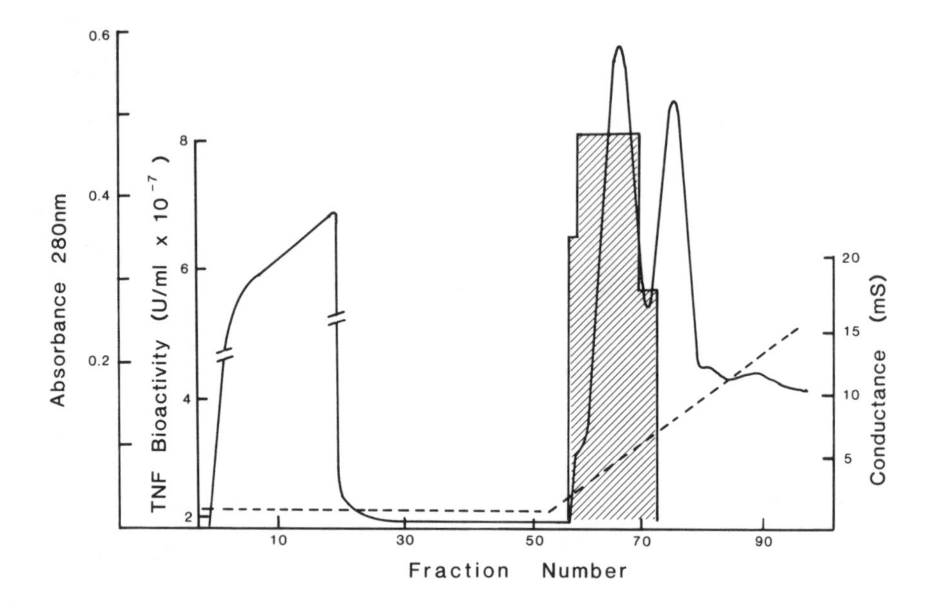

FIGURE 6. Anion Exchange Chromatography-Mono-Q/Pharmacia. Column is equilibrated with 0.1 M Tris HCl pH 7.0. Clarified bacterial disruptate is applied onto the column, and TNF eluted by a linear gradient of increasing NaCl concentration.

Hydrophobic Interaction Chromatography

The hydrophobic interaction matrix and the chromatographic conditions used for this separation is identical to that used for the separation of native TNF from tissue culture. TNF from either source elutes from this column under the same conditions.

Gel Filtration Chromatography

Gel filtration chromatographic step (Cellufine GH25) serves the purpose of desalting, where the phosphate buffer containing ammonium sulfate is exchanged for buffers suitable for further biochemical and biological characterization of the protein.

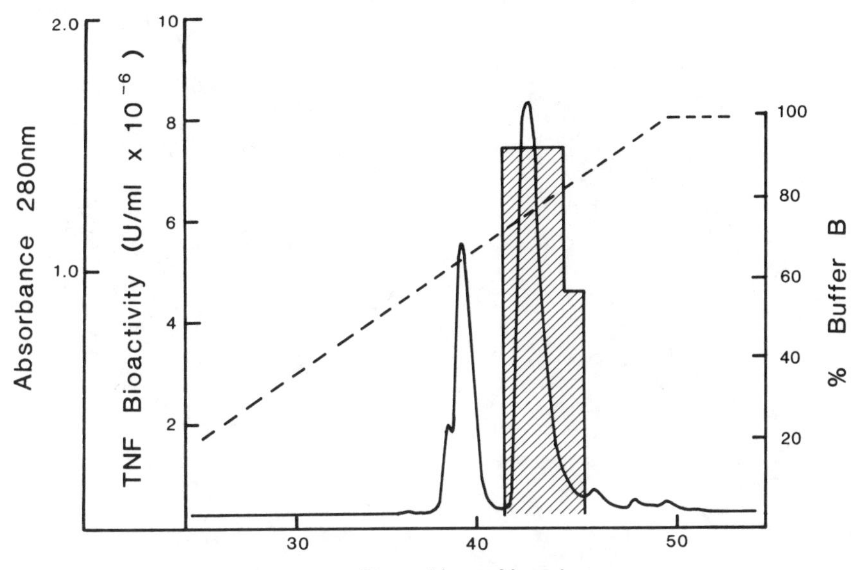

FIGURE 7. Hydrophobic Interaction Chromatography-TSK/Phenyl 5PW. Column is equilibrated with 0.1 M phosphate/1.8 M ammonium sulfate buffer pH 7.0. TNF solution containing 1.8 M ammonium sulfate is applied onto the column and eluted with a linear gradient of decreasing concentration of ammonium sulfate.

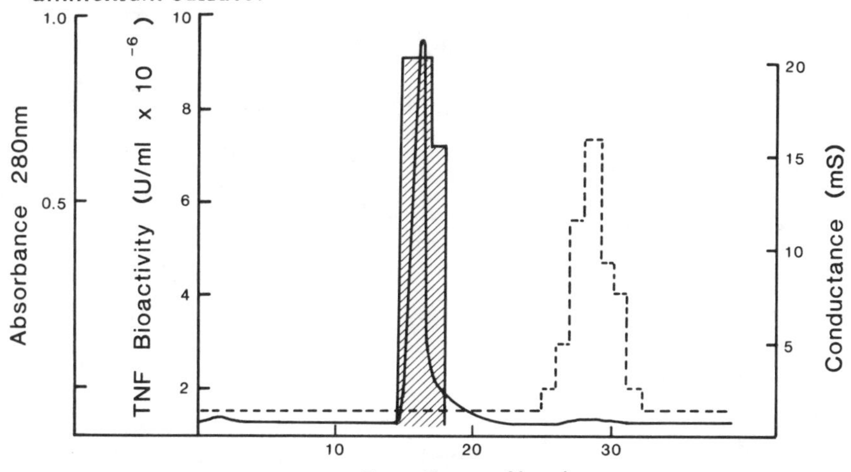

FIGURE 8. Gel Permeation Chromatography-Cellufine GH25. Column is equilibrated in 30 mM ammonium bicarbonate buffer pH 7.3. Column is developed in the same buffer.

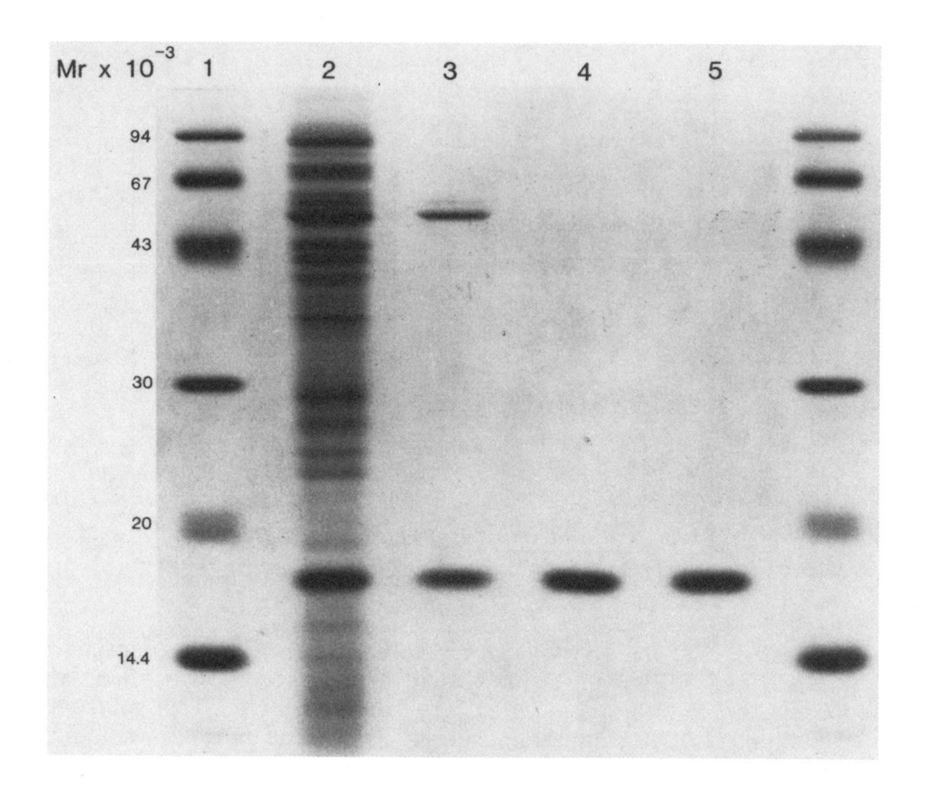

FIGURE 9. SDS-Polyacrylamide Gel.
Lane 1, Molecular weight marker (Pharmacia)
Lane 2, Total E. coli protein
Lane 3, Pooled Mono Q peak fractions
Lane 4, Pooled TSK/Phenyl peak fractions
Lane 5, Pooled GH-25 peak fractions

DISCUSSION

The native and the E. coli produced recombinant TNF is a 17,000 dalton protein, and contains a single intramolecular disulfide bond between cysteines 69 and 101. The amino terminal methionine is efficiently removed by E. coli (> 95% removal). TNF exists in aqueous solutions as a homodimer. There is no detectable intermolecular disulfide linkage or free thiol groups when the protein is released from the bacteria following cell disruption. These characteristics are identical to the native TNF molecule produced by tumor cell lines. We have applied a nearly identical purification protocol for the separation of the native and recombinant TNF from their respective contaminants. The purification of microgram amounts of native TNF approximately requires a 10,000 fold purification from liters of dilute tissue culture fluid, whereas the purification of miiligram amounts of recombinant TNF requires a mere 10 fold purification from less than a liter of bacterial culture. The major difference in the purification process for TNF protein from the two different sources is the overall yield. A greater than 40% yield is normally observed with the recombinant product, while a yield of 0.5% is generally observed for the native product. Our work on TNF demonstrates that nearly identical separation protocols can be used to obtain purified native TNF protein existing in dilute concentration or the recombinant TNF protein existing in a high concentration.

Protein Purification: Micro to Macro, pages 421–427

PURIFICATION OF RECOMBINANT HUMAN INTERLEUKIN 1ß PRODUCED FROM YEAST

M. Cristina Casagli, M. Giuseppina Borri, Cinzia D'Ettorre, Cosima Baldari, Cesira Galeotti, Paola Bossù, Paolo Ghiara and Guido Antoni

Sclavo Research Centre, Siena, Italy.

ABSTRACT A method is presented for the purification of recombinant human interleukin 1ß secreted from Sacch-romyces cerevisiae into the culture supernatant medium. After ten-fold concentration by ultrafiltration, the culture supernatant was chromatographed on a DEAE-Sepharose column and then further purified by chromatography on hydroxylapatite. The reverse-phase HPLC on C-18 Vydac column was selected to complete the purification of interleukin 1ß. The purified protein was electrophoresed on polyacrylamide gel and amino acid analysis was carried out. The different specific activity of native and recombinant human interleukin 1ß is discussed.

INTRODUCTION

Human interleukin 1ß (IL-1ß) is the predominant form of human IL-1, a term used to describe a protein mediator of several host defence responses.

IL-1ß has recently been produced in our laboratories by Saccharomyces cerevisiae transformed with a high copy number secretion vector into which the cDNA coding for aminoacids 121-269 of human IL-1ß have been cloned (1). This recombinant protein secreted into the culture medium is the mature form of IL-1ß, lacking the first four aminoacids and glycosylated at Asn 123, so that its apparent molecular weight determined by sodium dodecyl sulfate polyacrylamide gel electrophoresis (SDS-PAGE) is about 22,000 instead of the expected 17,400. The aim of the present study was to set up an efficient purification method for this protein. We also evaluated its biological activity in comparison with that described for the natural IL-1ß.

METHODS

Concentration of supernatant. One liter of crude yeast culture supernatant was concentrated by ultrafiltration with the Minitan System (Millipore) using a membrane with nominal molecular weight exclusion limit 10,000 and dialyzed against a 10 mM phosphate buffer, pH 7.5. The final volume was 100 ml.

SDS-PAGE. Electrophoresis was performed on a 15% acrylamide gel using Laemmli's procedure (2). Each sample was concentrated by precipitation with trichloroacetic acid (TCA) in the presence of sodium deoxycholate (3); the precipitated proteins were dissolved and heated at 100°C for 1 min in 30 mM Tris, 1% SDS and 0.18M ß-mercaptoethanol and applied to the gel.

Aminoacid analysis. Aminoacid analysis was performed by hydrolyzing the samples at 110°C for 24 h with 6N HCl containing 0.05% ß-mercaptoethanol and 0.05% phenol. A Chromakon 500 (Kontron) aminoacid analyzer was employed.

Protein determination. Proteins were determined by a modification of Lowry's procedure (3), after precipitation of the samples with TCA in the presence of sodium deoxycholate.

Carbohydrate determination. The analysis was carried out according to Svennerholm (4) using a glucose solution as standard.

Biological assay of IL-1. IL-1 activity was determined by its capacity to stimulate thymocyte proliferation in the presence of phytohemoagglutinin (5), using human recombinant IL-1ß from E. coli (Genzyme) as standard.

RESULTS

Crude concentrated yeast culture supernatant was first submitted to ion-exchange chromatography on a DEAE-Sepharose column (Fig. 1). Under the conditions used, IL-1ß was not adsorbed on DEAE-Sepharose and could be separated from the bound material which contained most of the contaminating proteins. Fractions were analyzed by SDS-PAGE and those containing IL-1ß were collected (peak A). The column regeneration was performed by washing with 1M NaCl.

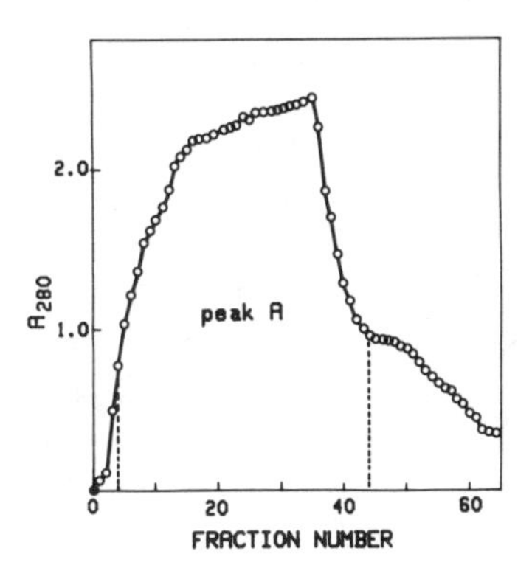

FIGURE 1. DEAE-Sepharose chromatography. 100 ml of concentrated supernatant was loaded on a 5x6.5 cm column equilibrated in 10 mM phosphate, pH 7.5. Flow rate was 30 ml/h.

Further purification and concentration of IL-1ß was achieved by hydroxylapatite (HTP) chromatography (Fig. 2). Peak A from DEAE-Sepharose chromatography was loaded directly onto the HTP column, and contaminating materials were removed from the column in the "flow through" (peak B), while IL-1ß was eluted in a sharp peak by increasing the phosphate buffer concentration to 0.1 M (peak C). Peak C was dialyzed against 10 mM ammonium bicarbonate and freeze-dried.

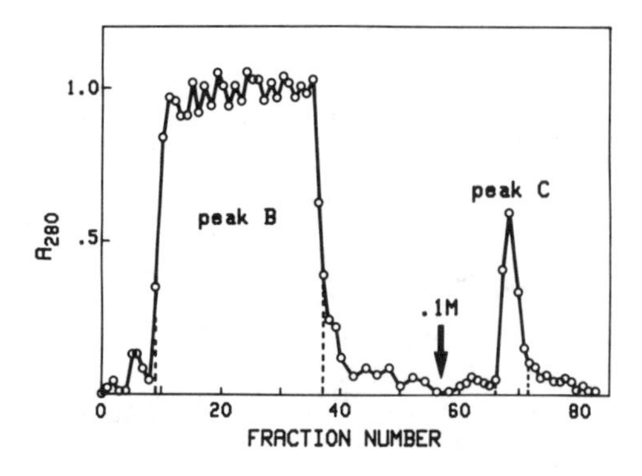

FIGURE 2. HTP chromatography. Peak A from DEAE chromatography (238 ml) was applied on a 2.5x16 cm hydroxylapatite column equilibrated in 10 mM phosphate, pH 7.5. The flow rate was 20 ml/h. Fractions were analyzed by SDS-PAGE.

After each chromatographic step, samples were analyzed for protein and carbohydrate concentrations (Table 1). These values and the SDS-PAGE analysis of the same samples (Fig. 3) show the results of IL-1ß purification achieved with the described steps. The DEAE-Sepharose chromatography is very efficient in separating IL-1ß from most of the extraneous proteins, whereas most of the contaminating carbohydrates were removed on the HTP column.

TABLE 1

SUMMARY OF PREPURIFICATION OF IL-1ß

Sample	Volume (ml)	Total protein (mg)	Total carbohydrate (mg)
Concentrated supernatant	100	73.0	946
DEAE-Seph., peak A	238	12.0	571
DEAE-Seph., 1M NaCl washing	100	23.0	49
HTP, peak B	225	5.2	472
HTP, peak C	43	6.4	50

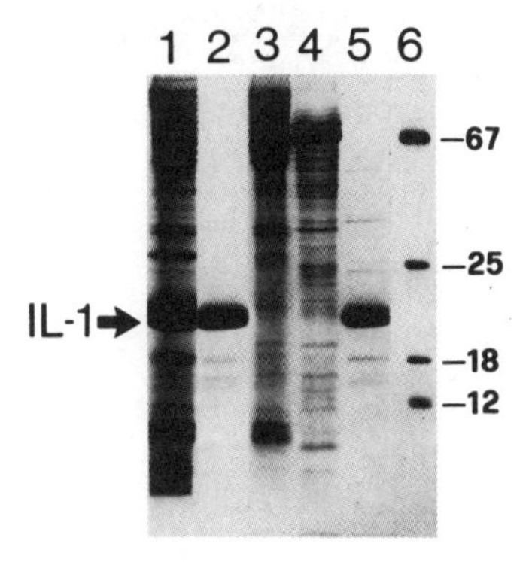

FIGURE 3. SDS-PAGE of samples obtained by DEAE and HTP chromatography. 1: concentrated supernatant; 2: peak A; 3: 1M NaCl washing; 4: peak B; 5: peak C; 6: molecular weight standards.

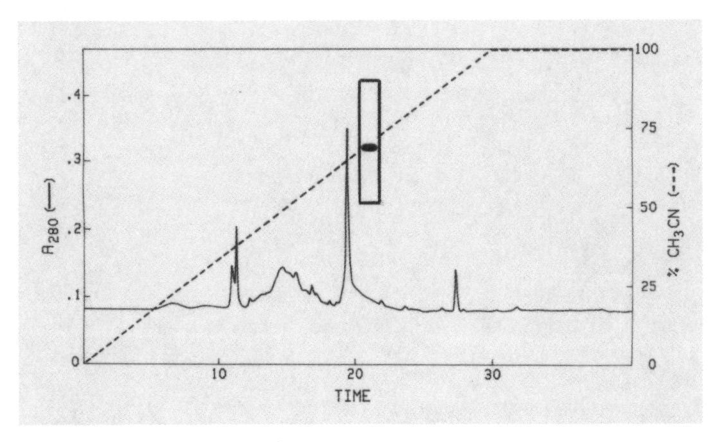

FIGURE 4. Reverse phase HPLC of IL-1ß. 15 mg of prepurified IL-1ß was loaded on a Vydac C18 column (1x25 cm) equilibrated in 0.1% trifluoroacetic acid (TFA) in water. The column was eluted with a 30 min linear gradient of 0.1% TFA in acetonitrile (0-100%) and subsequently washed for 10 min with the final solvent. Electrophoretic analysis of the main chromatographic peak is also shown.

It was possible to obtain a highly purified IL-1ß submitting the lyophylized product from HTP chromatography, containing about 10% of protein, to a final step on reverse-phase HPLC (Fig. 4).

Electrophoresis and aminoacid analysis (Table 2) of the main chromatographic peak showed that a homogeneous preparation of IL-1ß was obtained.

TABLE 2

AMINOACID COMPOSITION OF IL-1ß PURIFIED BY RP-HPLC

Aminoacid	Found	Expected
Asx	16.5	16
Thr	6.3	6
Ser	12.2	14
Glx	25.3	23
Pro	7.4	7
Gly	9.7	8
Ala	4.9	4
½-Cys	0.6	2
Val	9.2	10
Met	6.0	6
Ile	4.6	5
Leu	14.7	15
Tyr	3.9	4
Phe	8.3	9
Lys	14.3	15
His	1.0	2
Arg	2.5	2

Fractions from purification steps were tested for biological activity by the thymocyte proliferation assay. All the samples always showed very low activity; the final preparation of IL-1ß has a specific activity of 400 U/mg, much lower than that of the natural protein (10^8 U/mg)(6).

DISCUSSION

The procedure described in this report for the purification of recombinant IL-1ß is simple and yields a very homogeneous product. Unfortunately the purified protein has a very low specific activity in the biological assay. We checked whether the purification process adopted was responsible for a possible lack of biological activity. However, when a sample of a commercially available active protein produced in E. coli was subjected to the purification protocol described above, a quantitative recovery of biological activity was obtained (data not shown).

A possible cause of this low biological activity might be an incorrect folding of the protein molecule due to the lack of the first four residues or to the glycosylation in the position 123. Further studies will be necessary to confirm this hypothesis or to establish other causes for the low activity of the protein.

REFERENCES

1. Baldari C, Murray JAH, Ghiara P, Cesareni G, Galeotti C (1987). A novel leader peptide which allows efficient secretion of a fragment of human interleukin 1ß in Saccharomyces cerevisiae. EMBO J 6:229.
2. Laemmli VK (1970). Determination of protein molecular weight in polyacrylamide gels. Nature 277:680.
3. Bensadown A, Weinstein D (1976). Assay of proteins in the presence of interfering materials. Anal Biochem 70:241.
4. Svennerholm L (1956). The quantitative estimation of cerebrosides in nervous tissue. J Neurochem 1:42.
5. Gery I, Davies P, Derr J, Krett N, Barranger JA (1981). Relationship between production and release of lymphocyte-activating factor (interleukin 1) by murine macrophages. Cell Immunol 64:293.
6. Kronheim SR, March CJ, Erb SK, Conlon PJ, Mochizuki DY, Hopp TP (1985). Human Interleukin 1. Purification to homogeneity. J Exp Med 161:490

Protein Purification: Micro to Macro, pages 429–442

SOLUBILISATION, REFOLDING AND PURIFICATION OF EUKARYOTIC PROTEINS EXPRESSED IN *E. COLI*

Peter A. Lowe, Stephen K. Rhind,
Richard Sugrue and Fiona A.O. Marston

Celltech Ltd., 250 Bath Road,
Slough, SL1 4DY, U.K.

ABSTRACT The insolubility of many eukaryotic recombinant proteins expressed in *E.coli* (as N-terminal met-analogues or hybrid fusion proteins) presents both problems and opportunities for their purification. A brief review of solubilisation and refolding procedures, including fluorescence and light scattering will be outlined. These general considerations will be illustrated by reference to the laboratory scale production of the human peptide hormone calcitonin (hCT). hCT is a 32 amino acid hormone containing a 1-7 disulfide bridge and C-terminal proline amide residue. hCT was produced in *E.coli* as a precursor containing an additional C-terminal gly residue (hCTgly) fused onto the C-terminal region of the *E.coli* enzyme chloramphenicol acetyl transferase (CAT) via a proteolytically sensitive Arg residue (CATArghCTgly). The partially purified insoluble fusion protein was solubilised in 8M urea under reducing conditions and hCTgly liberated from the fusion by digestion with clostripain. hCTgly was partially purified by RP-HPLC and converted to hCT by enzymic amidation. Final purification of hCT was carried out by RP-HPLC.

INTRODUCTION

Many recombinant eukaryotic proteins are synthesised in *E. coli* in insoluble inactive forms as intracellular inclusion bodies (1).Furthermore, E. coli does not possess the enzymesystems required to reproduce the chemical modifications found in many eukaryotic proteins. These apparent deficiencies can be overcome for certain protein products by determination of appropriate solubilisation and refolding protocols and the use of selective chemical modification. A brief review will be presented of solubilisation and refolding methods together with the major variables to be considered in selecting conditions to maximise yield. Techniques for the analysis of the effectiveness of solubilisation and refolding procedures, including fluorescence, light scattering and bioactivity measurements will be outlined. These general considerations will be illustrated by reference to the laboratory scale production of the human peptide hormone calcitonin (hCT).

SOLUBILISATION

Intracellular expression of eukaryotic proteins in E. coli as hybrid fusions or N-terminal met-analogues can result in the deposition of the product in 'inclusion bodies'. Inclusion bodies contain the recombinant protein in an insoluble inactive form. It has recently been reported that inclusion body preparations also contain components of the E. coli transcription and translation machinery (2). Cell lysis followed by cycles of low speed centrifugation and washing with dilute aqueous detergent solutions is an effective method of purifying inclusion bodies and hence partially purifying the insoluble recombinant product. Purity levels in excess of 30% with respect to contaminating proteins may be obtained by this procedure. The recombinant protein in this form may be effectively solubilised by treatment with denaturants such as 8M urea, 6M guanidine HCl, detergents such as SDS or dilute alkaline solutions (pH) 10). In order to carry out efficient solubilisation and therefore maximise eventual product yield, several variables in the process of solubilisation may need to be considered. These include; protein purity, concentration of protein in denaturant, denaturant concentration, pH and other constituents of the solubilising denaturant (e.g. free thiols or thiol derivatising reagents). Finally, the time and mode of the solubilisation process should be examined. From studies with bovine prochymosin expressed in E. coli (3) it appeared that this protein was produced in vivo in a largely reduced form (the natural protein contains 3 disulphide bridges). Subsequent manipulations in air resulted in aberrant disulphide bridge formation. Hence, as outlined above, the solubilised recombinant protein may be reduced by the addition of free thiol groups or subjected to oxidative sulfitolysis followed by S-sulphonation (4).

REFOLDING

Since denaturants are used to effect solubilisation of the protein they must be subsequently removed to allow refolding of the protein. In theory this can be simply achieved by substitution of denaturant with a dilute aqueous solvent in which the protein product is known to be soluble and, where appropriate, bioactive. In practice, however, for solubilised recombinant proteins this approach often leads to precipitation, partial aggregation and a low recovery of correctly folded protein. Hence, in order to maximise product yield during refolding a number of variables may need to be considered. These include: Level of purity of the refolding protein, its concentration - this may be a particularly important consideration since the optimum concentration may be low (1mg/ml or less) higher concentrations favouring aggregation or precipitation, pH and other solvent constituents such as the presence or absence of exogenously added thiols (or reduced/oxidised thiol mixture). Finally, the constituents of the solvent chosen for refolding must be compatible with subsequent purification steps.

METHODS FOR THE ANALYSIS OF SOLUBILISATION AND REFOLDING PROCEDURES

Evaluation of the variables given above may produce an effective solubilisation and refolding protocol. This evaluation usually makes use of data on overall product recovery in terms of bioactivity, mass, aggregation state, structural integrity, etc. For example a procedure for prochymosin produced in E. coli was arrived at by determination of recovery of enzyme activity after variation of a number of the above considerations (5). Recently, we have attempted to monitor the process of solubilisation of prochymosin inclusion bodies and refolding of recombinant prochymosin by fluorescence and light scattering techniques in parallel with data on overall recovery of enzymically active product. Purified prochymosin inclusion bodies were used. However, it should be emphasised that the starting material was not homogeneous and was contaminated with E. coli proteins and cell debris. Hence, data obtained by these techniques must be treated with caution. Equilibration of prochymosin inclusion bodies with increasing concentrations of denaturant (urea, guanidine HCl or alkali) resulted in a progressive diminution in fluorescence intensity. At $\lambda EX = 290$ nm there was a shift in λEM from 330 nm to 348 nm in the case of urea and guanidine HCl. A summary of fluorescence data obtained is provided in Table 1.

TABLE 1
SUMMARY OF SOLUBILISATION OF PROCHYMOSIN INCLUSION BODIES WITH VARIOUS DENATURANTS AS MEASURED BY FLOURESCENCE INTENSITY

Denaturant	A	B	C	D
1) Urea	OM-8M	3.2M	6.0M	40%
2) Guanidine HCI	OM-6M	1.9M	3.5M	50%
3) Alkaline pH	pH8-pH13	pH11.2	pH12.8	55%

Table 1 Prochymosin inclusion bodies at 20 ug/ml protein at pH 8 (in the case of 1 and 2) were equilibrated under the above conditions in the stated denaturants. With λEX = 290 nm, increasing the concentration of denaturants resulted in a shift in λEM from 330 nm to 348 nm (350 nm in the case of alkaline pH.) A; range of denaturant concentrations utilised, B; molarity or pH at which 50% total decrease in fluorescence intensity observed, C; molarity or pH above which no further diminution in fluorescence intensity observed D; Approximate % fall in fluorescence intensity at concentration or pH given in C.

Similar experiments were carried out on prochymosin inclusion bodies using measurements of perpendicular scattered light signal intensity at λ = 460 nm. Increasing concentrations of urea, guanidine hydrochloride or increasing pH resulted in a progressive fall in scattered light signal intensity. A summary of the data obtained is provided in Table 2.

TABLE 2
SUMMARY OF SOLUBILISATION OF PROCHYMOSIN INCLUSION BODIES WITH VARIOUS DENATURANTS AS MEASURED BY PERPENDICULAR SCATTERED LIGHT SIGNAL INTENSITY

Denaturant	A	B	C	D
1) Urea	OM-8M	2.4M	5.0M	86%
2) Guanidine HCI	OM-6M	1.2M	3.5M	95%
3) Alkaline pH	pH8-pH13	pH11.0	pH = 12.5	68%

Table 2 Prochymosin inclusion bodies were equilibrated at pH = 8.0 (in the case of 1 and 2) in the stated denaturants. Perpendicular scattered light signal intensity was measured at λ =460nm. A; range of denaturant concentrations utilised, B; molarity or pH at which 50% total decrease in light scatter intensity observed. C; molarity or pH above which no further diminution in light scatter intensity observed D; approximate % fall in intensity at concentrations given in C.

With the reservations outlined above fluorescence and light scattering techniques can provide a convenient means of monitoring the process of inclusion body solubilisation. These methods can also be used to follow the process of refolding after removal of denaturant (data not shown). An increase in light scatter intensity on denaturant removal is indicative of reaggregation or precipitation. An unchanged light scatter intensity together with an increased fluorescence intensity signal may be indicative of refolding. The value of the fluorescence or light scattering measurements in determining effective solubilisation and refolding is, however, limited and should be carried out in concert with determinations of activity and mass recovery. For instance, in the case of prochymosin, effective solubilisation in 8M urea or 6M guanidine HCI followed by removal of denaturant with an equeous buffer at pH = 8 results in very low levels of activity recovery. (Light scatter measurements indicate precipitation on denaturant removal in the case of guanidine hydrochloride). However, removal of 8M urea by substitution with an aqueous solution at pH = 10.7 followed by lowering the pH to 8.0 results in recovery of activity (5).

Examples of some of the variables discussed above are given for the production of human calcitonin outlined below.

PRODUCTION OF HUMAN CALCITONIN (hCT) IN E. COLI

hCT is a calcium-lowering hormone with an important role in the long-term maintenance of skeletal mass and is used clinically in the treatment of Paget's desease of bone (6). The hormone consists of a 32 amino-acid polypeptide with a 1 - 7 disulfide bridge and C-terminal proline amide residue (Fig. 1). The production of recombinant hCT using *E.coli* has been described (7).

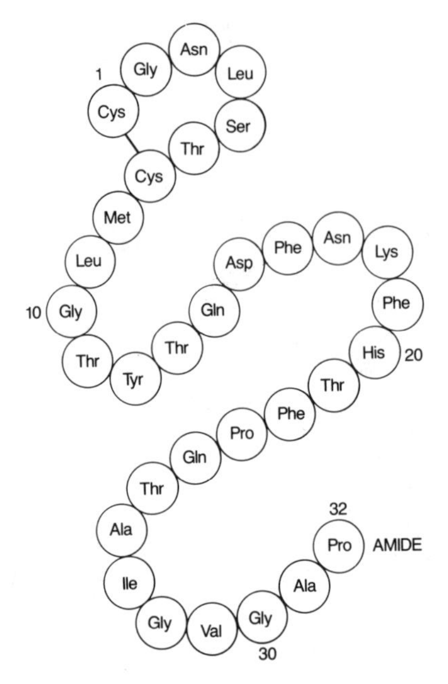

FIGURE 1 Primary Amino Acid Sequence of Human Calcitonin

In order to produce hCT, an analogue containing a C-terminal gly residue (hCTgly) was expressed in *E.coli* HB101 as a hybrid protein fused with the *E.coli* enzyme chloramphenicol acetyl transferase (CAT) as shown in Fig. 2. A proteinase cleavage site was introduced between CAT and hCTgly consisting of a clostripain sensitive Arg residue. The fusion protein CAThCTgly, Mr = approximately 25,000, was expressed as an insoluble product in the form of inclusion bodies. CAThCTgly represented approximately 15% of total *E.coli* protein.

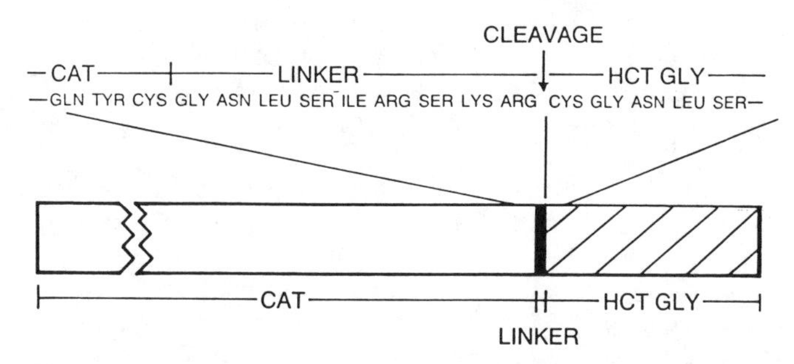

FIGURE 2 Structure of the CAThCTgly Fusion Protein

The CAThCTgly fusion protein is shown (not to scale) with the junction between CAT and hCTgly magnified. The fusion consists of the N-terminal 212 amino acids of CAT (natural CAT consists of 219 amino acids) fused to hCTgly via a linker sequence. The clostripain cleavage point for the generation of hCTgly from the fusion is indicated by an arrow. Details of expression plasmid construction for the expression of CAThCTgly are given in (7)

Production of hCT on a laboratory scale was carried out from 22g (wet weight) harvested E.coli cells. The cells were lysed, inclusion bodies isolated by low-speed centrifugation and washed in a soluble containing 0.05% (v/v) Trition X-100, 10mM EDTA (5). 176mg CAThCTgly was recovered at this stage with a purity of approximately 50% as determined by scanning of SDS polyacrylamide gels.

Solubilisation was achieved by homogenisation of the inclusion body preparation in 8M urea, 0.14M 2-mercaptoethanol. hCTgly was liberated from the fusion protein by clostripain cleavage at the Arg residue (indicated by an arrow in

Fig. 2). To achieve this the solubilised fusion protein solution was diluted to 4M urea and clostripain added in a ratio of one part by weight proteinase to 40 parts _E.coli_ protein. The

mixture was incubated at 37°C for 15 mins and liberation of hCTgly from the fusion protein monitored by analytical RP-HPLC. Trifluoraceticacid (TFA) was added to final concentration of 5% v/v and the incubation continued for 20 minutes at 4°C. Acid precipitated E.coli proteins were removed by centrifugation.

Acidified hCTgly in 4M urea was pumped onto a Synchropak RP-P (25cm 1.0cm) equilibrated in 30% (v/v) acetonitrile, 0.1% v/v TFA. Under these conditions hCTgly bound the column and was eluted with a gradient of a 30% v/v - 45% v/v acetonitrile, 0.1% (v/v) TFA in water (Fig. 3). hCTgly was detected in the chromatographic profile by radioimmunoassay or rechromatography of samples of appropriate fractions on analytical RP-HPLC using purified hCTgly as a reference standard. hCTgly was eluted in two peaks consisting of reduced (1,7cys) and 1,7 disulfide bridged oxidised forms. The two peaks were pooled, freeze dried and hCTgly redissolved at approximately 0.5 mg/ml in 50 mM Tris-HCl pH 8.5, 2mM EDTA. Under these conditions formation of the 1,7 disulfide bond occurred and was monitored by RP-analytical HPLC by a small shift in the hCTgly peak to a more hydrophobic region of the elution gradient. Oxidised hCTgly was purified by rechromatography as described above.

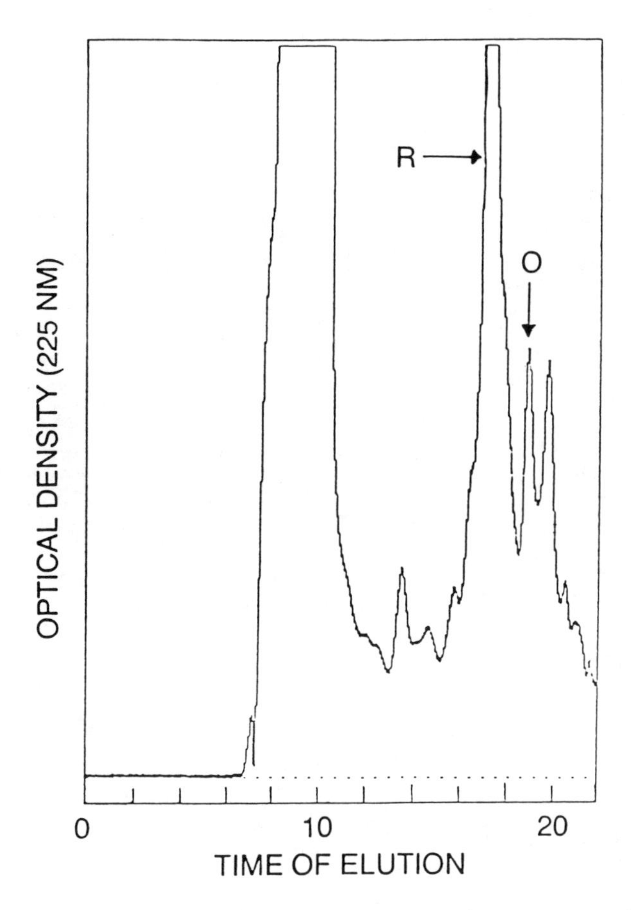

FIGURE 3 Purification of hCTgly by RP-HPLC

hCTgly liberated by clostripain digestion of CAThCTgly was partially purified by RP-HPLC. Column, Synchropak RP-P (25cm 1.0cm) equilibrated in 30% (v/v) acetonitrile, 0.1% (v/v) TFA in water. hCT was eluted with a gradient of 30% v/v - 45% v/v acetonitrile: 0.1% (v/v) TFA in water over a 20 min. period, flow rate 2 ml/min, 225 nm detection. Arrows, R and O indicate elution positions of reduced and oxidised hCTgly.

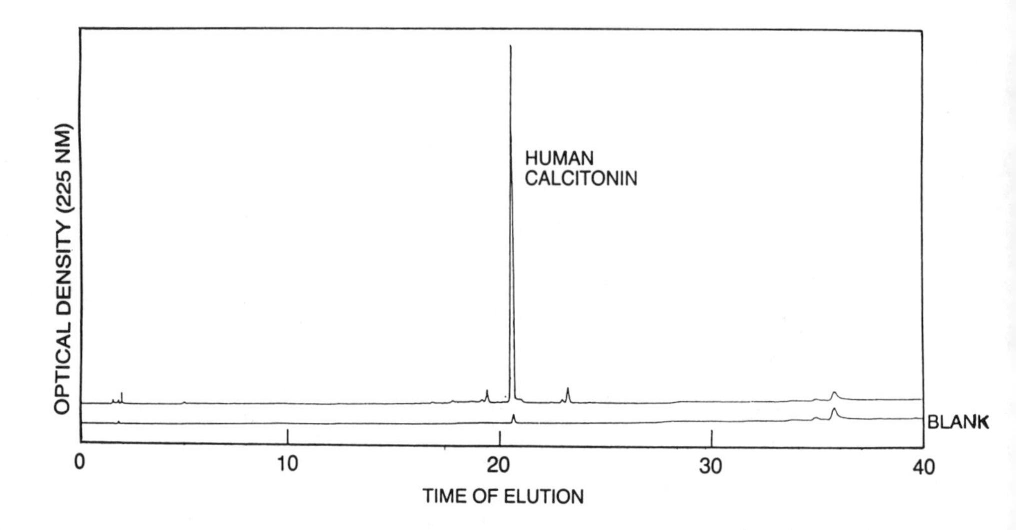

FIGURE 4 Analytical RP-HPLC of Human Calcitonin

hCT was separated from the constituents of the amidating enzyme mixture by repeated preparative RP-HPLC as described above. A sample of partially purified hCT is shown, subjected to analytical RP-HPLC (column, Vydac C18 15cm 0.46 cm, Gradient 10% v/v - 35% v/v acetonitrile in 20mM phosphate pH 6.7 over 40 min). The chromatographic profile of the analysis of a buffer blank is also shown.

Formation of the C-terminal proline amide of hCT from purified hCTgly was achieved by use of an amidating enzyme activity extracted from procine pituitary (8). Porcine pituitary amidating enzyme has a selective requirement for copper (II) ion as a co-factor and is stimulated by molecular oxygen and ascorbate (9, 10). Purification of hCT, produced by enzymic amidation, was achieved by RP-HPLC as described above (Fig. 4). The amino acid sequence of purified recombinant hCT was obtained by gas-phase microsequencing and FAB mass spectrometry and found to be that expected from (Fig. 1). Bioactivity of recombinant hCT was determined by an *in vivo* rat serum Ca^{2+} lowering assay (11) and was indistinguishable from that exhibited by chemically synthesis hCT (Sigma).

CONCLUSION

Although *E.coli* synthesises many recombinant proteins as denatured aggregates and does not possess the protein modification systems found in eukaryotes, this organism can be successfully utilised for the production of appropriate proteins by careful consideration of solubilisation, refolding, chemical modification and purification methodologies. This presentation has outlined some of these considerations in general and utilised our experience with the production of human calcitonin as an example. Production of hCT involved solubilisation of inclusion bodies containing the recombinant CAThCT hybrid, denaturant removal and chemical modification at the C-terminus by an enzymological method.

REFERENCES

1. Marston, FAO (1986) Biochem. J. 240, 1-12.

2. Hartley, D.L. and Kane, J.F. (1987) Abstracts of the Annual Meeting, American Society of Microbiology, pp226, pub. American Society of Microbiology.

3. Shoemaker, J.M., Brasnett, A.M. and Marston F.A.O. (1985) EMBO J 4, 775-780.

4. Jones, A.J.S., Lin, N.S.C., Olson, K.C., Shire, S.J., Pai, R-C and Wetzel, R.B. (1983) European Patent Application 0114806.

5. Marston, F.A.O., Lowe, P.A., Doel, M.T., Shoemaker, J.M., White, S., Angal, S. (1984) Biotechnology 2, 800-809.

6. Austin, L.A. and Heath III, M (1981) New England J. of Medicine, 304, 269-278.

7. Bennett, A.D., Rhind, S.K., Lowe, P.A., Hentschel, C.G.G. (1984) U.K. Patent Application GB 2140810A.

8. Bradbury, A.F., Finnie, M.D.A. and Smyth, D.G. (1982), Nature, 298, 686-688.

9. Bradbury, A.F., and Smyth, D.G. (1982) In: Peptides, 1982 (ed. K. Blaha and P. Mahou) Walker de Gruyter and Co., Berlin, pp. 381-386.

10. Eipper, B.A., Mains, R.E. and Glembotiski, G.G. (1983) Proc. Natl. Acad. Sci. U.S.A. 80, 5144-5148.

11. MacIntyre _et al_ (1980) in Handbuch der inneren Medizin Vl/1A. (Eds. Knochen _et al_), Springer, Verlag Berlin, 623-634.

Protein Purification: Micro to Macro, pages 443–458

SOLUBILIZATION OF INCLUSION BODY PROTEINS BY REVERSIBLE N-ACYLATION

Dante J. Marciani, Chung-Ho Hung, King-Lan Cheng[1], and Charlotte Kensil

Cambridge BioScience Corporation
365 Plantation Street
Worcester, MA 01605

ABSTRACT Recombinant proteins produced in E. coli, can be deposited inside the microorganism cell, as refractile inclusion bodies, i.e. clumps of insoluble protein. The inclusion bodies are separated from the host cell proteins through lysis of whole cells followed by centrifugation. Inclusion bodies are solubilized in the presence of guanidine hydrochloride or urea and 2-mercaptoethanol. The sulfhydryl groups of the protein(s) are protected by reversible or irreversible derivatization. The recombinant protein is dissociated by converting cationic amino groups to anionic carboxylic groups. These highly negative charged proteins will repel one another, reducing the tendency of recombinant proteins to aggregate. Introduction of the carboxyl groups is achieved by reacting the free amino groups of the protein with selected cyclic dicarboxylic acid anhydrides. This reaction leads to the formation of N-acyl groups, which can be deacylated by exposure of the modified protein to acid conditions. Following complete N-acylation, the modified proteins can be fractionated by use of conventional procedures, such as gel filtration, ion exchange chromatography, and others. The modified protein can be deblocked by exposing the protein to acid pH. The rate of this

[1]Present address: Department of Tumor Biology University of Texas System Cancer Center, M.D. Anderson Hospital, Houston, Texas 77030

reaction can be controlled by changes in pH and/or temperature. Following the removal of the N-acyl groups, sulfhydryls can be regenerated and if needed induce the formation of disulfide bonds.

INTRODUCTION

Recombinant DNA technology has permitted the expression of foreign proteins in different host cells, including bacterial, yeast, insect and mammalian. In some cases, high expression of recombinant proteins leads to the formation of intracellular aggregates, often referred to as "inclusion bodies" or "refractile bodies".

The inclusion bodies fall into two categories: (A) Paracrystalline arrays, in which the protein is presumably in a stable conformation, although not necessarily native; and (B) Amorphous aggregates containing partially and completely denaturated proteins, as well as aberrant proteins synthesized as a result of inaccurate translation. These aggregates of the heterologous protein may constitute a significant portion of the total cell protein.

Although inclusion bodies probably afford protection of the recombinant proteins against endogenous proteases, they present problems for the extraction and purification, due to the insolubility of these bodies in aqueous buffers. In many cases, denaturating agents have been used to solubilize and extract the protein from the inclusion bodies. These agents include denaturating agents, such as guanidine hydrochloride, urea, and detergents, e.g. sodium dodecyl sulfate (SDS), Triton X-100 and others. For proteins of biomedical interest, the use of these agents is, in many instances, undesirable. In some cases, the agent can irreversibly modify amino acid residues, i.e. high concentrations of urea at alkaline pH can carbamylate the amino groups of proteins. In other cases, it is difficult to completely remove the agent (e.g. detergents such as SDS) from the protein, particularly if the protein possess extensive hydrophobic domains.

Although strong denaturating agents, such as guanidine hydrochloride, may completely solubilize globular proteins, renaturation of the isolated protein may be difficult if not impossible. Moreover, solubilization of some amphipathic proteins with guanidine hydrochloride may result in extensive protein aggregation, a process that is more apparent at low concentrations of guanidine hydrochloride. As a consequence of this mechanism, an extensive insoluble aggregation of the recombinant proteins may be observed during the removal of the denaturant. The resultant soluble protein can be biologically inactive due to incorrect folding coupled to formation of discrete soluble aggregates.

Another problem in the purification of the recombinant proteins from inclusion bodies is the need to separate this protein from cellular contaminants. This association is perhaps mediated by ionic or hydrophobic interactions. Several approaches have been used to achieve these separations. In effect, Builder et al (2) and Olson et al (3) developed technology to solubilize the inclusion bodies using strong denaturating agents such as guanidine hydrochloride and ionic detergent such as sodium dodecyl sulfate. After solubilization, the recombinant proteins are refolded by exchange into weak denaturating agents such as urea in the presence of a reducing agent such as 2-mercaptoethanol. The recombinant proteins can then be isolated using ion exchange and gel permeation chromatography. Thus, in these methods, denaturants must be present throughout and subsequent to the process for recovering inclusion body proteins.

However, an alternative to denaturants or detergents is the N-acylation of amino groups by some acid anhydrides which has been used to dissociate and/or solubilize proteins from viruses and mammalian cells. Vande Woude and Bachrach (4) demonstrated that maleylation of the amino groups of Foot and Mouth Disease Virus protein solubilized it in aqueous nondenaturing solvents.

Eshhar et al (5) were able to solubilize 60-70% of membrane proteins from lymphocytes by citraconylation. Lundahl et al (6) selectively solubilized a portion of water-insoluble human erythrocyte membrane proteins by citraconylation. The N-acylated proteins were further separated on hydroxyapatite columns. The solubility of an unfolded protein increases, and aggregation is avoided, when amino groups are acylated with citraconic anhydride (7). Citraconylation of α-crystallin's lysine residues leads to complete disassociation of this protein. However, deacylation did not regenerate aggregates identical to the native ones (8).

A result of N-acylation of proteins is the disruption of their secondary and tertiary structure. Jackson et al (9) have demonstrated that maleylation of human plasma high density apolipoprotein-A-II causes a shift from a helical to a more disordered secondary structure. This change is concomitant with a decrease in immunoreactivity. Conformational transitions of immunoglobulin fragments induced by citraconylation has been described. A partial reversibility in conformation was achieved by high ionic strength (10). Brinegar and Kinsella (11) have described the formation of irreversible crosslinked dimers of B-lactoglobulin. This crosslinking arises from an S-alkylation reaction between a citraconylated amino group and free sulfhydryl groups.

This report describes the use of acylating agents for the solubilization and purification of recombinant proteins. The effect of side reactions and mechanisms for separation of the different N-acylated proteins are discussed.

METHODS

Acylation of amino groups in rec. proteins with citraconic anhydride was performed by adaptation of the methods described by Atassi and Habeeb (12). Sulfhydryl groups were modified irreversibly by alkylation or in a reversible fashion by disulfide exchange (13). Free amino groups in proteins were determined with trinitrobenzene sulfonic acid (TNBS) (14). Polyacrylamide gel electrophoresis was performed by the presence of SDS. Immunochemical reactivity of proteins was tested by enzyme linked immunoassay (ELISA) as described by Engvall (15). Chromatographic separations were performed by gel filtration or by medium-performance ion exchange chromatography (16). Recombinant proteins expressed in *E. coli* as "inclusion bodies" were processed as described in figure 1. The chemical modifications of very hydrophobic proteins, i.e. amphipathic proteins, were performed in the presence of appropriate denaturating agents. Reversal of the modifications were performed under conditions of pH, temperature, and ionic strength, selected to achieve maximum recovery of functional activity.

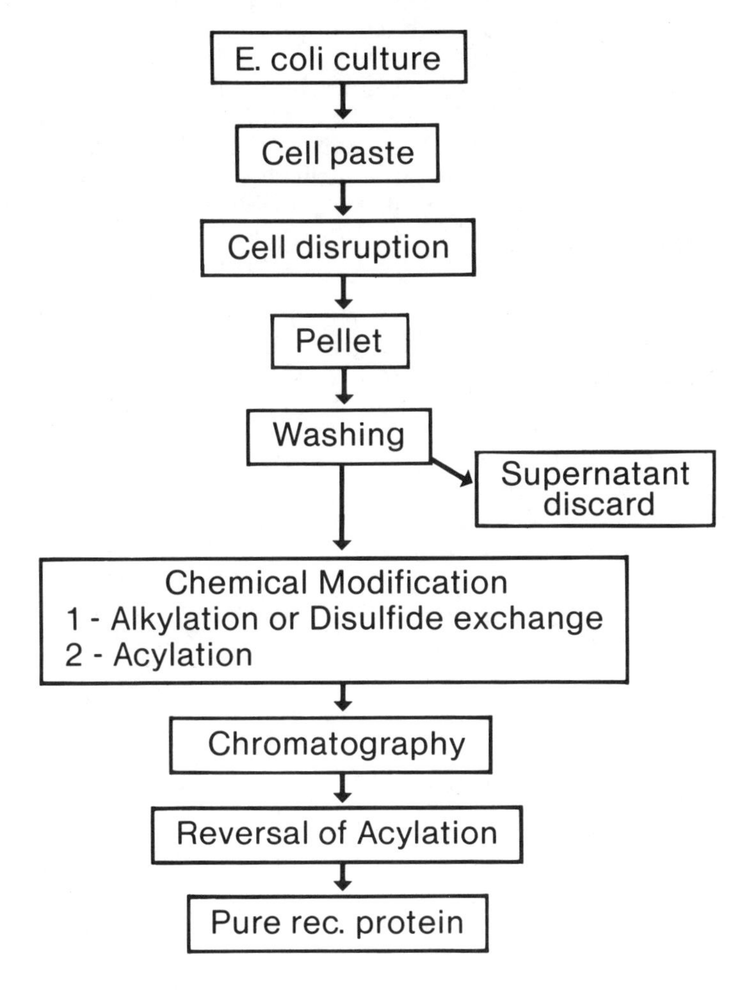
Processing of
Inclusion Bodies
E. coli culture
Cell paste
Cell disruption
Pellet
Washing
Supernatant
discard
Chemical Modification
1 - Alkylation or Disulfide exchange
2 - Acylation
Chromatography
Reversal of Acylation
Pure rec. protein

FIGURE 1.

RESULTS

Recombinant proteins of amphipathic nature can be solubilized, in many cases, by high concentrations of denaturating agents such as urea, Gu.HCl and SDS. Attempts to purify viral membrane or envelope proteins expressed in E. coli yielded unsatisfactory results. The proteins, due to their strong associative properties, tend to aggregate even in the presence of Gu.HCl or urea. Analysis of rec. envelope proteins from HIV, (17) (the AIDS etiological agent), by gel permeation chromatography in the presence of 6M Gu.HCl revealed a heterogenous aggregation of the polypeptide. Removal of the Gu.HCl from the sample resulted in the formation of visible aggregates. A more efficient solubilization and dissociation of these recombinant proteins was achieved by the use of the anionic detergent SDS. Gel permeation chromatography of these proteins in the presence of SDS separated the different polypeptides in their monomeric form. Unfortunately, the presence of SDS precluded the use of other chromatographic techniques such as ion exchange, affinity and others. Moreover, this detergent binds strongly to proteins, making its complete removal extremely difficult. Weak-denaturating agents, such as urea and non-ionic detergents, failed to completely solubilize these recombinant proteins. Attempts to transfer these proteins from strong denaturating conditions, 6M Gu.HCl, to weak denaturants resulted in significant losses of protein due to precipitation.

Citraconylation

Solubilization of the rec. proteins from the inclusion bodies was achieved by citraconylation of the protein in the presence of 7-8M urea or 6M Gu.HCl at pH 8-10. N-acylation was performed by reacting the protein with 40-60 fold molar excess of the citraconic anhydride. The modified rec. protein from HIV envelope remained soluble upon removal of denaturant, by dialysis into 50mM Na. borate pH 9. A somewhat similar behavior was observed with the rec. envelope protein from FeLV.

Gel Filtration

The rec. proteins modified by citraconylation were studied by gel filtration. The untreated HIV rec. envelope protein from inclusion bodies, in the presence of 6M Gu.HCl and 2-mercaptoethanol, showed different degrees of aggregation (Fig 2A). The content of monomeric form in this preparation was only a fraction of the total rec. protein. This distribution of aggregates indicates strong residual forces, presumably hydrophobic interactions, that can exist even in the presence of Gu.HCl. Citraconylated HIV envelope protein analyzed by gel filtration on Ultrogel AcA 34 in the presence of 50mM Na.borate pH 9 revealed a homogenous population (Fig 2B). The molecular weight of the N-acylated protein was estimated to be around 34,000 from comparison with globular protein standards. This molecular weight is similar to that calculated from the amino acid sequence, around 35,000 and confirmed by SDS-polyacrylamide gel electrophoresis. These results indicate that introduction of citraconyl groups in this polypeptide is enough to achieve complete dissociation into soluble monomers.

A different behavior is observed with a rec. envelope protein derived from another retrovirus, FeLV. This amphipathic protein when dissolved in urea, without amino groups modification, tends to stay in an aggregated state. Citraconylation of this protein results in increased solubility and dissociation of the aggregated form. However, an apparent equilibrium between monomeric and polymeric forms is detected by gel filtration. Comparison of the elution profiles of the N-acylated rec. protein at two different concentrations, indicates an increase in the polymeric form at the higher protein concentrations (Fig 3). Citraconylation of the amino groups leads to dissociation of the aggregates. However, at increasing concentrations of modified protein a discrete association of the protein molecules occur, in a fashion similar to that observed with detergents. This mechanism explains the bimodal distribution observed by gel filtration (Fig 3).

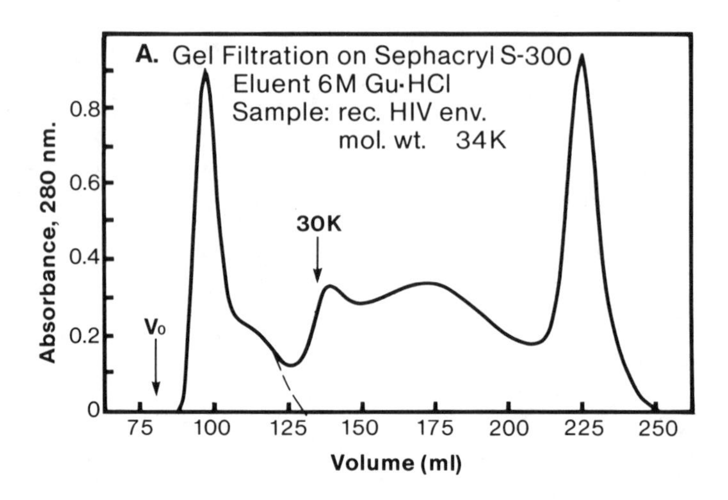

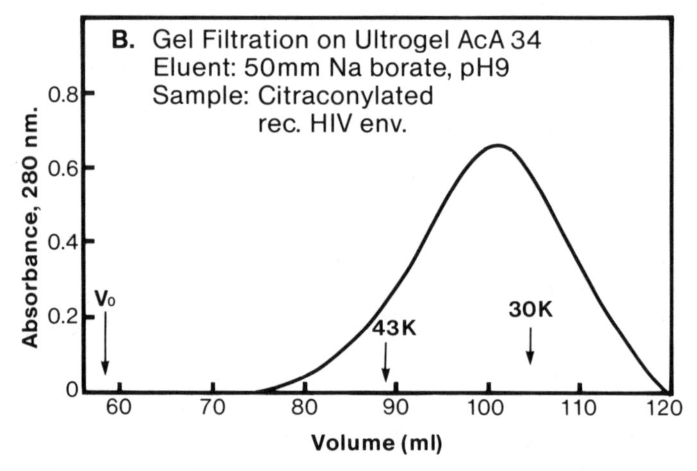

FIGURE 2. Effect of citraconylation on the associative properties of the rec. HIV envelope. (A) Gel filtration of inclusion bodies solubilized in 6M Gu.HCl, pH 9, containing 0.1% dithiotreitol (DDT). HIV protein was eluted between 90 and 125 ml. (B) Gel filtration of citraconylated rec. protein in 50mM Na borate, indicates no aggregates. The 43K and 30K indicate mol. wt. standards.

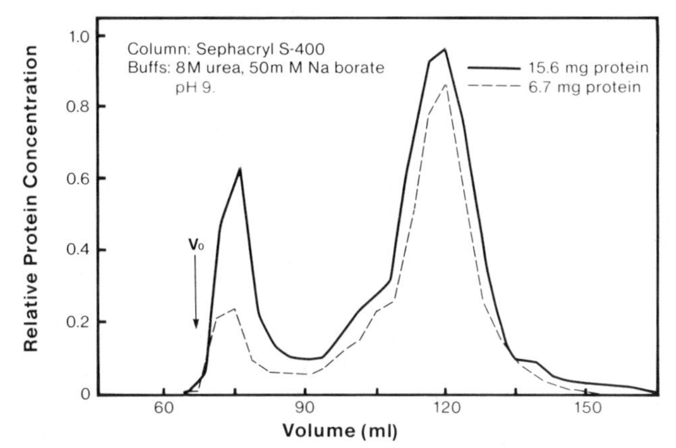

FIGURE 3. Associative properties of a citraconylated amphipathic rec. protein, FeLV envelope. The N-acylated protein, with alkylated sulfhydryl groups, was analyzed by gel filtration in urea. Increasing protein concentration resulted in an increase in the formation of discrete aggregates. Gel filtration was performed on a Sephacryl S-400 column, 1.5 x 90 cm.

Ion-Exchange Chromatography

N-acylated proteins are polyanions that can be separated by ion-exchange chromatography. Elution profiles of citraconylated inclusion bodies containing HIV envelope protein are shown in figure 4. The sample concentration has an effect on the elution characteristics. Decreasing the sample concentration resulted in a lengthened elution time. This shift in the elution profile was not correlated with the salt concentration of the linear gradient (Fig 4). Prolonged elution at constant salt concentration, resulted in patterns somewhat similar to those presented here, with peaks broader than those obtained in the presence of a salt gradient. Analysis by SDS-polyacrylamide gel electrophoresis of individual samples along the elution pattern indicated a direct correlation between the molecular weight of the N-acylated proteins and its retention time. The chromatographic characteristics for these modified proteins indicate that the basic mechanisms for salt elution do not apply here and suggests a "displacement" process (18).

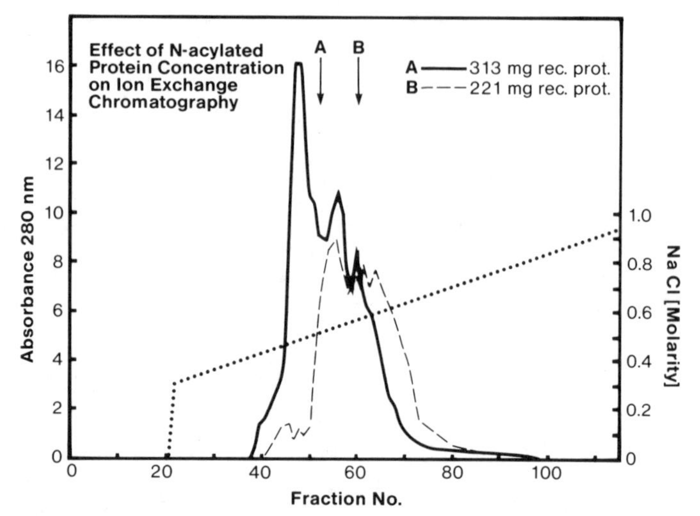

FIGURE 4. Ion exchange chromatography of citraconylated rec. HIV envelope inclusion bodies. The citraconylated proteins in 50mM Na borate, pH 9, were applied to a 80 ml. TSK-DEAE column and eluted with a NaCl linear gradient from 0.3 to 1M. Individual fractions were analyzed by SDS-PAGE and Western Blot.

Properties of Decytraconylated Proteins

Citraconyl groups were easily removed at pH values below 4. In the present case, the pH values selected for the deacylation were 2.5 and 3, as shown in figure 5. Attempts to deblock the proteins at pH values around 4, were complicated due to the low solubility exhibited by the modified proteins around that pH. Citraconylation decreased the apparent pI of the proteins to around 4 instead of the pI 10.5, calculated from the protein amino acid sequence. Therefore, isoelectric precipitation was avoided by working outside the range of insolubility.

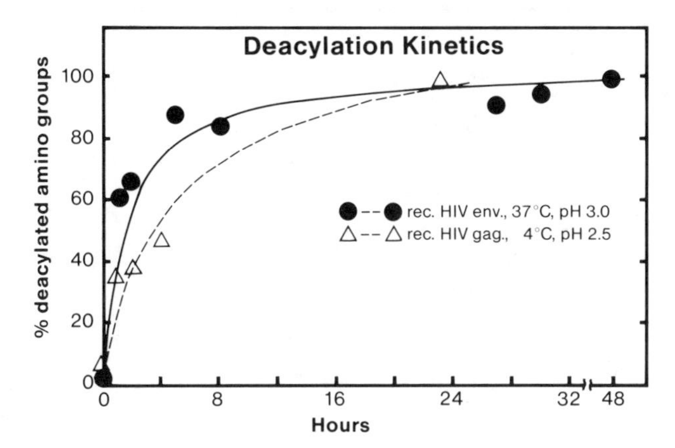

FIGURE 5. Deacylation of blocked amino groups in citraconylated rec. proteins, HIV env (●) and HIV gag (Δ). Proteins were subjected to acid pH at the temperature indicated. At different times the free amino groups were determined by TNBS.

A quantitive estimate of the effects of the chemical modification on the immunological properties of the rec. proteins was made using the ELISA procedure. The rec. HIV envelope protein, untreated, N-acylated and deacylated, was tested for its immunoreactivity with high titer human sera against HIV. Figure 6 shows that citraconylation of the protein eliminated reactivity with the human serum, except at high protein concentrations. After removal of the citraconyl groups, the protein regained its full immunological properties and was indistinguishable from the untreated protein.

Additional evidence that rec. proteins purified by use of an N-acylated intermediate phase can recover all their immunological properties is shown in figure 7. Mice immunized with rec. envelope from FeLV, isolated by the citraconylation procedure, presented a better humoral immunoresponse than those immunized with protein isolated by more classical procedures. The immunological response was tested against the FeLV itself. These results suggest that the deacylated rec. protein resembles the native protein to a higher degree than the rec. protein isolated by other procedures.

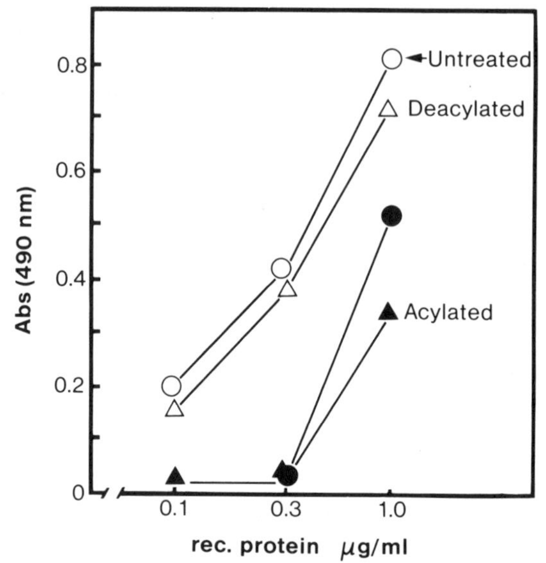

FIGURE 6. Effect of citraconylation on the immunochemical properties of rec. HIV envelope. Citraconylated (●▲), decitraconylated (△), and untreated (O), proteins were tested by ELISA against a high titer human serum reacting with HIV.

DISCUSSION

Strong denaturating agents, such as guanidine HCl and SDS are required to obtain initial solubilization of inclusion body rec. proteins, and thereafter weak denaturants can maintain solubility. In order to overcome the need of denaturants to maintain solubility, a large negative charge can be introduced into the rec. protein by a citraconylation reaction. The negatively charged carboxyl groups will induce electrostatic repulsions of enough magnitude to counteract the noncovalent attractive forces holding the polypeptide chains together. Therefore, citraconylation is accompanied by dissociation of the rec. polypeptides and a more expanded structure of the polypeptide chains.

Citraconylated rec. proteins are very stable in aqueous solutions at pH values around 8-9.5. These water soluble forms can be purified by different chromatographic

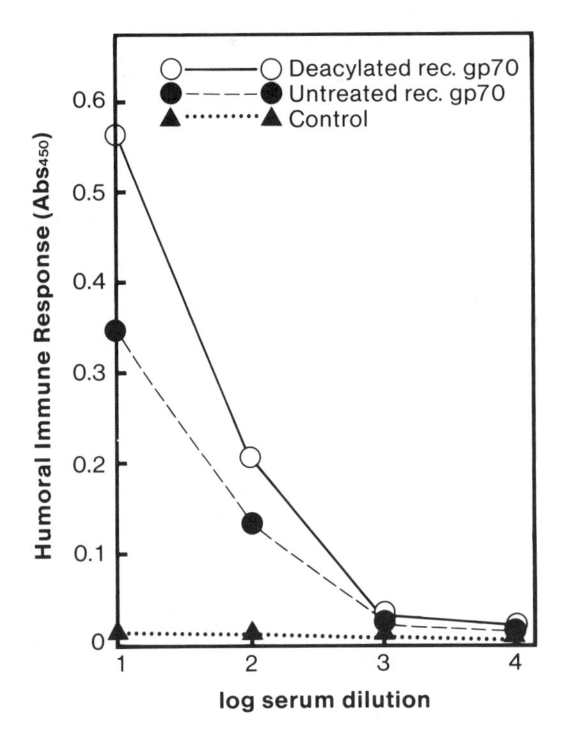

FIGURE 7. Effect of citraconylation-decitraconylation on the humoral immunoresponse of immunized animals. Mice were immunized with FeLV envelope protein prepared through a reversible citraconylation step (○), or prepared without a N-acylation step (●), plus adjuvants. The controls were injected with the adjuvant alone. The immunoresponse was tested by measuring the ELISA response toward FeLV of serial dilutions of sera.

procedures. Citraconic anhydride can also react with free sulfhydryl groups resulting in S-alkylation. Moreover, residual citraconylated amino groups can participate in an S-alkylation reaction with a different polypeptide chain to form an irreversible cross-linked dimer (11). In the event that sulfhydryl groups are essential for biological activity or to form disulfide bonds, the -SH groups should be protected by a reversible modification such as disulfide exchange. If the sulfhydryl groups are not critical for refolding of the polypeptide or its biological activity, irreversible blocking is recommended.

Citraconylation is reversible under mild acid conditions, and upon deacylation the properties of the native protein should be mostly restored. The reversibility of N-acylation by cyclic acid anhydrides depends on the presence of a protonated carboxyl group specifically oriented to participate as an intramolecular catalyst of the hydrolytic reaction. The carbon-carbon doubled bond present in these reagents maintains the terminal carboxyl group in the spatial orientation suitable for deacylation (13). Succinic anhydride, which is devoid of the C-C double bond, forms groups that are quite stable under acidic conditions.

Citraconylation of the rec. proteins leads to an increase in their hydrodynamic volume that will depend on the extent of the interacting hydrophobic domains and their distribution. The hydrodynamic volume of the citraconylated rec. HIV envelope protein corresponds to that of a globular protein of similar molecular weight. The extensive distribution of hydrophobic interactions occuring in the protein limits the degree of expansion due to electrostatic repulsions. The behavior of the rec. FeLV envelope protein, an amphipathic structure with assymetric distribution of hydrophobic regions, is different.

Citraconylation of this protein resulted in an increased solubility when exposed to weak denaturants. However, increase in the protein concentration resulted in an enhancement of an aggregation process. Gel filtration showed the existance of two populations, monomeric forms and discrete aggregates. This aggregation can be facilitated through the hydrophobic tail of the molecule, in a fashion similar to the micelle formation observed with detergents.

Substitution of the protein amino groups by negatively charged carboxyl groups, a result from N-acylation, changes its electrochemical properties from amphoteric to polyanionic. The negative charge density of these polyelectrolytes will be somewhat similar, as a result of the frequency of occurrence for lysine in different proteins. The close similarities in intrinsic charge for the different modified proteins reduces the ability of ion exchange chromatography to resolve these polyanions. A more suitable separation mechanism will be one based on displacement (18,19). A soluble polyanion having a number of negative charges per molecule to match those on an anion exchanger being used by a given protein, can be expected to compete with that protein for binding. If the polyanion can offer additional negative charges, it is likely that it will move the protein ahead of it on a column. A series of polyanions, such as citraconylated proteins, should form a displacement train based on the differences in the number of effective negative charges for molecule. The citraconylated proteins will find a position in that train according to their own abilities to compete for binding to the exchanger. As a consequence, for a set of polyanions with similar intrinsic charge, the binding strength should be proportional to the molecular weight of the polypeptide. This mechanism of displacement is not dependent on changes of ionic strength, as is in the classical elution chromatography.

N-acylation offers a viable alternative for solubilization of rec. proteins. The reversibility of the modification may be used in the process of refolding of rec. proteins to achieve functional molecules.

REFERENCES

1. Old, and Primrose (1985) "Principles of Gene Manipulation". Oxford Blackwell Scientific Publisher, p. 289.
2. Builder, S.E. Ogez, J.R. (1985) Purification and Activity Assurance of Precipitated Heterologous Proteins, U.S. Patent 4511502.
3. Olson, N.C. Pai, D.C. (1985) Purification and Activity Assurance of Precipitated Heterologous Proteins, U.S. Patent 451103.
4. Vande Woude GF, Bachrach HL (1971). Number and molecular weights of Foot-and Mouth Disease virus capsid proteins and the effects of maleylation. J. Virology, 7:250.
5. Eshhar, Z. Gafni, M. Givol, D. and Sela, M. (1971) Solubilization of Lymphocyte and Thymocyte Antigens by a Reversible Chemical Modification. Eur. J. Immunol. 1:323.
6. Lundall P. (1975) Proteins selectively released from water-extracted human erythrocyte membranes upon citraconylation or maleylation. Biochem. Biophy Acta 379:304.
7. Light A. (1985) Protein solubility, protein modifications and protein folding. BioTechniques 3:298.
8. Bindels JG, Misdom LW, Hoenders HJ. (1985) The reaction of citraconic anhydride with bovine α-crystalline lysine residues. Biochem. Biophys. Acta. 828:255.
9. Jackson RL, Mao SJT, Gotto AM (1974) Effects of maleylation on the lipid-binding and immunochemical properties

of human plasma high density apolipoprotein A-II. Biochem. Biophys. Res. Common 61:1317.
10. Nakagawa Y, Liaw WY, Hunt A.H. Jirgensons B (1974) Conformational transitions of immunoglobulin fragments by citraconylation of their lysine side chains. Immunochemistry 11:483.
11. Brinegar AC, Kinsella J.E. (1981) Reversible modification of lysine in B-lactoglobulin using citraconic anhydride Int. J. Peptide Protein Res. 18:18.
12. Atassi, MZ, Habeeb AFSA (1972) Reaction of proteins with citraconic anhydride. Methods in Enzymol. 25:546.
13. Glazer AN, Delanger RJ, Sigman D.S. (1975) Chemical modification of proteins. In Work TS, Work E (eds). "Laboratory Techniques in Biochemistry and Molecular Biology" New York: American Elsevier, p. 101.
14. Habeeb AFSA. (1966). Determination of free amino groups in proteins by Trinitrobenzenesulfonic acid. Anal. Biochem. 14:328.
15. Engvall E, (1980) Enzyme Immunoassay ELISA and EMIT. Methods in Enzymol. 70:419.
16. Katz Y, Nakamura K, Hashimoto T, (1982) Characterization of TSK-gel DEAE-Toyopearl 650 ion exchanger. J. Chromatogr. 245:193.
17. Thorn RM, Beltz GA, Hung CH, Fallis BF, Feldhouse SW, Cheng KL, Marciani D.J. (1987). An enzyme immunoassay using a novel recombinant polypeptide to detect human immunodeficiency virus (HIV) env antibody, J. Clin Microbiology (In press).
18. Peterson EA, Torres Ar,(1984) Displacement Chromatography of proteins. Methods in Enzymol. 104:113.
19. Peterson EA, (1978) Ion-exchange displacement chromatography of serum proteins, using carboxymethyldextrans as displacers. Anal. Biochem. 90:767.

Protein Purification: Micro to Macro, pages 459–473

MAMMALIAN CELLS HAVE MULTIPLE PATHWAYS FOR DEGRADING PROTEINS[1]

Lloyd Waxman[2] and Julie M. Fagan[3]

[2]Department of Biological Chemistry, Merck Sharp & Dohme, West Point, PA 19486
[3]Department of Animal Sciences, Rutgers University, New Brunswick, NJ 08903

ABSTRACT Protein degradation in mammalian cells can take place in the cytoplasm as well as in the lysosome. One cytoplasmic pathway requires energy and involves the conjugation of ubiquitin to proteins. This modification signals their degradation by an ATP-requiring protease. A variety of other proteases have been isolated from the cytoplasm, some of which may participate in energy-independent pathways. The features of a protein which govern its half-life and its site of degradation within the cell are not known. Recent experiments are reviewed which attempt to address this important question.

INTRODUCTION

Lysosomes have been regarded as the main site for breaking down proteins in mammalian cells, primarily because of the high concentration of acid hydrolases within these organelles. However, with the use of inhibitors of lysosome function and lysosomal proteases, it is now clear that the degradation of proteins to their constituent amino acids can also occur in the cytoplasm. In this paper we will summarize some of the properties of several newly discovered cytosolic proteases active at neutral pH. We will also review

[1] This work was supported by research grants from the Muscular Dystrophy Association, Rutgers University and the New Jersey Agricultural Experiment Station.

some current ideas on what features of a protein affect its susceptibility to proteolysis and hence its half-life.

ATP-DEPENDENT PROTEOLYSIS

An intriguing aspect of intracellular protein breakdown in animal cells as well as in bacteria is the absolute requirement for energy (1). Although ATP is thought to be required to maintain the intracellular pH of the lysosome, at least part of this energy requirement for proteolysis must be nonlysosomal since neither reticulocytes nor bacteria contain these organelles. In reticulocytes, a cytoplasmic protein degradative system which requires ATP is believed to play a major role in the degradation of abnormal and short-lived proteins (2,3). According to the current model, the hydrolysis of ATP provides the energy to covalently link ubiquitin, a 76 amino acid peptide, to free amino groups on proteins. Much is known about the molecular biology of this highly conserved peptide, but its roles in a variety of apparently unrelated biological systems remain a puzzle (Table 1). Ubiquitin conjugation is thought to render proteins more susceptible to proteolysis (Figure 1). More recently, Hershko et al. suggested that the conjugation of ubiquitin to the α-amino group, and not the ε–amino group on lysyl residues marked proteins for degradation in reticulocytes (4). Other experiments have shown that the enzyme (E_3) which transfers ubiquitin to proteins (see Figure 1) binds better to proteins with free α-amino groups than to those having blocked amino termini (5).

In contrast, in Dictyostelium calmodulin which has an acetylated α-amino group, the conjugation of ubiquitin to its single ε-amino group still results in ATP-stimulated proteolysis (6). We recently examined the breakdown in reticulocyte extracts of bovine kappa-casein which also has a blocked amino terminus (pyrrolidone-carboxylic acid). Its rate of proteolysis by the ATP-dependent pathway was found to be five times greater than that for α or β casein, and equal to that for lysozyme, a model substrate for the ATP-dependent proteolytic system. The results of these experiments suggest that conjugation to the α-amino group on some proteins cannot be the signal for their degradation.

TABLE 1
PROPERTIES OF UBIQUITIN

1. Essential component in the ATP-dependent proteolytic system of eukaryotes, including yeast and plants (7).
2. Attached to some molecules of histone H2A and H2B (8,9), associated with transcription of active genes (10).
3. Found bound to some plasma membrane receptors (11).
4. Synthesized as a polyubiquitin precursor of identical repeating units (12).
5. Component of paired helical filaments in brains of patients with Alzheimer's disease (13).
6. Heat shock protein (14).

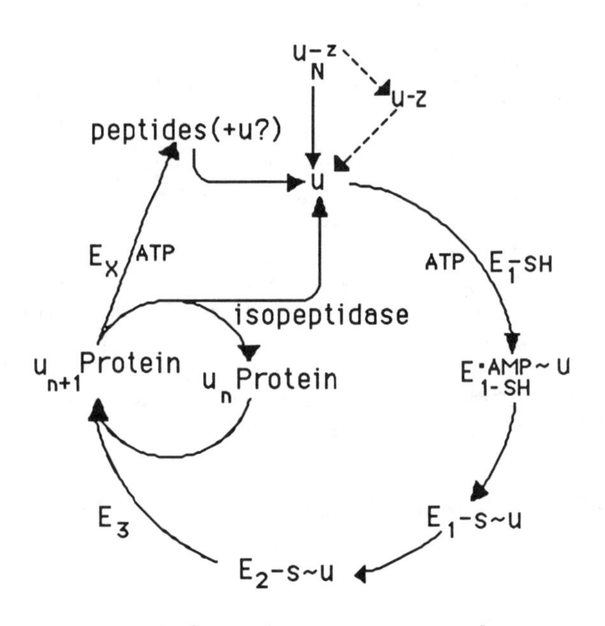

Figure 1. A model of ubiquitin metabolism in mammalian cells (14). This multi-component pathway consists of: 1) the heat-stable polypeptide ubiquitin (u) which is synthesized as a polyubiquitin precursor (u_N), 2) three factors, E_1, E_2, and E_3, which are required to covalently transfer ubiquitin to amino groups on proteins, 3) an ATP-requiring protease (E_x) which can degrade ubiquitin-protein conjugates and 4) multiple enzymes called isopeptidases which play a role in regenerating free ubiquitin from proteins or peptide fragments.

In addition to the participation of ATP in the conjugation of ubiquitin to proteins, ATP is required for the degradation of ubiquitin-protein conjugates (15). Recently, we and others have identified a high molecular weight (1500 kD) ATP-requiring protease which will degrade proteins to which ubiquitin is attached, but will not degrade proteins which lack this modification (16,17). We have also found evidence for this ATP-dependent pathway in muscle and liver and have been able to isolate a similar ATP-requiring protease from tissue extracts (18). It seems likely, therefore, that all mammalian cells contain this degradative pathway, but its importance *in vivo* compared to other degradative mechanisms has not been established. Strong genetic support for an essential role for the pathway comes from studies with a mouse cell line which contains a temperature-sensitive mutation in the enzyme E_1. In these cells the breakdown of analog-containing abnormal proteins is completely blocked at the non-permissive temperature (19).

Several other ATP-requiring proteases have been found in mammalian cells. An ATP-dependent ubiquitin-independent protease has been described in erythroleukemia cells, but its function is not known (20,21). Mitochondria also contain an ATP-dependent ubiquitin-independent protease similar to the ATP-dependent enzyme purified from bacteria (22). It probably functions to degrade abnormal and damaged proteins in the mitochondrial matrix.

OTHER PROTEASES IN MAMMALIAN CELLS

Several proteases have been purified and characterized from the cytosol of a variety of cells and tissues (Table 2). Unfortunately, it has been difficult to determine their physiological role since mutants lacking active proteases have not been identified nor are there specific inhibitors available.

Probably the best studied cytoplasmic enzymes are the calcium-activated thiol proteases (calpains). Much is known about the biochemistry of these enzymes including their sequence specificity and mechanism of action (23). An important unresolved question has been whether or not there is enough free calcium in the cell (except under pathological conditions) to permit activation of these enzymes. There is now some evidence that calpains associate with membranes in the presence of calcium (24) and that in the presence of phospholipids there is a decrease in the calcium requirement for their autoproteolytic activation (25). To determine the importance of the calpains in protein breakdown *in vivo*, these proteases can be blocked by the nontoxic thiol protease inhibitors, leupeptin (26) or the epoxide E-64 (27). Except for neutrophils and platelets, there is no strong evidence to indicate that these proteases are active under physiological conditions.

TABLE 2
PROPERTIES OF MAMMALIAN CYTOPLASMIC PROTEASES

	Type	Mol Wt (kD)	In vivo inhibitors
Conjugate-degrading protease	?	1500	?
Large alkaline multi-functional protease	Serine/ Thiol	720	?
Calpains	Thiol	110	E-64, leupeptin
Insulinases (1)	Metallo	300	?
(2)	Serine	115	CBZ-Pro-Pro aldehyde

A large 720kD multifunctional alkaline protease has been purified from many mammalian tissues, including muscle (28), liver (29), lung (30), pituitary (31) and red blood cells (32). It is currently believed to have multiple copies of three active sites of differing specificity; one which hydrolyzes fluorometric peptides cleaved by trypsin, one with a specificity for fluorometric peptides cleaved by chymotrypsin, and a third which degrades proteins (e.g. BSA and casein) whose specificity is not clear. Based on inhibitor studies, it is still not resolved whether this is a thiol or serine-type protease. Its function in the cell is not known since there are no specific inhibitors available which can be used *in vivo*.

Two other major proteases, termed "insulinases", degrade polypeptides the size of insulin (~6kD) or smaller (33). One insulinase is a metalloenzyme, a trimer of molecular weight 300,000. Little is known about its specificity or chemistry (34,35). The second insulinase is a well studied enzyme also known as proline endopeptidase (36,37). It is a serine-type protease extremely sensitive to DFP. *In vivo*, it can be inhibited by peptides containing proline aldehyde (38). Although it has been suggested to be the enzyme responsible for degrading neuropeptides *in vivo*, there is little data to support this suggestion (38).

All mammalian cells contain a variety of metallo-aminopeptidases of varying specificity. One major function is to perform the final step in the degradation of proteins, the generation of amino acids from peptides (39). Many of these are inhibited by bestatin and amastatin (40).

Under certain conditions, mammalian cells also secrete proteases. They are commonly encountered in culture media of fibroblasts and keratinocytes in a latent form and probably require proteolysis for activation. Their secretion is stimulated by a variety of growth factors as well as tumor promotors. One group of secreted proteases is similar to the lysosomal thiol protease, cathepsin L. This enzyme is a major secretory product in some transformed cells (41-43) and possibly plays a role in metastasis where it may degrade components of the extracellular matrix to facilitate tissue invasion by cancer cells. These enzymes are strongly inhibited by E-64, leupeptin, and the polypeptide thiol protease inhibitor, cystatin. A second major class of secreted proteases, including collagenases, gelatinases, and proteoglycanases, contain zinc and require calcium for optimal activity (44). *In vivo*, these proteases may be important in the destruction of connective tissue in diseases such as rheumatoid arthritis. Metal chelators, such as o-phenanthroline and EDTA, will block their activity. Inhibition of secreted proteases may be particularly important when a cloned protein is engineered to be secreted into the culture medium.

A ROLE FOR ATP-INDEPENDENT PROTEOLYSIS

In addition to the cytosolic ATP-dependent pathway, mammalian cells also contain an ATP-independent pathway for degrading some short lived proteins, proteins present in excess, and certain kinds of abnormal proteins. For example, the breakdown of the short-lived protein ornithine decarboxylase (ODC) in cultured cells (45) and the degradation of excess spectrin chains during the assembly of the cytoskeleton in chicken red cells (46) occurs by a nonlysosomal ATP-independent degradative mechanism. In erythroid cells, hemoglobin damaged by oxidants is also rapidly degraded intracellularly by an ATP- and ubiquitin-independent proteolytic pathway (47). The proteases responsible for the intracellular degradation of oxidant-damaged proteins have not yet been identified. In reticulocyte extracts, the breakdown of the free α and β chains of hemoglobin is dependent on ATP (Table 3). However, exposure of these proteins to oxidants known to damage the heme group renders the chains susceptible to breakdown by an ATP-independent process. Extracts from erythrocytes, which have lost the ATP-dependent degradative system during the course of maturation from reticulocytes, cannot degrade

TABLE 3
EFFECT OF PHENYLHYDRAZINE ON PROTEOLYSIS OF HEMOGLOBIN AND GLOBIN α AND β CHAINS BY RETICULOCYTE LYSATES[a]

Substrate	Protein Degradation (nmoles Ala/2h) Substrate Pretreatment			
	None		Phenylhydrazine	
	-ATP	+ATP	-ATP	+ATP
Hemoglobin	0	0	10.4	10.4
α chain	0.4	9.6	21.1	21.6
β chain	0.2	5.0	14.8	15.1
α globin chain	1.4	4.5	1.2	5.2
β globin chain	0.5	6.4	1.1	6.8

[a]Dialyzed reticulocyte lysates were incubated in the presence of one of the following substrates: purified hemoglobin, the indivdual α and β chains, or the individual chains from which heme had been removed. Each substrate was also treated with phenylhydrazine, which damages the heme moiety. Proteolysis was measured by the release of alanine from each substrate using a sensitive fluorometric assay (47).

native α or β hemoglobin chains but hydrolyze the oxidant-treated chains almost as well as reticulocyte extracts. After removing the heme group, the α and β globin chains exposed to phenylhydrazine are no longer degraded by an ATP-independent mechanism. Under these conditions the globin chains appear to be degraded by an ATP-dependent degradative process at a rate similar to the untreated globin chains.

EVIDENCE THAT PROTEINS MAY CONTAIN SEQUENCES WHICH MARK THEM FOR PROTEOLYSIS

Over the years, some attempts have been made to correlate the rate of turnover of a protein and the properties of a protein, including its size, isoelectric point, hydrophobicity, thermal and proteolytic susceptibility *in vitro* , and the oxidation state of its thiol groups. Recently, three novel approaches suggest that the information controlling a protein's half-life may be built into its primary sequence.

Proteins with very short half-lives (<2h) were shown to contain amino acid sequences rich in proline (P), glutamic acid (E), serine (S), threonine (T) and to a lesser extent aspartic acid (D), called PEST regions (48). More than ten rapidly degraded proteins have been shown to contain PEST regions, but only 3 of 35 proteins with half-lives from 20-200h contain a PEST sequence. PEST sequences are identified by examining stretches of >10 amino acids flanked by positive residues. This algorithm combines the mole percent of P,E,S,T and D, with the hydrophobicity of the sequence. The latter factor is included in the calculation because hydropathy plots of PEST sequences include stretches of extreme hydrophilicity followed or preceded by a hydrophobic peak (48). Some examples of proteins which contain such sequences are shown in Table 4. Kappa-casein, a protein which we found to be a good substrate for the reticulocyte ATP-dependent proteolytic system also has a strong PEST sequence. This method of analysis does not specify the site of degradation of the protein in the cell.

If the PEST hypothesis proves to be a general rule, then attaching such sequences to stable proteins by chemical or genetic means should produce rapidly degraded hybrid proteins. Similarly, removal of PEST sequences should stabilize proteins within animal cells.

TABLE 4
EXAMPLES OF PEST SEQUENCES

Protein	PEST Score	Residues	Sequence	Half-Life (Hours)
β casein	+6.6	1-25	RELEELNVPGEIVESLSSSEESITR	2-5
Tyrosine amino-transferase	+0.3	382-395	HFPEFENDVEFTER	0.5
C-FOS	+10.1	128-139	KVEQLSPEEEEK	0.5
K casein	+ 4.5	116-169	KTEIPTINTIASGEPTSTPTIEAVE STVATLEASPEVIESPPEINTVQV TSTAV	---

Needless to say, confirmation of such a recognition mechanism would find wide application for the expression of proteins in mammalian cells.

Varshavsky and colleagues have also suggested that the half-life of a protein may be dependent on its primary sequence. Their hypothesis, "the N-end rule", proposes that proteins which have either Arg, Lys, Asp, Glu, Phe, Leu, Ile, Tyr, Gln at the N-terminal will have short half-lives (Table 5) (49). This conclusion was based on experiments in which ubiquitin was fused to β-galactosidase. When the plasmid vector containing this construct was expressed in yeast, ubiquitin was found to be rapidly cleaved from the α-amino end of β-galactosidase. By site-directed mutagenesis, different amino acids were inserted at the amino terminus of β-galactosidase which were then exposed upon removal of ubiquitin. This minor change resulted in proteins that varied widely in their stability depending on the amino acid exposed (Table 5) (49).

These experiments suggest that 1) ubiquitin does not actually mark proteins for degradation, 2) ubiquitin is not added *in vivo* to amino termini of proteins since it would be immediately hydrolyzed, and 3) ubiquitin might be added to the ε-amino groups of lysine residues to help destabilize proteins that are already marked for destruction. The importance of the N-terminal amino acid in determining the half-life of a protein might explain why bacteria can degrade proteins although they have no ubiquitin.

TABLE 5
PROTEIN HALF-LIFE IS DETERMINED BY ITS N-TERMINAL AMINO ACID

$t_{1/2}$ of X-β gal	Residue X in ub-X-β gal					
> 20 hours	Met	Ser	Ala	Thr	Val	Gly
~ 30 minutes	Ile	Glu				
~ 10 minutes	Tyr	Gln				
~ 3 minutes	Phe	Leu	Asp	Lys		
~ 2 minutes	Arg					

Of 208 intracellular proteins whose amino termini and half-lives are known, no exceptions were found to the N-end rule (49). However, this rule has only been shown to be valid for long-lived proteins. We examined the sequences of several short-lived proteins (derived from their cDNA's) and predicted their amino terminal acids (Table 6). These rapidly degraded proteins do not appear to follow the N-end rule. In the case of RNAse A, the N-terminal lysine residue should signal its rapid breakdown based on this rule. However, its experimentally determined half-life is ~100h (50).

TABLE 6
SOME EXCEPTIONS TO THE N-END RULE[a]

Protein	Amino-Terminal Sequences	Half-Life (hours)
RNAse	K E T A A A K F E R	100
ODC	M S S F T K D E F D	0.5
C-FOS	M M F S G F N A D Y	0.5
P53	M T A M E E S Q S D	0.5
P_{730}	M S S S R P A S S S	1.0

[a] Except for RNAse, sequence data was obtained from the cDNA. Based on several studies (51,52), the amino terminal of ODC and P730 should be Ac-Ser; C-FOS, Met-Met or Ac-Met-Met; and P53, Ac-Thr.

A very different signal may govern intracellular lysosomal proteolysis. Dice and colleagues have suggested that the enhanced uptake of certain proteins by the lysosome during nutrient deprivation is due to the recognition of specific amino acid sequences within these proteins. Residues 7-11 in the S-peptide of RNAse A, Lys-Phe-Glu-Arg-Gln, were shown to be an important signal for the enhanced lysosomal breakdown of RNAse A following serum deprivation. (50,53,54). To obtain further evidence for this model, the S-peptide was covalently coupled to lysozyme and the A-chain of insulin, proteins which normally do not respond to serum removal when microinjected. These hybrid molecules were now degraded 2-fold faster upon nutrient removal. Moreover, five cellular proteins whose rates of degradation are known to be enhanced during serum withdrawal also appear to contain related sequences.

CONCLUSIONS

Our understanding of protein breakdown in mammalian cells has made great progress over the last few years. A number of new proteases have been identified whose properties are unusually complex but whose chemistry and regulation are not known. In addition to the lysosome, there appear to be both energy-requiring and energy-independent cytoplasmic degradative pathways. What determines which route a protein takes is still not clear, although several new hypotheses have emerged which can be easily tested with the aid of site-directed mutagenesis, the facile synthesis of model peptides, and the ability to microinject model proteins into cells. Some of these signals are probably built into the primary sequence, perhaps analogous to the "signals" that govern compartmentalization of cellular proteins. Clearly, knowledge of these rules and the properties of the degradative systems within mammalian cells as well as in yeast and plants will enable us to use these cells more efficiently for expression of cloned proteins.

REFERENCES

1. Goldberg AL, St John AC (1976). Intracellular protein degradation in mammalian and bacterial cells. Ann Rev Biochem 45:747.
2. Hershko A, Ciechanover A (1982). Mechanisms of intracellular protein breakdown. Ann Rev Biochem 51:335.
3. Ciechanover A (1987). Mechanisms of energy-dependent intracellular protein breakdown. The ubiquitin-mediated proteolytic pathway. In Glaumann H, Ballard FJ (eds): "Lysosomes: Their Role in Protein Breakdown," London: Academic Press, p 561.
4. Hershko A, Heller H, Eytan E, Kaklij G, Rose IA (1984). Role of the α-amino group of protein in ubiquitin-mediated protein breakdown. Proc Natl Acad Sci USA 81:7021.
5. Hershko A, Heller H, Eytan E, Reiss Y (1986). The protein substrate binding site of the ubiquitin-protein ligase system. J Biol Chem 261:11992.
6. Gregori L, Marriott D, West CM, Chau V (1986). Specific recognition of calmodulin from Dictyostelium discoideum by the ATP, ubiquitin-dependent degradative pathway. J Biol Chem 261:5232.
7 Shanklin J, Jabben M, Vierstra RD (1987). Red light-induced formation of ubiquitin-phytochrome conjugates: Identification of possible intermediates of phytochrome degradation. Proc Natl Acad Sci USA 84:359.
8. Busch H, Goldknopf IL (1981). Ubiquitin-protein conjugates. Mol Cell Biochem 40:173.
9. West MHP, Bonner WM (1980). Histone 2B can be modified by the attachment of ubiquitin. Nucleic Acids Res 8:467.
10. Levinger L, Varshavsky A (1982). Selective arrangement of ubiquitinated and D1 protein-containing nucleosomes within the Drosophila genome. Cell 28:375.
11. Siegelman M, Bond M, Gallatin WM, St John T, Smith H, Fried VA, Weissman IL (1986). A putative lymphocyte homing receptor is a ubiquitinated branched-chain glycoprotein: Additional cell surface proteins also appear ubiquitinated. Science 231:823.
12. Ozkaynak E, Finley D, Varshavsky A (1984). The yeast ubiquitin gene: head-to-tail repeats encoding a polyubiquitin precursor protein. Nature 312:663.
13. Mori H, Kondo J, Ihara Y (1987). Ubiquitin is a component of paired helical filaments in Alzheimer's disease. Science 235:1641.
14. Finley D, Varshavsky A (1985). The ubiquitin system: functions and mechanism. Trends Biochem Sci 10:343.
15. Hershko A, Leshinsky E, Ganoth D, Heller H (1984). ATP-dependent degradation of ubiquitin-protein conjugates. Proc Natl Acad Sci USA 81:1619.
16. Hough R, Pratt G, Rechsteiner M (1986). Ubiquitin-lysozyme conjugates; identification and characterization of an ATP-dependent protease from rabbit reticulocytes. J Biol Chem 261:2400.

17. Waxman L, Fagan JM, Goldberg AL (1987). Demonstration of two distinct high molecular weight proteases in rabbit reticulocytes, one of which degrades ubiquitin conjugates. J Biol Chem 262:2451.
18. Fagan JM, Waxman L, Goldberg AL (1987). Skeletal muscle and liver contain a soluble ATP+ubiquitin-dependent proteolytic system. Biochem J 243:335.
19. Ciechanover A, Finley D, Varshavsky A (1984) Ubiquitin dependence of selective protein degradation demonstrated in the mammalian cell cycle mutant ts 85. Cell 37:57.
20. Rieder RF, Ibrahim A, Etlinger JD (1985). A particle-associated ATP-dependent proteolytic activity in erythroleukemia cells. J Biol Chem 260:2015.
21. Waxman L, Fagan JM, Tanaka K, Goldberg AL (1985). A soluble ATP-dependent system for protein degradation from murine erythroleukemia cells. J Biol Chem 260:11994.
22. Goldberg AL (1987). The ATP-dependent pathway for protein breakdown in mitochondria. In Glaumann H, Ballard FJ (eds): "Lysosomes: Their Role in Protein Breakdown," London: Academic Press, p 715.
23. Suzuki K (1987). Calcium activated neutral protease: domain structure and activity regulation. Trends in Biochemical Sciences 12:103.
24. Gopalakrishna R, Barsky SH (1986). Hydrophobic associations of calpains with subcellular organelles: compartmentalization of calpains and the endogenous inhibitor. J Biol Chem 261:13936.
25. Coolican SA, Hathaway DR (1984). Effect of L-α-phosphatidylinositol on a vascular smooth muscle Ca^{2+}-dependent protease. J Biol Chem 259:11627.
26. Umezawa H (1982). Low molecular weight inhibitors of microbial origin. Ann Rev Microbiol 36:75.
27. Barrett AJ, Kembhavi AA, Brown MA, Kirschke H, Knight CG, Tamai M, Hanada K (1982). L-trans-epoxysuccinyl-leucylamido (4-guanidino) butane (E-64) and its analogues as inhibitors of cysteine proteinases including cathepsins B, H, and L. Biochem J 201:189.
28. Dahlmann B, Kuehn L, Rutschmann M, Reinauer H (1985). Purification and characterization of a multicatalytic high-molecular-mass proteinase from rat skeletal muscle. Biochem J 228:161.
29. Tanaka K, Ii K, Ichihara A, Waxman L, Goldberg AL (1986). A high molecular weight protease in the cytosol of rat liver. J Biol Chem 261:15197.
30. Zolfaghari R, Baker CRF Jr, Canizaro PC, Amirgholami A, Behal, FJ (1987). A high-molecular-mass neutral endopeptidase-24.5 from human lung. Biochem J 241:129.
31. Wilk S, Orlowski M (1983). Evidence that pituitary cation-sensitive neutral endopeptidase is a multicatalytic protease complex. J Neurochem 35:842.

32. McGuire MJ, DeMartino GN (1986). Purification and characterization of a high molecular weight protease (macropain) from human erythrocytes. Biochim Biophys Acta 873:279.
33. Kirschner RJ, Goldberg AL (1981). Nonlysosomal insulin-degrading proteinases in mammalian cells. Methods Enzymol 80:702.
34. Kirschner RJ, Goldberg AL (1983). A high molecular weight metallo-endoprotease from the cytosol of mammalian cells. J Biol Chem 258:967.
35. Shii K, Baba S, Yokono K, Roth RA (1985). Covalent linkage of ^{125}I-insulin to a cytosolic insulin-degrading enzyme. J Biol Chem 260:6503.
36. Yoshimoto T, Orlowski RC, Walter R (1977). Postproline cleaving enzyme: identification as serine protease using active site specific inhibitors. Biochemistry 16:2942.
37. Andrews PC, Hines CM, Dixon JE (1980). Characterization of proline endopeptidase from rat brain. Biochemistry 19:5494.
38. Friedman TC, Orlowski M, Wilk S (1984). Prolyl endopeptidase: inhibition *in vivo* by N-benzyloxycarbonyl-prolyl-prolinal. J Neurochem 42:237.
39. Botbol V, Scornik OA (1985). Peptide intermediates in the degradation of cellular protein. In Khairallah EA, Bond JS, Bird JWC (eds): "Intracellular Protein Catabolism," New York: Academic Press, p 573.
40. Wilkes SH, Prescott JM (1985). The slow tight binding of bestatin and amastatin to aminopeptidases. J Biol Chem 260:13154.
41. Falanga A, Gordon SG (1985). Isolation and characterization of cancer procoagluant : a cysteine proteinase from malignant tissue. Biochemistry 24:5558.
42. Gal S, Gottesman MM (1986) The major excreted protein of transformed fibroblasts is an activatable acid-protease. J Biol Chem 261:1760.
43. Denhardt DT, Hamilton RT, Parfett CLJ, Edwards DR, St Pierre R, Waterhouse P, Nilsen-Hamilton M (1986). Close relationships of the major excreted proteins of transformed murine fibroblasts to thiol-dependent cathepsins. Cancer Research 46:4590.
44. Barrett AJ, Saklatvala J (1985). Proteinases in joint disease. In Kelley WN, Harris ED Jr, Ruddy S, Sledge CB (eds): "Textbook of Rheumatology" Philadelphia: WB Saunders Co, p 182.
45. Glass JR, Gerner EWJ (1987). Spermidine mediates degradation of ornithine decarboxylase by a non-lysosomal, ubiquitin-independent mechanism. J Cell Physiol 130:133.
46. Woods CM, Lazarides E (1985). Degradation of unassembled α– and β–spectrin by distinct intracellular pathways: regulation of spectrin topogenesis by β–spectrin degradation. Cell 40:959.
47. Fagan JM, Waxman L, Goldberg AL (1986). Red blood cells contain a pathway for the degradation of oxidant-damaged hemoglobin that does not require ATP or ubiquitin. J Biol Chem 261:5705.
48. Rogers S, Wells R, Rechsteiner M (1986) Amino acid sequences common to rapidly degraded proteins: the PEST hypothesis. Science 254:364.
49. Bachmair A, Finley D, Varshavsky A (1986). *In vivo* half-life of a protein is a function of its amino-terminal residue. Science 234:179.

50. Backer JM, Bourret L, Dice JF (1983). Regulation of catabolism of microinjected ribonuclease A requires the amino-terminal 20 amino acids. Proc Natl Acad Sci USA 80:2166.
51. Flinta C, Persson B, Jornvall H, von Heijne G (1986). Sequence determinants of cytosolic N-terminal protein processing. Eur J Biochem 154:193.
52. Boissel J-P, Kasper TJ, Shah SC, Malone JI, Bunn HF (1985). Amino-terminal processing of proteins: Hemoglobin South Florida, a variant with retention of initiator methionine and N^{α}-acetylation. Proc Natl Acad Sci USA 82:8448.
53. Dice JF, Chiang H-L, Spencer EP , Backer JM (1986). Regulation of catabolism of microinjected ribonuclease A. J Biol Chem 261:6853.
54. Backer JM, Dice JF (1986). Covalent linkage of ribonuclease S-peptide to microinjected proteins causes their intracellular degradation to be enhanced during serum withdrawal. Proc Natl Acad Sci USA 83:5830.

Protein Purification: Micro to Macro, pages 475–489

SCALING UP PROTEIN PURIFICATION

E. N. Lightfoot, S. J. Gibbs, M. C. M. Cockrem
and A. M. Athalye

Department of Chemical Engineering, University of Wisconsin
Madison, Wisconsin 53706

ABSTRACT. Examination of economic data suggests that recovery of valuable solutes from dilute solution is dominated by the costs of processing large masses of unwanted material. This suggestion is confirmed by examination of the most widely used current processing techniques, and a general strategy is suggested for reducing recovery costs. Important to this strategy is rapid volume reduction. Means are also suggested for reducing the costs of individual operations. As a specific example, the design of large-scale liquid chromatographic operations is considered from the points of view of process simulation and parameter estimation.

INTRODUCTION

Development of economical and effective separation equipment and processes for preparation of high-value proteins has been identified (1) as critical in the commercialization of biotechnology. However, significant interest in these problems has arisen only since the first cloning of genes in 1973 (2). Since both protein properties and process requirements tend to differ from those of more familiar commercial products and processes, it is necessary for the chemical engineering community to take a comprehensive approach, encompassing problem definition, elaboration of problem solving strategies, data acquisition and both equipment and process design. We examine these aspects of protein purification via representative examples.

A STRATEGY FOR RECOVERY FROM DILUTE SOLUTION

The Importance of Problem Definition

Here we demonstrate the utility of defining separations problems in operational terms, and we begin by noting that concentration of desired products from dilute feeds frequently dominates overall separation cost. We focus our attention on such processes, to identify their salient characteristics and to suggest strategies for improving equipment and process design. In this way we define a compact family of processing problems closely enough related to justify grouping them together and important enough to make such a grouping useful. We then ask the basic questions: How do process economics scale with the concentration of the desired product in process feeds, and why? How can one improve equipment effectiveness and process economics? We suggest that other classes of protein separations can be identified in similar ways, but we do not pursue that suggestion here.

Classification of Separation Processes

Our first task is to narrow the scope of analysis. We begin by recognizing that protein separation costs generally depend only upon the composition of the feed and product streams. We suggest that protein separations can be classified into three limiting types. i) Concentration: increasing the concentration of a single valuable substance originally present in small amount in a mixture of undesired species. This is the situation of primary interest here. ii) Fractionation: separating two or more closely related species of comparable value, all initially present in economically significant amounts. iii) Purification: removal of a single undesired impurity from a mixture in which it occurs at a low initial concentration. This classification appears to be compatible with that of J.-C. Janson (3). We now contrast these three common limiting situations from an economic standpoint.

We begin by looking at Fig. 1, commonly called a Sherwood plot, taken in this case from Nystrom as reported by Dwyer (4). Shown is an empirically observed relation between selling price per unit mass of various fermentation derived chemicals and their concentrations in the material from which they were produced. Despite the wide variety of

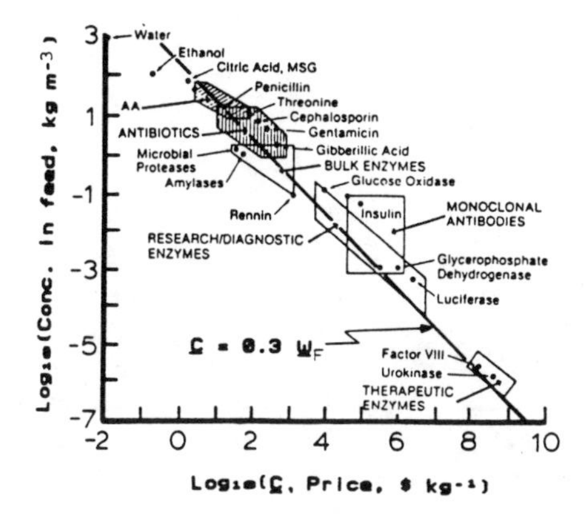

FIG. 1. Costs of materials handling: a Sherwood plot.

materials shown we find a good correlation, and similar behavior is observed by others (5). These data may be correlated by the simple equation:

$$C = KW_F \tag{1}$$

where C = cost per kilogram of desired material ($/kg), W_F = mass of inert material in the feed per kilogram of desired material (kg/kg), and K = a universal constant. This simple correlation, shown superimposed on the data, has two important characteristics. i) It suggests that processing cost is independent of the amount of the desired species present. ii) The correlation does not contain product composition. The first of these characteristics is examined in detail below, but here we concentrate on purification, where costs commonly correlate according to the relation

$$C = K' \ln[W_F/W_P] \tag{2}$$

in which F and P refer to feed and product conditions, respectively. Since the logarithmic dependence on concentration is much weaker than linear, purification costs tend to be small unless W_P becomes very large. This is indicated in Fig. 2 where these two kinds of behavior are contrasted for a hypothetical situation: $K = K' = 0.01$, and W_P of 10^{-4}, corresponding to 99.99% product purity. We would thus expect concentration costs to dominate unless very high purity is needed.

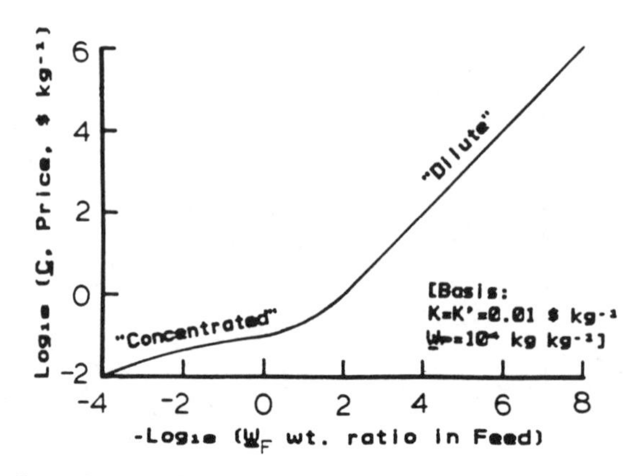

FIG. 2. Comparison of recovery costs: dilute vs. concentrated solutions.

Fractionation costs appear much more difficult to generalize, and we shall only note that they tend to scale differently with feed concentration than do concentration costs.

We find that concentration does exhibit distinct economic characteristics, and we are led to the definition:

> **Concentration, or recovery from dilute solution, comprises those processes in which processing costs depend only upon the composition of the feed solution and are independent of the amount of the potentially desirable species present.**

We have already discussed this definition (5), and we now summarize its significance to equipment and process design. Consider for example the following common operations.

Filtration and Centrifugation. Filtration and centrifugation are common early processing steps, and applications include removal of unwanted particulates, recovery of desirable particulates and removal of water and other solvents.

For these processes, cost relates only to the volume of suspending liquid and to the concentration and mechanical characteristics of solids present. It is not appreciably affected by the amount of product present, whether dissolved or suspended, if these are a minor constituent. Moreover, for individual classes of products at least, for example those produced from genetically altered **E. coli**, filtration characteristics are insensitive to the specific fermentation.

Filtration and centrifugation costs should then be proportional to the ratio of inerts to potential product, and independent of the degree of purification later desired and thus behave consistently with the Sherwood plot and the explanation given for its cost/concentration relation given above. It may also be noted that these processes are transport limited: processing cost is dominated by fluid flow considerations.

Liquid Extraction. Liquid extraction, another favorite early processing step seems at first sight, much more responsive to the amount of valuable material present: it is in principle possible to use highly selective solvents and thus to obtain rapid volume reduction proportionate to concentration of desired solutes. In practice, however, this is not normally true. Available liquid extractors perform well only if the ratio of extract and raffinate streams is kept within narrow limits: examination of process data shows that volume reductions on extraction from dilute solution are typically five to ten fold, and that they rarely exceed this latter limit, irrespective of the solute concentration of the feed. Processing costs, including solvent losses, are then dependent only upon the amount of inert material present, for example water, and the process is transport limited. If solute activity coefficients are increased, as by addition of sodium or ammonium sulfate, this cost, including that of waste disposal, is again related to the mass of inerts rather than of desired solutes. Once again we find the Sherwood-plot economics justified.

Precipitation. We now consider direct precipitation of the desired material from solution, a technique which includes both sulfate precipitation and crystallization. We see immediately that sulfate precipitation and any crystallization technique requiring a change in bulk solution thermodynamics give a cost picture which tends to be dominated by the amount of inerts present. However, filtration cost is now related to the amount of product obtained, and the cost-concentration relations begin to become more favorable to processing of dilute solutions, especially if a reversible, stoichiometric precipitant is used. We thus find here our first exception to Sherwood-plot economics.

Sorption Processes. In fixed-bed sorption processes, the restriction on stream-rate ratios characteristic of liquid extractors does not exist, and it is possible to find highly selective sorbents. One can often produce processes with reduced domination by costs of handling inerts.

However, there are productivity limiting factors here as well which tend to relate to the mass of inerts: hydrodynamic resistance and bed saturation. In preparative LC for example, column productivity is often limited by high pressure drops and crushing strength of the packing. In Janson's terminology, the dynamic capacity tends to be low. Moreover, equilibrium bed capacity may be determined by sorption of inert materials which must be removed later. Both these situations tend again to produce Sherwood-plot economics.

Separation Sequencing

The above discussion makes it clear that the sheer volume, or mass, of a dilute solution is a key characteristic, and that rapid volume reduction is important to economic design. It may also be seen that solute concentration and activity become independent parameters from an economic point of view and that the necessity of efficiently processing the entire feed stream is the most important single problem facing the designer.

It is therefore useful to plot the course of solute recovery on an activity/concentration diagram such as that shown in Fig. 3. Here a feed of low activity and concentration is to be converted to a nearly pure product approaching an activity of unity. One may take any of a large number of paths across this diagram, but the two shown are of particular interest: (a) The classic approach is to first increase the activity, to position I on the figure, and then to increase the concentration. A familiar example is addition to the feed of $(NH_4)_2SO_4$, to bring solute activity up to, or even above, unity. The concentration can then be completed in any of several ways: i) Extract with even a poor solvent and back to water. ii) Precipitate, as by cooling or just allowing the solution to stand. Such a procedure offers the advantages of simplicity: one need not find a highly selective solvent or precipitant. However, one must process large volumes to the end of the process, and use large quantities of salt. This salt in turn must either be recovered or disposed of in some acceptable, usually expensive, way. (b) A potentially more economic approach is first to concentrate the solute from the feed, to intermediate E in the figure, and then to increase its partial molal free energy. A simple example is the adsorption of a negatively charged protein on an

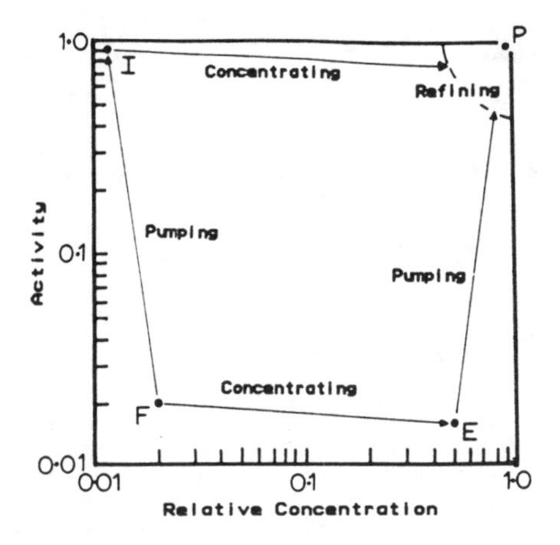

FIG. 3. Process sequencing: activity-concentration diagram

anion exchanging resin, followed by elution with another anion, for example a simple inorganic salt. Such a process offers the advantage of rapid volume reduction and is thus inherently superior to the first approach. However, it requires a greater knowledge of system chemistry and more ingenuity.

Improving Equipment Performance

Processing of dilute solutions tends to be dominated by fluid mechanics problems: obtaining high volumetric throughputs and uniform flow distribution at acceptable pressure drops. However, one must also obtain high volumetric mass transfer rates and minimize dispersion, and it is therefore important to understand the different effects of geometry and flow regimes on momentum and mass transfer.

Flow dominated systems: removal and/or concentration of particulates. If removal of particulates is by filtration, the primary problem is the slowness of percolation through a particulate bed, described by the Blake-Kozeny equation in which flow rate is heavily dependent on particle diameter and void fraction. If removal is by centrifugation, the problem is the low terminal velocity of small particles. In both cases, permissible velocity increases with the square of particle diameter.

Combined flow and mass transfer: transport capacity.

Spherical Bead Annular Cylinder

FIG. 4. A sketch of two packing geometries.

Almost all recovery processes require effective mass transfer between two partially miscible phases, and normally the degree of segregation is insufficient to permit batch contacting. One must then normally use a close approach to counterflow, and this requires high volumetric mass-transfer coefficients, low dispersion in the flow direction and good uniform lateral distribution. These conditions usually demand long narrow columns of small packing or sorbent particles, and these typically have limited flow capacity.

Fixed-bed sorbers operate at very low Reynolds number (Re), where one may assume creeping flow. It is important to minimize momentum transfer relative to mass transfer, and it is convenient to define the ratio of these quantities in terms of Nusselt number for mass transfer and the momentum transfer analog Ref/2, where f is the friction factor. There are many possible situations, but we consider here those, quite common in practice, where mass-transfer resistance is predominantly in the sorbent. We compare a spherical packing with one of the most easily realized form-drag free geometries, bundles of parallel hollow cylinders. Such a system is sketched in Fig. 4, and for simplicity we assume the active portion of the cylindrical solid phase to be a thin shell within which curvature can be neglected. Friction factors for these two systems are readily available, and we calculate them for the spherical packing using a void fraction of 0.4. Then:

$$Ref/2 \approx 211 \text{ (sphere)}; \approx 8 \text{ (cylinder)} \qquad 3,4$$

Asymptotic mass transfer coefficients can be calculated using the zeroth approximation of Reis et al. (6), and corresponding Nusselt numbers are obtained using the sphere and cylinder diameters as reference lengths. We thus obtain:

$$Nu_{asym} = 10 \quad \text{(sphere)}; \quad = 3\,D_{cyl}/\delta \quad \text{(cylinder)} \qquad 5,6$$

It follows that

$$[Ref/2]/Nu_{asym} = 21 \ \text{(sphere)}; = (8/3)(\delta/D_{cyl}) \ \text{(cylinder)} \qquad 7,8$$

Then the resistance ratio for spheres is about one to two orders of magnitude higher than that for cylinders, depending upon the thickness of the active layer δ in the latter. This geometry offers the additional advantage of mechanical support by an imbedding matrix.

SCALING-UP LIQUID CHROMATOGRAPHY

Fixed granular beds are used for recovery and purification of biologically active proteins both because they can afford a large volume reduction, and because they can perform highly selective separations. Liquid chromatography will serve as the basis for our discussion.

Process Modelling

Use of fixed beds for adsorption and chromatographic processes has a long history and there is a wealth of literature (7,8). We will focus on pseudo-continuum models (9-13), which are most frequently formulated by neglecting radial non-uniformities within the bed and beginning with the pseudo-continuum equation of continuity:

$$\varepsilon_b\left(\frac{\partial c_f}{\partial t} + v\,\frac{\partial c_f}{\partial z} - \mathcal{E}\,\frac{\partial^2 c_f}{\partial z^2}\right) = -(1 - \varepsilon_b)\,\frac{\partial c_b}{\partial t} \qquad 9$$

where c_f = solute concentration in the fluid phase, c_b = average solute concentration in the sorbent phase, z = axial coordinate down the bed, v = average fluid velocity, $\mathcal{E}$ = an effective fluid phase dispersion coefficient which takes into account radial non-uniformities in the fluid velocity, ε_b = the fractional interparticle void volume of the bed, and t = time.

We desire the effluent solute concentration in order to gauge column performance, and a number of solutions have been found for various approximations to the boundary conditions at the ends of the column and for the transport and equilibrium behavior of the solute. For complex equilibrium and transport behavior within the particles, numerical methods such as finite differences (14) or orthogonal collocation (15) are generally required, but they pose no severe computational difficulties for description in terms of known parameters.

We now demonstrate the success of process modelling for correlation of experimental data, using as an example, gradient elution ion-exchange chromatography of a 40,000 MW ligninase (11). We assume a linear isotherm because of the dilute solutions involved in our experiments, and consider three phenomena in modelling transport of the protein from the fluid to within the anion-exchange particle: (i) diffusion through a liquid film surrounding the particle, (ii) diffusion through the pore liquid and, (iii) adsorption on the pore wall. We consider the sorbent particles to be spherical and thus write

$$-(1-\varepsilon_b)\frac{\partial c_b}{\partial t} = k_c a_b \varepsilon_p (c_p|_{r=R} - c_f); \quad \frac{\partial c_s}{\partial t} = K_{ad}(c_p - c_s/K_D) \qquad 10,11$$

$$\varepsilon_p \frac{\partial c_p}{\partial t} = \varepsilon_p \mathcal{D}_p \frac{1}{r^2}\frac{\partial}{\partial r}\left(r^2 \frac{\partial c_p}{\partial r}\right) - (1-\varepsilon_p)\frac{\partial c_s}{\partial t}; \quad \frac{\partial c_{p,s}}{\partial r}\Big|_{r=0} = 0 \qquad 12,13$$

where c_p = solute concentration in liquid filling pores, c_s = solute concentration adsorbed on pore walls, ε_p = fractional pore volume of particle, $\mathcal{D}_p$ = diffusivity of solute in pores, K_{ad} = first order rate constant for adsorption, and K_D = equilibrium distribution coefficient = $[c_s/c_p]_{eqbm}$, and a_b = the surface area for mass-transfer per unit column volume, and k_c = boundary-layer mass-transfer coefficient. Note that we have neglected the effects of surface diffusion of the adsorbed protein and have assumed the boundary-layer mass-transfer coefficient to be uniform over the surface of the particle.

For Eq. 9, we use the commonly assumed long column boundary conditions, $\lim_{z\to\pm\infty} \partial c_f/\partial z = 0$. Others have been considered (16) but the solution of Eq. 9 does not appear to be sensitive to the choice of boundary conditions for the long columns of experimental interest (11). The increased computational effort required for other boundary conditions is not justified in view of other modelling ambiguities such as the nature of the variation of the boundary-layer mass-transfer resistance over the surface of a packing particle (17).

Further, we assume the column to be initially free of solute and the feed to the column to be an impulse function:

$$c_b(\tau=0,z)=0; \quad c_f(\tau=0,z) = 0; \quad c_f(\tau,z=0)=\delta(\tau) \qquad 14,15,16$$

A convenient, approximate solution for these equations has been developed (6,18) which allows for temporal, but not spatial, dependence of distribution equilibria. The lowest order approximation, which appears to be adequate

Axial Dispersion

$$T_1 = (\mathcal{E}/v_o^2)\left(\frac{\alpha\epsilon_b}{1-\epsilon_b+\alpha\epsilon_b}\right)$$

Convective Mass-Transfer Resistance

$$T_2 = (R/k_c)\left(\frac{(1-\epsilon_b)\epsilon_b^2\alpha}{(1-\epsilon_b+\alpha\epsilon_b)^3 3\epsilon_p}\right)$$

Intraparticle Diffusion

$$T_3 = \left(\frac{R^2}{\alpha\epsilon_p\mathcal{D}_p+(1-\alpha\epsilon_p)\mathcal{D}_s}\right)\left(\frac{(1-\epsilon_p)\alpha^2\epsilon_b^3}{(1-\epsilon_b+\alpha\epsilon_b)^3 15\epsilon_p}\right)$$

Adsorption Kinetics

$$T_4 = (1/K_{ad})\left(\frac{(1-\alpha\epsilon_p)^2(1-\epsilon_b)\epsilon_b^2\alpha}{(1-\epsilon_b+\alpha\epsilon_b)^3(1-\epsilon_p)}\right)$$

$$u(\alpha) = \frac{\alpha\epsilon_b}{1-\epsilon_b+\alpha\epsilon_b} \qquad \alpha = \frac{1}{\epsilon_p+(1-\epsilon_p)K_D}$$

FIG. 5.

Contributions of Individual Dispersive Mechanisms to Peak Variance

(long time asymptotes)

for many conditions of experimental interest (11), gives for the fluid phase solute concentration,

$$c_f = \frac{c_0(R^2/\mathcal{E})}{2\sqrt{\pi\int_0^t T_{tot}\,dt}}\exp\left[\frac{-(z/v_0-\int_0^t u(\alpha)dt)^2}{4\int_0^t T_{tot}\,dt}\right] \qquad 17$$

with $u(\alpha)$ = fraction of solute in moving fluid phase, T_{tot} = sum of time constants T_i for individual dispersive phenomena, c_0 = a reference concentration, and R = the particle radius, where the time constants T_i, corresponding to the individual dispersive phenomena, and other symbols are defined in Fig. 5. It is the sum of these time constants that defines the width of a solute peak, and it is only this sum that can be determined from the analysis of a single chromatogram.

We now explore the utility of this solution for correlating data from isocratic elution and for extrapolation from isocratic to gradient elution. In Fig. 6, we show the comparison of the experimentally measured enzyme concentration at the exit of a laboratory column, Pharmacia's MonoQ, to a least squares fit of the above model to the experimental data. The values of the distribution coefficient K_D and the sum of the time constants T_{tot} calcu-

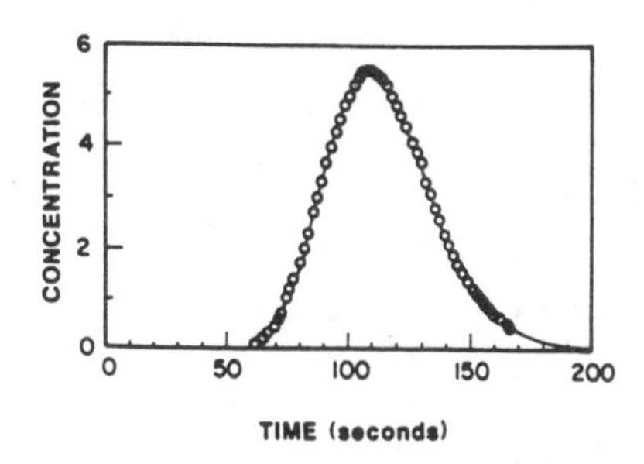

FIG. 6. Isocratic elution

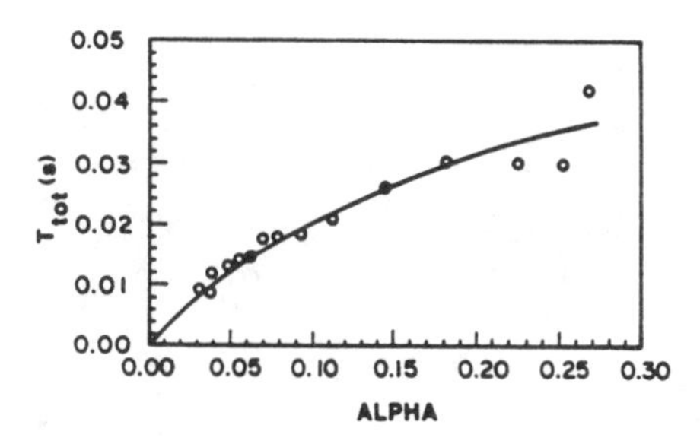

FIG. 8. System response time

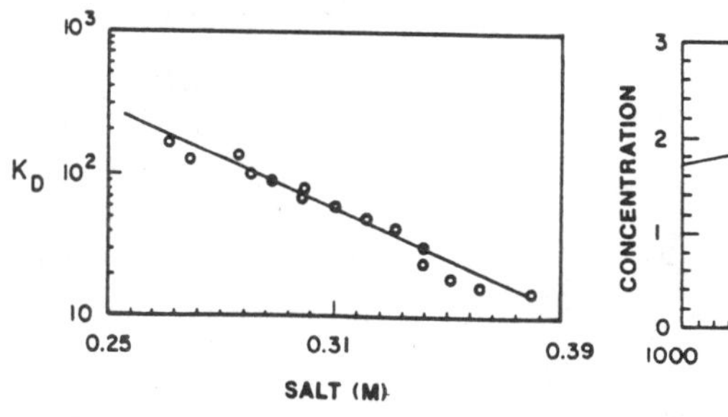

FIG. 7. Equilibrium distribution

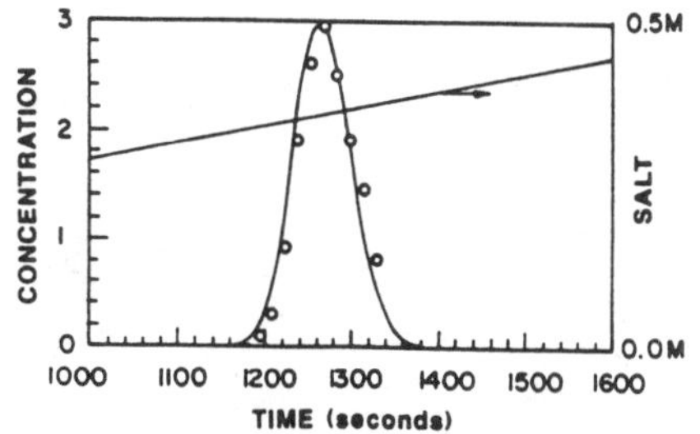

FIG. 9. Gradient elution

lated by assuming $\varepsilon_b = 0.4$ and $\varepsilon_p = 0.8$ are 23.9 and 0.031 s respectively. Additional experiments at constant flow rate and different eluant salt concentrations allow the determination of the dependence of K_D and T_{tot} on the salt concentration in the eluant. These results are shown in Figs. 7 and 8. In Fig. 9 we see an excellent agreement between the model predictions using Figs. 7 and 8 and experimental gradient elution data.

Equipment Characterization and Parameter Estimation

The successes of various modelling efforts (11-13,19), upon first examination, appear to provide a firm basis for the design of large-scale separations equipment using information obtained from laboratory experiments. However, the performance of large-scale equipment frequently does not meet design expectations. There are many possible explanations for this failure: flow maldistribution in the larger diameter columns commonly used, sensitivity of equipment performance to variations in the composition of the feed and/or eluant, and mechanical degradation of the sorbent material, leading to both poor mass-transfer and poor flow distribution. There are several scale-dependent factors which have not been analyzed simply because they are important only in large columns and so cannot be adequately studied in small column experiments. Convective dispersion caused by non-uniform packing density, inadequate header design and viscosity-induced dynamic instabilities such as fingering, can cause poor performance.

The scope of reliable designs predictable from measures of the total mass-transfer resistance, or T_{tot}, is limited by the reliability of the extrapolation of the magnitude of this parameter to the conditions of interest, for example high solute loadings, different sorbent geometry and internal morphology, and different flow rates and pressures. A robust and optimized design then requires estimates of the individual parameters governing solute sorbent equilibria, mass-transfer characteristics, and macroscopic flow behavior and axial dispersivities.

Moreover, the estimation of individual transport properties from measures of overall column performance or measurement of a total mass-transfer resistance in batch experiments is even more difficult because of the number of complex phenomena governing solute transport. Batch experiments are surely easier to interpret than column experiments, but may be difficult to perform for the small particles of interest in HPLC which exhibit very short time scales of equilibration.

Transport phenomena in porous sorbents are complex and we must consider the level of modelling detail required to predict intra-particle behavior for conditions of interest. Although it has been recognized that adsorption kinetics are important in the description of protein transport in affinity sorbents (20-22), it remains to be demonstrated whether adsorption kinetics or surface diffusion play a significant role in protein transport in other sorbents. Some work (10) suggests that both adsorption kinetics and surface diffusion may play an important role in the transport of glucose and fructose in ion-exchangers. Therefore, it is not unlikely that these phenomena may play an important role in the transport of proteins in porous sorbents.

For the purposes of our discussion, we define an effective diffusion coefficient $\mathcal{D}_{eff}$, as that which would be observed if the sorbent were modelled as a homogeneous particle in which the solute migrates by Fickian diffusion. Also, we consider fast adsorption kinetics and a linear isotherm, $c_s = K_D c_p$. Then, we can write the effective diffusion coefficient as

$$\mathcal{D}_{eff} = \frac{\varepsilon_p \mathcal{D}_p + (1-\varepsilon_p)\mathcal{D}_s K_D}{\varepsilon_p + (1-\varepsilon_p)K_D} \qquad 18$$

Therefore we note that in this simple example, the effective diffusion coefficient depends on the distribution equilibria. For finite adsorption kinetics and non-linear isotherm, the situation is more complex (23).

This type of analysis illustrates the complexity in interpreting measures of an overall particle mass-transfer resistance to estimate individual parameters and suggests caution in extrapolating measures of an effective diffusivity to systems with different solvent chemistry (and hence different distribution equilibria), different sorption ligand density (and hence different surface diffusivity), or different particle size (and thus a different contribution of sorption kinetics to the total mass transfer resistance).

The primary problem we are faced with in designing large-scale fixed bed separations equipment is one of parameter estimation and equipment characterization, not process simulation. We are therefore focusing in our laboratory on developing methods for more direct determination of protein transport proterties within sorbent particles by pulsed field-gradient NMR, which allows the observation of molecular motion at equilibrium (24), and of the flow behavior of protein solutions within packed columns by means of NMR imaging techniques, which can give three dimensional characterizations of protein profiles in time.

REFERENCES

1. National Academy of Sciences Committee on Separations Science and Technology. Report in progress.
2. Commercial Biotechnology: An International Assessment, NAS-NAE-IOM, Report of the Research Briefing Panel on Chemical and Process Engineering for Biotechnology (1984). In "Research Briefings", Washington, DC:

National Academy Press, p 31.
3. Jansson J-C, Hedman P (1982). In Fiechter A (ed): "Advances in Biochemical Engineering", Springer, pp 43-49; --,-- (1986). Industrial Scale Chromatography of Proteins, paper 42a, Annual Meeting, AIChE, Miami.
4. Nystrom JM, Product Purification and Downstream Processing (1984). 5th Biennial Executive Forum, AD Little, Boston, MA, 3-6 June. Cost data reproduced in Dwyer JL, p 597 (Nov 1984).
5. Lightfoot EN, Cockrem MCM. What Are Dilute Solutions. Accepted, Separation Science and Technology.
6. Reis JFG, Lightfoot EN, Noble PT, Chaing AS (1979). Sep Sci Technol 14:367.
7. Giddings JC (1965). "Dynamics of Chromatography." New York: Marcel Dekker, vol 1; Hougen OA, Watson KM (1943). "Chemical Process Principles." New York: Wiley, part III.
8. Danckwerts PV (1953). Chem Eng Sci 2:1.
9. Kirkby NF, Slater NKH, Weisenberger KH, Addo-Yobbo F, Doulia D (1986). Chem Eng Sci 41(8):2005.
10. Ghim YS, Chang HN (1982). Ind Eng Chem Fundam 21:369.
11. Gibbs SJ, Lightfoot EN (1986). Ind Eng Chem Fundam 25:490.
12. Graham EE, Fook CF (1982). AIChE J 28(2):245.
13. Pinto NG, Graham EE (1987). Reactive Polymers 5:49; Tsou H, Graham EE (1985). AIChE J 31(12):1959.
14. Mansour A, von Rosenberg DU, Sylvester ND (1982). AIChE J 28(5):765.
15. Raghavan NS, Ruthven DM (1983). AIChE J 29(6):922.
16. Parulekar SJ, Ramakrishna D (1984). Chem Eng Sci 39(11):1571.
17. Gibbs SJ, Karrila SJ, Lightfoot EN. Submitted to Biochem Eng J.
18. Gill WN, Sankarasubramanian R (1970). Proc Roy Soc London A316:341.
19. Arnold FH, Blanch HW, Wilke CR (1985). Chem Eng J 30:B25.
20. Arnold FH, Blanch HW (1986). J Chromatogr 355:13.
21. Chase HA (1984). J Chromatogr 297:179.
22. Arve BH, Liapis AI (1987). AIChE J 33(2):179.
23. Aris RA (1983). Ind Eng Chem Fundam 22(1):150.
24. Tanner JE, Stejskal BO (1968). J Chem Phys 49(4):1768.

Protein Purification: Micro to Macro, pages 491–500

LARGE-SCALE PRODUCTION OF RECOMBINANT PROTEIN A AND ITS USE IN PURIFYING MONOCLONAL ANTIBODIES

R. Love, A. Profy, B. Kaylos, M. Belew, W. Herlihy

Repligen Corporation
One Kendall Square, Building 700
Cambridge, MA 02139

ABSTRACT

Protein A is a cell wall component of *Staphylococcus aureus* which binds a wide range of IgG subtypes through the Fc domain. Immobilized protein A has been evaluated for *ex vivo* removal of IgG and immune complexes as a therapy for a variety of autoimmune disorders and cancers. Commercial use of protein A will require low cost hundred kilogram production of therapeutic-grade protein. We evaluate here two alternative techniques for large-scale purification of recombinant Protein A from *E. coli*. We also present a novel method for coupling protein A to agarose beads and discuss its particular suitability for the affinity purification of mouse antibodies of subtype IgG1.

THERAPEUTIC APPLICATIONS OF PROTEIN A

Plasmapheresis is a well-established therapy for the treatment of a range of autoimmune disorders including myasthenia gravis and systemic Lupus erythematosus (SLE) (1). In this therapy, a portion of the patient's blood is separated *ex vivo* into plasma and cellular fractions with a plasma separator. The plasma fraction, which contains the pathogenic immune complexes, is discarded, and a plasma replacement solution of albumin/saline or fresh frozen plasma is recombined with the cellular fraction and returned to the patient. In 1986, an estimated 200,000 plasmapheresis treatments were performed. Although this therapy is successful in alleviating the symptoms of autoimmune disease, it suffers from the risk of using potentially contaminated blood products. In addition, the patient is deprived of a significant fraction of essential plasma components (clotting factors, transferrin, complement system, etc.) when the discarded plasma is replaced with an albumin/saline solution. An alternative to discarding patient plasma is to pass it over a solid support containing immobilized protein A which will selectively bind and remove IgG and immune complexes. This procedure, which is typically carried out in a continuous, on-line mode, has been shown to remove antibodies against factor IX from a patient suffering from hemophilia B (2). In this application, two columns containing 600 mg of protein A were employed. Plasmapheresis utilizing immobilized protein A columns has also been used in cancer therapy, although the mechanism of the observed effect remains to be elucidated (3). Widespread use will require a high-volume, low-cost source of therapeutic-quality protein A. Assuming the preferred case of a single-use cartridge containing protein A, hundred kilogram quantities will be required annually.

LARGE-SCALE PRODUCTION AND PURIFICATION OF RECOMBINANT PROTEIN A (rPROTEIN A TM)

Prior to 1982, the only commercial source of protein A was methicillin-resistant, secreting stains of Staph. aureus. This source suffers from potential problems, such as large-scale use of a pathogen, possible contamination with staphylococcal enterotoxins, and moderate productivity. To circumvent these problems, scientists at Repligen cloned the gene for protein A and expressed it in E. coli (4). Fermentative growth of the recombinant organism has been performed on a 500-liter scale.

Purification schemes leading to an rProtein A ™ product were developed on the basis of three criteria. First, a purity of 98% was considered sufficient, since the product would be immobilized in a device rather than introduced directly into the patient. However, reduction of E. coli-derived endotoxin to a level of 1 E. U./mg was essential. Second, each step in the process was required to achieve a throughput of >1 kg protein A per batch. Third, the process was required to support substantially reduced costs.

Several purification schemes were developed for a standard batch size of 20 kg E. coli (wet cell weight). One of these is illustrated in Scheme 1. Cells were lysed in 0.5% Triton X-100 with a bead mill, and the lysate was clarified by hollow-fiber filtration. Purification was achieved by chromatography on two sequential ion-exchange columns: 1) a DEAE-cellulose DE-52 (Whatman) column at pH 8.3, and 2) a DEAE-Sepharose Fast Flow (Pharmacia) column at pH 8.1. Both columns were eluted with a salt gradient (0-200 mM KCl), and fractions were analyzed by sodium dodecyl sulfate-polyacrylamide gel electrophoresis (SDS-PAGE). Pooled protein A-containing fractions were concentrated and passed over a 250-ml immobilized polymyxin B column to remove endotoxin contaminants. After polymyxin B chromatography, the protein A product was filter sterilized and lyophilized.

Although the purification process shown in Scheme 1 was successfully scaled up to produce 50-g lot sizes of rProtein A ™, several problems were encountered. Gradient elution of the ion-exchange columns was time and labor intensive, and separation of the protein A from contaminants was not reproducible. Furthermore, the DE-52 column could not be regenerated by stripping in situ with 0.1 N NaOH due to channel formation. The material therefore had to be unpacked after each run, regenerated in a batch process, and repacked. This constraint reduced the practical size limit of this column to 25 l. In order to maintain a large batch size,

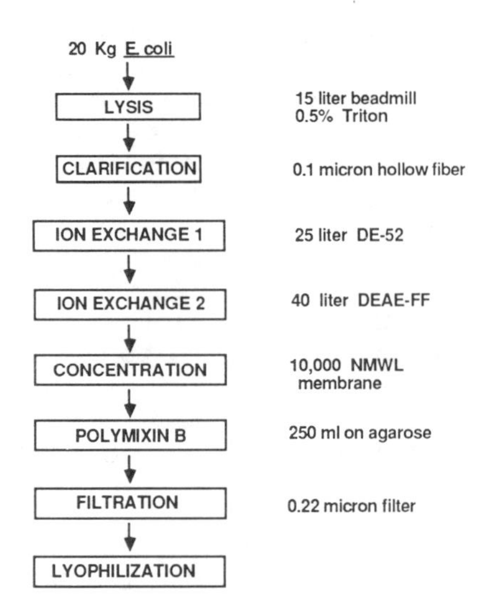

SCHEME 1. Early process for rProtein A ™ production.

the eluates from three column runs were pooled before chromatography on the 40-l DEAE-Sepharose Fast Flow column. This column could be loaded and eluted rapidly, and regenerated with 0.1 N NaOH *in situ*.

A more serious problem with the process shown in Scheme 1 was the variable removal of endotoxin. The final rProtein A ™ product typically contained between 20 and 200 E. U./mg, much greater than the desired 1 E. U./mg. Although immobilized polymyxin B has occasionally been effective in removal of endotoxins (5), many batches of rProtein A ™ seem to contain endotoxin which does not bind to this matrix. It is known that polymyxin B does not bind all types of endotoxin (6,7), and this could explain our results. Alternatively, failure to completely remove endotoxin may be due to a specific interaction between the endotoxin and protein A.

This is supported by the fact that other proteins that we have purified by similar techniques contain significantly lower levels of endotoxin.

In designing an improved purification, we attempted to make use of the properties of rProtein A ™ that most distinguish it from the E. coli contaminants. We previously established that rProtein A ™ which had been eluted from reverse-phase HPLC columns with organic solvents retained full activity after the organic solvents were removed. Secondly, we observed that rProtein A ™ was stable to prolonged heating at 80°C. Finally, it was found that endotoxin could be removed by binding rProtein A ™ to a DEAE column and washing the column with 1% Triton X-100. The protein A, after subsequent elution, contained <10% of the starting endotoxin. These observations formed the basis for the improved purification method diagramed in Scheme 2.

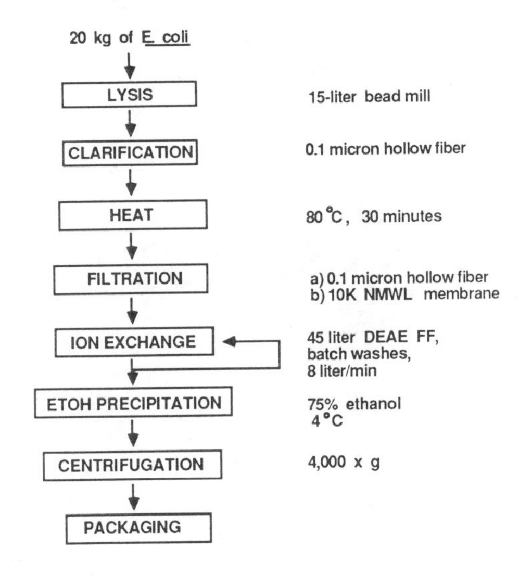

SCHEME 2. Improved process for rProtein A ™ production.

After lysis and clarification, the lysate supernatant was heated to 80°C for 30 minutes, resulting in precipitation of many E. coli proteins but quantitative recovery of rProtein A ™ (Figure 1). In order to achieve optimal precipitation, Triton X-100 was omitted from the lysis buffer. After filtering to remove suspended particulate material, the supernatant was diafiltered with a 10,000 NMWL membrane to reduce the ionic strength and to remove low molecular weight contaminants which could interfere with the capacity of the subsequent ion-exchange column. DEAE-Sepharose Fast Flow ion-exchange chromatography was carried out in a batch mode employing a 45-l column with a 15-cm bed height. This column was washed and eluted at a flow rate of 8 liters/minute. Endotoxin was removed at this step by binding the protein to the gel in 25 mM Tris-HCl (pH = 8.3) and then washing the bound protein with the same buffer containing 1% Triton X-100. The detergent was then removed with a 25 mM Tris-HCl wash, and the

FIGURE 1. Purification achieved by improved rProtein A ™ production process. SDS-PAGE was performed on 1) recombinant cells, 2) clarified cell lysate, 3) supernatant from heat treatment, and 4) ion-exchange column eluate.

product was eluted with a Tris buffer containing 150 mM KCl. If insufficient endotoxin was removed, the ion-exchange step could be repeated. The protein A was further purified and concentrated by precipitation with 75% ethanol at 4°C. The purified rProtein A ™ product was shown to be >98% pure by HPLC analysis, and endotoxin levels were <1 E. U./mg.

The improved purification procedure outlined in Scheme 2 satisfied our requirements for the production of rProtein A ™. We currently use the process to prepare the product in 250-g batches, and have the potential for hundreds of kilograms of annual production at significantly reduced costs. This process illustrates the power of combining efficient recombinant gene expression with cost-effective precipitation and batch purification steps.

ANTIBODY PURIFICATION USING IMMOBILIZED rPROTEIN A ™

Agarose-immobilized protein A has been used for a number of immunochemical procedures, including the purification of IgGs by affinity chromatography (8). In this procedure, an IgG-containing solution such as serum, ascites fluid, or tissue culture supernatant is applied to a column of protein A-agarose, and the contaminants are washed away at neutral or higher pH. The bound antibodies are then eluted by lowering the pH of the eluent. Product purity is typically better than 95%, which compares favorably with purities obtained by ion-exchange chromatography (9).

Since monoclonal antibodies may find widespread use as _in vivo_ diagnostic or therapeutic agents, affinity chromatography on immobilized protein A is being evaluated as an industrial-scale purification process. Concerns have been raised, however, about the stability of conventional CNBr-immobilized protein A, and about the low binding capacity for monoclonal antibodies of mouse subclass IgG1.

We have developed a procedure for coupling rProtein A to crosslinked 4% agarose through a stable amide bond (Scheme 3). The agarose is first treated with epichlorohydrin and the resulting oxirane groups are reacted with arginine. The arginine carboxylic acid groups are then activated by conversion to the N-hydroxysuccinimide esters which, in turn, are reacted with rProtein A ™ to form amide bonds with protein A amino groups. Unreacted esters are blocked with ethanolamine, and the gel is thoroughly washed to remove non-covalently bound rProtein A.

Immobilized rProtein A ™ prepared by this procedure binds mouse IgG1 monoclonal antibodies more effectively than conventional CNBr-immobilized protein A. In Table 1, the binding capacities (mg IgG bound per ml gel) are compared. Although the capacity of Immobilized rProtein A ™ is only moderately higher for human and mouse polyclonal IgGs, capacities for each of three IgG1 monoclonals are several fold higher. The underlying chemical basis for the uniquely high IgG1 binding capacity shown by Immobilized rProtein A ™ is not understood, but is currently under investigation.

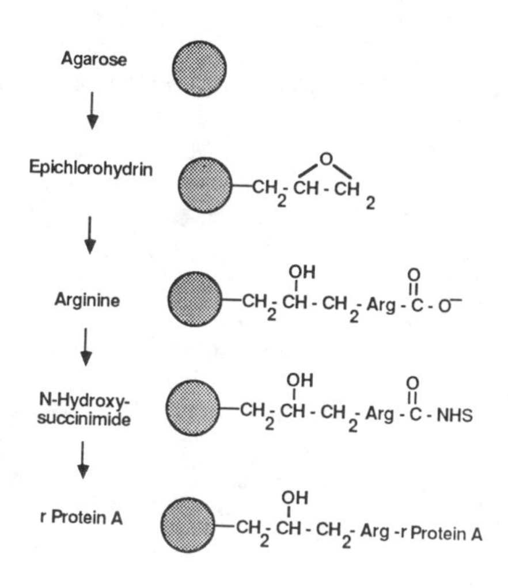

SCHEME 3. Process for Production of Immobilized rProtein A ™

TABLE 1
IgG BINDING CAPACITIES FOR IMMOBILIZED rPROTEIN A ™ AND CNBr-IMMOBILIZED PROTEIN A

IgG Binding (mg/ml)	Immobilized rProtein A	CNBr	%Increase
Human IgG	20	17	18%
Mouse IgG	17	11	55%
Mouse IgG1			
103-13	7	2	250%
103-23	6	1	500%
103-28	8	1	700%

REFERENCES

1. Nose Y, Malchesky MS, Smith JW, and Krakauer RS (Eds.) (1982) Plasmapheresis : Therapeutic Applications and New Techniques. Raven Press, New York.
2. Nilsson IM, Jonsson S, Sundqvist S-B, Ahlberg A, and Bergentz SE (1981), Blood, 58:38-43.
3. Fer MF and Oldham RK (1985) in Contemporary Topics in Immunobiology Vol. 15 : Immune Complexes and Human Cancer (Salinas FA and Hanna MG, Jr., Eds.) pp. 257-276 Plenum Press, New York.
4. Colbert D, Anilionis A, Gelep P, Farley J and Breyer R (1984) J. Biol. Resp. Modif. 3:255-259.
5. Issekutz AC (1983) J. Immunol. Methods 61:275-281.
6. Maitra SK, Yoshikawa TT, Guze LB and Schotz MC (1981) J. Clin. Microbiol. 13:49-53.

7. Kluger MJ, Singer R and Eiger SM (1985), J. Immunol. Methods 83:201-207.
8. Langone JJ (1982), J. Immunol. Methods 55:277-296.
9. Manil L, Motte P, Pernas P, Troalen F, Bohblon C and Bellet D (1986), J. Immunol. Methods 90:25-37.

Index

0 00 02 0245853 5
MIDDLEBURY COLLEGE